How to use this book

These pages show you all the different features you will find in your Student Book. Each feature is designed to support and develop the skills you will need for your examinations, as well as foster and stimulate your interest in chemistry.

Chapter openers

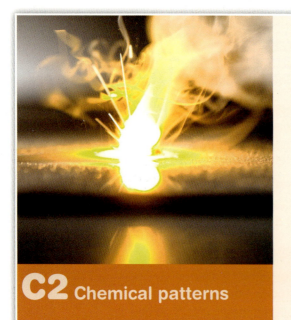

C2 Chemical patterns

Why study chemical patterns?

Copper makes ideal water pipes but another metal, potassium, reacts violently with water. Why do these elements behave so differently? How can we predict their properties? Scientists use what they know about elements and compounds to choose substances with the best properties for particular purposes. They use their knowledge of chemical reactions to work out how to make these substances.

What you already know

- Substances are made of tiny particles, called atoms. Every element has its own type of atom and its own chemical symbol.
- Compounds are made up of atoms of two or more elements that are joined together. You can use chemical formulae to represent compounds.
- During a chemical reaction, atoms are rearranged. The atoms in the reactants are joined together differently to the atoms in the products.
- In a chemical reaction, the total mass of the reactants is the same as the total mass of the products. This is the law of conservation of mass.
- Equations are used to represent chemical reactions. The formulae of the reactants are on the left of the arrow, and the formulae of the products are on the right of the arrow.
- In metal displacement reactions, more reactive metals displace less reactive metals from their compounds.
- Catalysts speed up some chemical reactions. The catalyst is not used up in the reaction.
- The Periodic Table lists all the elements, roughly in order of relative atomic mass.
- In the Periodic Table, each vertical column is called a group of elements. Elements in the same group have similar properties. The horizontal rows are called periods. There is a gradual change in properties from left to right of a period. You can use the position of an element in the Periodic Table to predict its properties.
- In the Periodic Table, the elements on the left of the stepped line are metals, and the elements on the right of the stepped line are non-metals.
- Most metals are [...] heat and elect[...]
- Most non-met[...] that are in the [...] conductors of [...]

The Science
In this chapter you will discover how the tiny particles that make up atoms give elements their properties. You will use the Periodic Table to find patterns in the physical and chemical properties of elements, and to make predictions about their reactions.

You will also explain some strange observations. Why does salt solution conduct electricity, when solid salt is an insulator? And why do some elements react violently while others do not react at all?

Ideas about Science
This chapter traces the development of two vital ideas in chemistry – [...] and the Periodic Table.

You will start by finding out about atoms from the earliest thoughts of Greek philosophers to the discovery of even tinier subatomic particles continue today.

Next you will learn about orga[...] elements. How did a Russian c[...] leap of imagination help to ma[...] sense of other chemists' data a[...] lead to the development of the [...] Periodic Table?

Why study
This explains why what you are about to learn is useful to scientists as well as to you.

The Science
This summarises the science in the chapter you're about to study.

Ideas about Science
This provides some ideas about how science explanations are developed and the impact of science and technology on society, in the context of the science in the chapter.

What you already know
This is a summary of the things you've already learnt that will come up again in the chapter. Check through them in advance and see if there is anything that you need to recap.

Main spreads

Key words
- electrolysis
- reducing agent
- oxidation
- bioleaching
- displacement reaction
- phytoextraction

Using carbon to extract metals
Many metals can be extracted from their ores by heating with carbon. This works for some metals but not for very reactive metals. Reactive metals, such as aluminium, are extracted by **electrolysis**.

Carbon works well to extract metals from ores that contain metal oxides, but some metal ores contain metal sulfides. Zinc ores include zinc blende (ZnS), zinc spar ($ZnCO_3$), and zincite (ZnO). Zinc blende, which is zinc sulfide (ZnS), is the main ore for zinc extraction. Sulfide ores can be heated in air so that the sulfide compounds become oxides, for example, turning zinc sulfide into zinc oxide.

To extract zinc from zinc oxide, the oxygen is removed so that ZnO becomes Zn. This process of removing the oxygen is called reduction. Carbon is used as a **reducing agent** to remove oxygen.

zinc oxide + carbon → zinc + carbon monoxide

Zn loses O to C and gets reduced

$$ZnO + C \rightarrow Zn + CO$$

C takes O from Zn and gets oxidised

Further **oxidation** of the carbon monoxide forms carbon dioxide, which is less harmful.

Carbon is often used as a reducing agent to extract metals because:
- Carbon, in the form of coke, can be made cheaply from coal.
- At high temperatures, carbon has a strong tendency to react with oxygen, so it is a good reducing agent.
- The carbon monoxide formed is a gas, so it is not left behind to make the zinc impure.

Synoptic link
You can learn more about oxidation in C1.1 How has the Earth's atmosphere changed over time, and why?

Carbon can also be used to extract iron and copper. These reactions can be summarised as:

iron oxide + carbon → iron + carbon dioxide
copper oxide + carbon → copper + carbon dioxide

Green meth[...]
Scientists are de[...] waste heaps, wh[...] methods are 'gr[...] much lower ene[...] on the environ[...] are toxic so th[...] resource. At the[...] far too small to [...] important as ou[...]

Synoptic link
[...]can learn more about [...]rolysis in C3.3 What are [...]rolytes and what happens during [...]rolysis?

Key words
Key words boxes highlight words that are useful to know and may be important to understand for your exams. You can look for these words in the text or check the glossary to see what they mean.

Worked example
Worked examples provide step-by-step instructions for how to carry out calculations and analyse data.

C2.4 How are equations [...] chemical reactions

A: Writing formulae

Every element and compound has its own **chemical formula**. Chemists work out chemical formulae by experiments, and you can look them up in data tables.

The formula of sodium chloride is $NaCl$ because there is one sodium ion, Na^+, for every chloride ion, Cl^-. There are no molecules in sodium chloride, only ions.

Not all ions have single positive or negative charges. The formula of lead bromide is $PbBr_2$. In this compound there are two bromide ions [...] every one lead ion. All compounds are overall electrically neutral, so [...] charge on a lead ion must be twice that on a bromide ion. A bromid[...] ion has a single negative charge, Br^-, and a lead ion has a double pos[...] charge, Pb^{2+}.

The diagram shows the charges on some simple ions. If you know [...] charges of the ions in a compound, you can work out its formula.

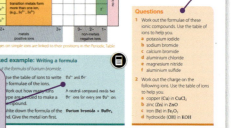

The charges on simple ions are linked to their positions in the Periodic Table.

Worked example: Writing a formula
Work out the formula of barium bromide.

Use the table of ions to write [...] Ba^{2+} and Br^- [...] the formulae of the ions.

Work out how many ions [...] A neutral compound needs two [...] type are needed to make a [...] Br^- ions for every one Ba^{2+} ion. [...]compound.

Write down the formula of the [...] Barium bromide = $BaBr_2$ [...]nd. Give the metal ion first.

Find out about
Every section in the book has a list of the key points that are explored in the section.

Find out about
- how to write formulae of ionic compounds

Questions
Every section has questions that you can use to see if you've understood the content of the section.

Questions
1 Work out the formulae of these ionic compounds. Use the table of ions to help you.
   a potassium iodide
   b sodium bromide
   c calcium bromide
   d aluminium chloride
   e magnesium nitride
   f aluminium sulfide
2 Work out the charge on the following ions. Use the table of ions to help you.
   a copper (Cu) in $CuCl_2$
   b zinc (Zn) in ZnO
   c iron (Fe) in Fe_2O_3
   d hydroxide (OH) in KOH

[...]ow are equations used to represent chemical reactions? 73

TWENTY FIRST CENTURY
sci@nce

GC
Ch

Project Director
Mary Whitehouse

Project Manager
Alistair Moore

Editors
Mutlu Cukurova
Helen Harden
Maria Turkenburg

Authors
Maureen Borley
Helen Harden
Philippa Gardom Hulme
Emma Palmer
Ann Tiernan
Dorothy Warren
Lynda Dunlop

OXFORD
UNIVERSITY PRESS

Contents

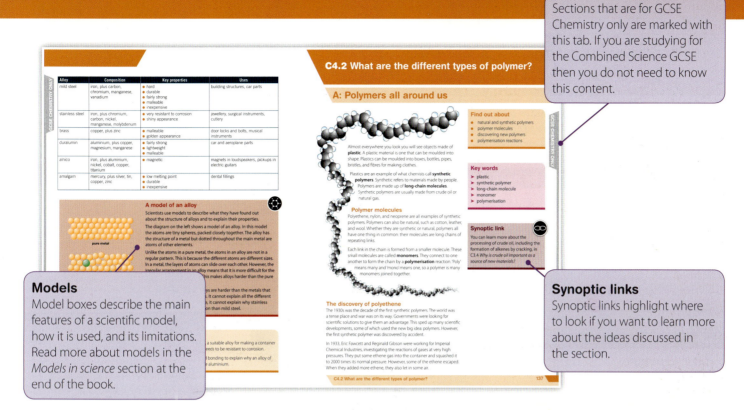

Sections that are for GCSE Chemistry only are marked with this tab. If you are studying for the Combined Science GCSE then you do not need to know this content.

Models
Model boxes describe the main features of a scientific model, how it is used, and its limitations. Read more about models in the *Models in science* section at the end of the book.

Synoptic links
Synoptic links highlight where to look if you want to learn more about the ideas discussed in the section.

Science explanations and Ideas about Science

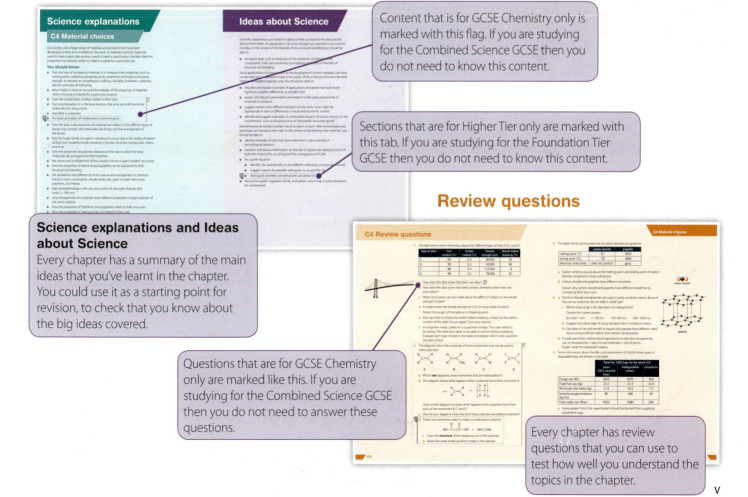

Content that is for GCSE Chemistry only is marked with this flag. If you are studying for the Combined Science GCSE then you do not need to know this content.

Science explanations and Ideas about Science
Every chapter has a summary of the main ideas that you've learnt in the chapter. You could use it as a starting point for revision, to check that you know about the big ideas covered.

Sections that are for Higher Tier only are marked with this tab. If you are studying for the Foundation Tier GCSE then you do not need to know this content.

Review questions

Questions that are for GCSE Chemistry only are marked like this. If you are studying for the Combined Science GCSE then you do not need to answer these questions.

Every chapter has review questions that you can use to test how well you understand the topics in the chapter.

v

Structure of assessment

There will be two examination papers for GCSE Chemistry.

Paper 1: Breadth in Chemistry

This paper covers all the Chemistry chapters. It contains short answer questions worth up to three marks, including problem solving, calculations, and questions about practical work.

Time	Marks available	Percentage of GCSE
1 hour 45 minutes	90	50%

Paper 2: Depth in Chemistry

This paper covers all the Chemistry chapters. It contains structured questions covering both theory and practical work, including questions requiring an extended written response.

Time	Marks available	Percentage of GCSE
1 hour 45 minutes	90	50%

Top tips

In each examination paper, you will have to:

- demonstrate your knowledge and understanding of scientific ideas, techniques, and procedures

- apply your knowledge and understanding to new and familiar contexts

- interpret and evaluate information and data, make judgements and conclusions, and evaluate scientific procedures and suggest how to improve them.

Some questions will use contexts that you are not familiar with, including examples of science from the real world, issues from the news, and reports of scientific investigations. Remember that although the context may be different, the science is the same. The questions are designed so that you can answer them if you combine your own knowledge and understanding with the information given in the question. You should:

- think about how the context is similar to something you have learnt about

- look for information in the question that suggests how you can relate what you know to the new context.

Kerboodle

This book is also supported by Kerboodle, offering unrivalled digital support for building your practical, maths, and literacy skills.

If your schools subscribes to Kerboodle, you will also find a wealth of additional resources to help you with your studies and with revision:

- Animations, videos, and podcasts
- Webquests
- Activities for every assessable learning outcome
- Maths skills activities and worksheets
- Literacy skills activities and worksheets
- Interactive quizzes that give question-by-question feedback
- Ideas about Science case studies

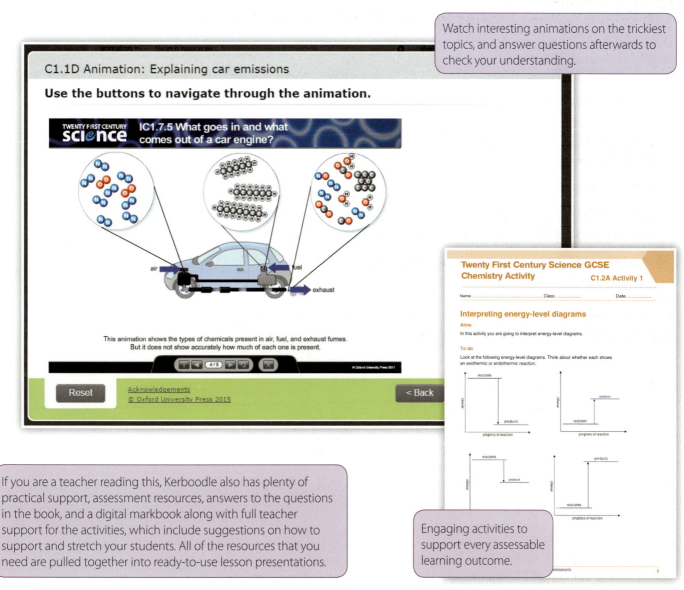

Watch interesting animations on the trickiest topics, and answer questions afterwards to check your understanding.

If you are a teacher reading this, Kerboodle also has plenty of practical support, assessment resources, answers to the questions in the book, and a digital markbook along with full teacher support for the activities, which include suggestions on how to support and stretch your students. All of the resources that you need are pulled together into ready-to-use lesson presentations.

Engaging activities to support every assessable learning outcome.

C1 Air and water

Why study air and water?

Do you believe everything you read online? Do you make decisions based on headlines? Do you trust what others tell you, or do you prefer to make up your own mind?

A knowledge of the science behind air quality and climate change, as well as a greater understanding of the data produced by scientists, will give you the confidence to start to make your own decisions and make a difference to the world.

What you already know

- The properties of different states of matter can be explained in terms of the particle model.
- The changes of state can be explained in terms of the particle model.
- The differences between atoms, elements, and compounds.
- The chemical symbols for elements and formulae for compounds.
- Matter is conserved during changes of state and chemical reactions.
- Filtration and distillation are simple techniques for separating mixtures.
- Chemical reactions are the rearrangement of atoms.
- Chemical reactions are represented using formulae and equations.
- Catalysts speed up chemical reactions.
- Energy changes take place during changes of state.
- There are two types of reaction: exothermic and endothermic.
- The carbon cycle describes how carbon is recycled.
- The atmosphere contains mainly nitrogen and oxygen, with small amounts of other gases.
- Carbon dioxide is produced by human activity and has an impact on climate change.

The Science

This chapter gives you the opportunity to use your current science knowledge to develop your GCSE understanding of the atmosphere, air quality, climate change, and how to ensure a supply of potable (drinking) water.

You will also be introduced to key ideas, such as balancing equations. This will support your learning through the rest of the course.

Ideas about Science

In this chapter you will:

- interpret data, allowing for uncertainties in measurements
- consider the difference between correlation and cause
- find out how scientists peer review each other's work and how evidence may, or may not, change the accepted explanation.

C1.1 How has the Earth's atmosphere changed over time, and why?

A: What types of substance are found in the atmosphere?

Find out about

- the atmosphere
- the mixture of elements and compounds found in the atmosphere
- how the particle model can help to explain the properties of elements and compounds
- how substances enter and leave the atmosphere
- physical and chemical changes

oxygen molecule containing two oxygen atoms

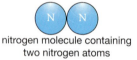

nitrogen molecule containing two nitrogen atoms

Most of the atmosphere consists of nitrogen molecules and oxygen molecules.

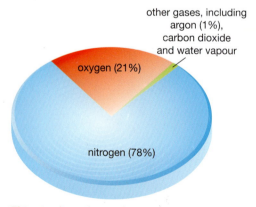

other gases, including argon (1%), carbon dioxide and water vapour

oxygen (21%)

nitrogen (78%)

This pie chart shows the proportion of different elements and compounds in the air.

The atmosphere

The Earth's **atmosphere** forms a layer of air around our planet. The atmosphere is essential to life on Earth. Without it we would have no oxygen to breathe and temperatures would be too cold for life.

The Earth from space. White clouds formed from condensed droplets of water vapour can be seen in the atmosphere.

Elements, mixtures, and compounds

The air in the atmosphere consists mainly of two elements: nitrogen and oxygen. An **element** is made of only one type of **atom**.

Air also contains compounds. A **compound** is made from two or more elements that are chemically combined, for example, water vapour (H_2O) and carbon dioxide (CO_2). The compound carbon dioxide is made from a combination of the elements carbon and oxygen.

The atmosphere is a **mixture** of elements and compounds because the different substances are not chemically combined. The atmosphere is about 100 km thick, but three quarters of its mass is within the first 11 km.

The water in our oceans and rivers is a compound made from the elements hydrogen and oxygen. Other compounds are dissolved in the water, which provides a habitat for a huge variety of living organisms. Creating water that is clean and safe to drink is the focus of C1.4.

Explaining properties of substances

The element gold is sometimes found in rock as a solid nugget. The particle model can be used to explain why a gold nugget keeps its shape. The particles in pure gold are gold atoms. Individual gold atoms do not have the properties of solid gold. When gold is heated it changes state and becomes a liquid. The particle model can explain why it can now flow and fill the shape of a mould.

The element oxygen is found in the atmosphere as a gas. The particle model represents each particle of oxygen as a sphere but in reality oxygen exists as molecules. A **molecule** is a group of atoms (two or more) that are chemically combined. Each oxygen molecule is made of two atoms of oxygen chemically joined together. The particle model can explain how oxygen spreads to fill an available space.

Key words

> atmosphere
> element
> atom
> compound
> mixture
> molecule

The particles in a solid remain in a fixed position and only vibrate about the same point.

The particles in a liquid can move around.

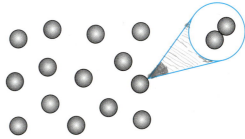

The particles in a gas can move about in three dimensions. The particle model shows them as single spheres but the particles could be molecules made from two (or more) atoms that are joined.

A simple particle model of matter

The particle model gives us a way to visualise the arrangement and movement of particles in solids, liquids, and gases. This can be used to explain the different properties of each state. In this simple model:

- All matter is made of very tiny particles.
- There are attractive forces between particles.
- In the solid state, the particles are close together and unable to move away from their neighbours.
- In the liquid state, the particles are also close together but can slide past each other.
- In the gas state, the particles are further apart and can move freely.

H The model makes some simplifications, such as:

- The particles are pictured as spheres even though they may be molecules with a different shape.
- The spheres are drawn the same size even though they may really have different sizes and masses.
- When drawn, the gaps between the particles in a gas must enable a diagram to fit on a page when in fact they are much larger relative to the size of the particles.

These simplifications mean that there are limitations to what the basic model can explain.

C1.1 How has the Earth's atmosphere changed over time, and why?

5

Physical changes

On Earth, the compound H_2O can be found in the liquid state (water), solid state (ice), and gas state (water vapour). When H_2O **evaporates** it moves from the oceans and rivers to the atmosphere. The reverse process occurs when H_2O, in the form of water vapour, **condenses** and falls as rain.

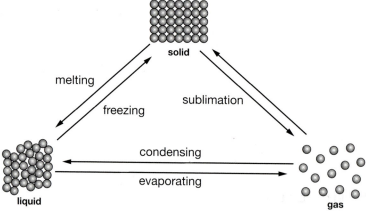

A diagram to show changes of state.

Even though H_2O changes its name in each state, it is still the same substance. Each process in the diagram above is an example of a **physical change**.

Chemical changes

Coal is mined from the rocky surface of the Earth. When coal burns, the carbon atoms in the coal chemically combine with oxygen atoms in the air. The carbon dioxide that is produced is in the gas state and enters the atmosphere. It has completely different properties to both carbon and oxygen. It is a new substance. This process is an example of a **chemical change**.

The burning of fuels and how this affects our atmosphere – and indeed our whole planet – is explored later in the chapter.

Questions

1 Use the particle model to explain why:
   a molten lava can flow down a volcano
   b lava stops flowing when it cools and solidifies.

H 2 Explain why the particle model cannot explain why:
   a a gold bar is much heavier than an aluminium bar of the same size
   b gold and iron have different melting points.

3 Are the following chemical or physical changes? Explain your answers.
   a Water vapour in your bathroom condensing on the window.
   b Iron exposed to oxygen becoming coated in rust.

4 When methane gas burns, carbon dioxide and water are formed. Explain why both these substances enter the atmosphere.

B: What determines the state of a substance on Earth?

Using data to predict the state of a substance

You may only think of carbon dioxide as a gas, but on the planet Mars the wintertime temperatures at the poles are so low that carbon dioxide freezes. When sunlight returns in the spring this carbon dioxide changes back to a gas without turning into a liquid first. This is another type of physical change, known as **sublimation**.

You use methane gas each time you light a Bunsen burner but on Titan, a moon of Saturn, lakes have been detected that are thought to contain liquid methane.

You can use **melting point** and **boiling point** data to work out the state of a substance at a given temperature. The Worked example shows you how to do this by drawing a thermometer. As you get more confident, you should be able to visualise the thermometer in your head.

Worked example: Predict the state of a substance from data

The melting point of methane is −182 °C and its boiling point is −161 °C.

What is the state of methane at −180 °C?

Step 1: Draw a thermometer and mark on the melting point (MP) and boiling point (BP) temperatures.

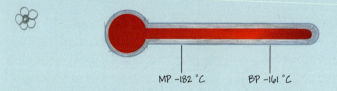

Step 2: Mark on the thermometer the zones where the substance will be in the gas state, the liquid state, and the solid state.

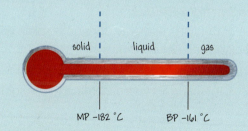

Step 3: Mark on the thermometer −180 °C.

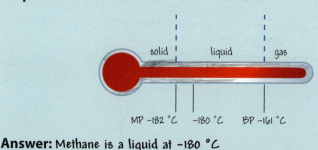

Answer: Methane is a liquid at −180 °C

Find out about

- using data to predict the state of a substance at a given temperature
- forces of attraction between particles
- limitations of the basic particle model

Key words

➤ sublimation
➤ melting point
➤ boiling point

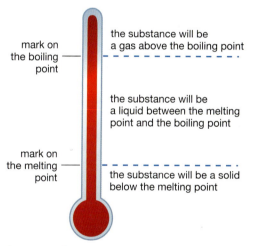

Drawing a thermometer will help you visualise the relationship between boiling point, melting point, and the states of a substance.

Why do substances have different boiling points?

Particles within a substance have a force of attraction between them that varies from substance to substance. The strength of the force of attraction between particles depends upon the atoms and molecules that make up the substance.

When a liquid is heated, the particles move faster because they have more kinetic (movement) energy. Eventually they have enough energy to overcome the attractive forces between particles and can move apart. The substance is then in the gas state.

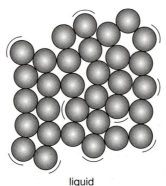

liquid

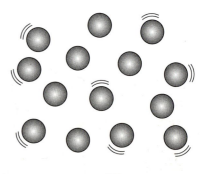

gas

Particles in the liquid have less kinetic energy and move more slowly. They do not have enough kinetic energy to overcome the attractive forces between the particles.

Particles in the gas have more kinetic energy and move faster. They have enough kinetic energy to overcome the attractive forces between the particles.

Water is a special case

Water molecules are particularly strongly attracted to each other. This makes the boiling point of water much higher than that of other substances with molecules of similar size and shape. Water is liquid at temperatures where similar compounds would already be in the gas state. Without the strong attraction between water molecules, the water on Earth would evaporate and conditions would no longer be suitable for life.

Improving the particle model

The basic version of the particle model does not explain why substances have different melting and boiling points.

The model can be improved by adding the idea that the forces of attraction between the particles vary according to the substance.

Questions

1   The melting point of hydrogen sulfide (H_2S) is –85 °C and its boiling point is –61 °C .
    a   What is the state of hydrogen sulfide at 25 °C?
    b   What is the state of hydrogen sufide at –70 °C?

2   The boiling point of water is (H_2O) 100 °C. State whether water or hydrogen sulfide has a stronger force of attraction between its molecules. Explain your answer.

C: How did our atmosphere form?

Early Earth

The early Earth was a violent place, constantly bombarded by meteors and covered with active volcanoes. The atmosphere consisted mainly of carbon dioxide and water vapour, with smaller amounts of other gases. These gases probably came from volcanoes, which still belch out a variety of gases today. If it had stayed that way, Earth would have been as inhospitable as our neighbouring planet Venus.

Find out about

- the composition of gases in the Earth's earliest atmosphere
- evidence for an increase in oxygen in the atmosphere
- processes by which carbon dioxide may be removed from the atmosphere
- why scientific research is peer reviewed before publication

This computer generated image of the surface of Venus is based upon radar data collected by the Magellan spacecraft. The temperature of the surface of Venus is almost 500 °C. Its atmosphere is largely carbon dioxide and its clouds are formed of sulfuric acid.

On Earth temperatures began to cool, causing the water vapour to gradually condense and form oceans.

Exactly what happened next is less certain. Scientists are still investigating when and how the Earth's atmosphere changed. Why did oxygen levels increase and what processes caused carbon dioxide levels to decrease?

Finding out about the formation of Earth's atmosphere

Scientists cannot directly identify or measure the gases in Earth's ancient atmosphere; they have to use indirect evidence instead.

Collecting data about gases emitted from modern volcanoes and gases that are found in the atmosphere of other planets can help scientists better understand Earth's very early atmosphere.

Scientists are able to date rocks and fossils. The chemical make-up of rocks can give clues about the composition of the atmosphere when the rocks were formed. Fossils provide information about the earliest forms of life.

An artist's impression of early Earth.

Air bubbles trapped in ice cores drilled in Antarctica allow scientists to analyse the composition of the air from hundreds of thousands of years ago. Ice cores do not go back far enough in time to provide evidence of the earliest atmosphere. In the beginning, the Earth's surface was too hot for water to stay in the solid state.

Explaining the increase in oxygen levels

Scientists generally agree that oxygen levels rose dramatically about 2.3 billion years ago.

Recent evidence suggests that oxygen was first released long before this by photosynthesising bacteria. Just as in plants today, these bacteria used energy from sunlight to convert carbon dioxide from the air into oxygen. This reaction also uses water and produces glucose.

This early oxygen is thought to have reacted with other substances before it could enter the atmosphere. Later, plants developed, which resulted in the release of larger quantities of oxygen. Atmospheric levels of oxygen only increased when oxygen was being produced at a higher rate than it could react with other substances on Earth.

Scientists constantly discuss and evaluate each other's ideas and explanations. As more evidence is discovered, theories on how the atmosphere formed change.

Decrease in carbon dioxide levels

The decrease in carbon dioxide levels can be explained by the **carbon cycle**. Carbon dioxide is removed from the atmosphere as it dissolves in the oceans. Photosynthesis uses carbon dioxide and water to produce oxygen and glucose, so an increase in plant life substantially lowers carbon dioxide levels in the atmosphere. When some of these plants died they formed fossil fuels, locking carbon in the ground for millions of years.

Carbon dioxide is also locked up in the ground by the formation of sedimentary rock, such as limestone (formed from the calcium carbonate shells of ancient sea creatures).

It is the burning of fossil fuels that is now rapidly adding carbon dioxide back into the atmosphere. We will return to this later in the chapter.

Iron pyrite is made of iron sulfide, which only forms if there is no oxygen present. If oxygen is present then iron sulfate would be formed instead.

Red iron oxide rocks only form if there is oxygen present.

Sharing research

Scientists collaborate to share their ideas and to check each other's results and methods. One way scientists do this is by publishing their research in **peer-reviewed** journals. Your peer group is made up of other students of a similar age. In the scientific world, peers are other scientists.

All new research papers must be peer reviewed (evaluated by other scientists) to make sure that they are:

- Valid – do the results measure what they say they do, were the methods used (e.g., in data collection and analysis) appropriate, and was the research design correct?

- Original – are the results new? Has the work of other scientists been properly credited?

- Significant – are the research findings important?

Once research has been published other scientists can read it, and either test it by trying to reproduce the research themselves or use it to inform and inspire further research. If the new evidence contradicts the current most accepted theory this gives rise to very robust scientific debate.

Cutting-edge research

Scientists are searching for evidence of the earliest photosynthesising bacteria.

For example, publication of the discovery of bacteria-like fossils in 3.5 billion-year-old samples of rock was a major breakthrough.

However, when another research team analysed similar rocks using a different technique, they claimed the shapes were made of a crystalline material.

So, without the research being reproduced by other scientists, the scientific community will not accept the discovery.

Scientists study ecosystems today that are similar to those found on Earth 3 billion years ago. These pools in Mexico contain over 70 species of bacteria found nowhere else on Earth, including types of photosynthesising bacteria.

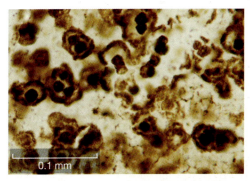

0.1 mm

A microscope view of a thin section of rock containing some of the earliest life forms found (about 2 billion years old). These were discovered in ancient rocks which contain iron oxide.

Questions

1 No samples of iron pyrite (**FeS**) have been dated as less than 2 billion years old. Rocks made of iron oxide (**FeO**) have been dated to 2 billion years ago.
   a What does this tell you about when oxygen first appeared in the Earth's atmosphere?
   b What discovery would give you further information about when oxygen first appeared?

2 By what process did carbon dioxide transfer from the atmosphere to the plants that then formed coal?

3 Some scientists have developed a method for detecting organic material (material from living organisms) in rocks. How could this help scientists discover when oxygen was first produced on Earth?

D: How do human activities affect the atmosphere?

Find out about

- how air pollutants are formed
- the problems different air pollutants cause

Key words

➤ air pollutants
➤ oxidation
➤ combustion
➤ hydrocarbon
➤ particulate
➤ impurities
➤ incomplete combustion
➤ emission

Synoptic link

You can learn more about oxides and oxidation in C3.2A *Extracting metals from ores*.

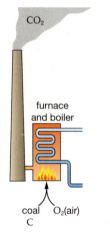

When coal (carbon) reacts with oxygen, carbon dioxide is produced.

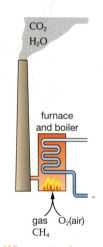

When natural gas (CH_4) reacts with oxygen, carbon dioxide and water vapour are produced.

Air pollution

Almost every time you are driven somewhere by car or when you switch on the lights at home, gases are released into the atmosphere. They are released by the car or the power station where electricity is generated. These chemical compounds are released as a result of fossil fuels reacting with oxygen in the air. Some of the substances produced are harmful to the environment and to humans, they are called **air pollutants**.

Types of reactions

Reactions where oxygen is gained by a reactant are called **oxidation** reactions. Some chemicals (e.g., fossil fuels) react rapidly with oxygen to release energy and often light. This type of oxidation reaction is called **combustion** (in everyday life this is often referred to as burning). Oxygen is necessary for combustion and combustion can be used to identify the presence of oxygen.

Synoptic link

You can learn more about how to safely test for oxygen and carbon dioxide in C8L *Identifying products using gas tests*.

Air pollution from power stations

Most power stations are fuelled by coal or natural gas.

When coal (which is made of carbon atoms) combusts, carbon dioxide is formed in the gas state and enters the atmosphere.

Natural gas, petrol, diesel, and fuel oil are all mainly made up of **hydrocarbon** molecules. These are molecules made up of carbon and hydrogen only. When these fuels react with oxygen, carbon dioxide and water vapour are formed. These gases also enter the atmosphere.

If coal or natural gas combust with insufficient oxygen (incomplete combustion), carbon monoxide (**CO**) is produced as well as particles of unburnt carbon known as **particulates**.

If the fuel contains sulfur **impurities** then sulfur dioxide (SO_2) will be released.

Finally, the high temperatures inside the furnace cause some of the nitrogen from the air to react to produce nitrogen monoxide (**NO**). This is another example of a combustion reaction.

NO is then oxidised further to form nitrogen dioxide (NO_2). This is an oxidation reaction. Together, **NO** and NO_2 are referred to as nitrogen oxides (**NOx**).

Air pollution from vehicles

Vehicle engines burn petrol or diesel, which are both types of hydrocarbon.

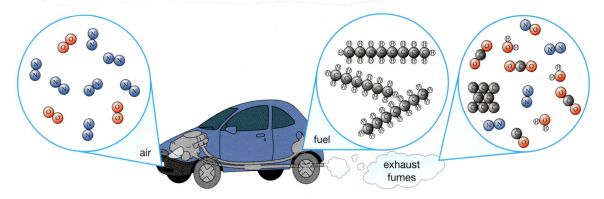

Fuel, and oxygen and nitrogen from the air enter the engine. Carbon dioxide, water vapour, carbon monoxide, and nitrogen monoxide are produced. The formation of carbon and carbon monoxide is referred to as **incomplete combustion**.

The effect of air pollutants

The **emission** of substances into the atmosphere can affect both our health and the environment.

Particulates (tiny bits of solid suspended in the air)	Particulates stick to surfaces and make them dirty. They can be breathed into your lungs and cause breathing problems, and can make asthma worse.
Carbon monoxide (CO)	CO is a very poisonous gas. It blocks oxygen from being carried in the blood, which can make people's existing heart conditions worse.
Nitrogen oxides (NO and NO_2)	NO reacts quickly with oxygen in the air to form a secondary pollutant, NO_2. This happens within a few metres of the vehicle's exhaust pipe. NO_2 can cause breathing problems and make asthma worse.
Sulfur dioxide (SO_2)	SO_2 (and NO_2) reacts with water and oxygen to produce acid rain. This can damage buildings, and harm trees and plants.
Water (H_2O)	H_2O is harmless and not classed as a pollutant.

Questions

1 Why is low-sulfur diesel better for air quality?

2 Carbon dioxide is produced when fuels burn. It is used during photosynthesis and is essential for plant growth. Human activities are causing an increase in atmospheric carbon dioxide levels, which could lead to climate change.
   Do you think carbon dioxide is an atmospheric pollutant? Give reasons for your answer.

3 Why do you think NO_2 is called a secondary pollutant?

E: How are scientists working to reduce air pollution?

Find out about

- how air pollution is measured
- how data is checked and used
- what we can all do to reduce air pollution
- how scientists are designing methods to reduce the emission of air pollutants by cars and power stations

Key words

➤ true value
➤ accuracy
➤ outlier
➤ mean value
➤ range

Local authorities have a responsibility to monitor the quality of air in their streets.

Measuring air pollution

Some people suffer from asthma or hay fever. They may be able to feel when the air quality is poor. But most people do not know whether the air quality is good or bad.

There are automatic instruments situated around the country that collect air samples, and measure and record the concentration of a range of pollutants. The media summarise the data in reports that may give the day's overall air quality on a number scale or describe it as low, medium, or high quality.

🟠 molecules of pollutant

🔵 other molecules in air

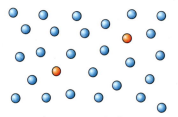

lower concentration

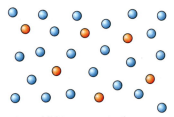

higher concentration

A low concentration of pollutants. There are very few pollutant molecules in a certain volume of air. This is an indication of good air quality.

A high concentration of pollutants. There is a large number of pollutant molecules in a certain volume of air. This shows that the air quality is poor.

Concentration is the amount of pollutant in a certain volume of air. It is important to note that the air molecules are normally much more spread out than shown in the diagrams.

Making measurements

If you measure the concentration of nitrogen dioxide in a sample of air several times, you will probably get different results. This may be because:

- you used the equipment differently
- there were differences in the equipment itself.

If you take just one reading, you cannot be sure it is accurate. So it is better to take several measurements. Then you can use them to estimate the true value.

The **true value** is what the measurement should really be. The **accuracy** of a result is how close it is to the true value.

How can you make sure your data is accurate?

The Worked example below shows what you should do to get a measurement of the level of nitrogen dioxide that is as accurate as possible.

Worked example: Handling a data set to find the best estimate

A student makes 10 measurements of the concentration of NO_2 in a sample of air. The values recorded, in parts per billion (ppb), were 18.8, 19.1, 18.9, 19.4, 19.0, 19.2, 19.1, 19.0, 18.3 and 19.3.

What is the student's best estimate of the true value?

Step 1: The measurements (10 in this case) are called the data set. To look for any patterns in the results, plot the values on a number line.

This shows that the 18.3 ppb measurement is very different from the others. A result that is very different from the others is called an **outlier**.

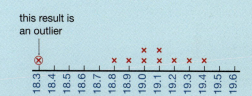

this result is an outlier

The value 18.3 ppb looks very different from the others, which are clustered together. Perhaps a mistake was made, as these were all from the same sample.

Step 2: Decide whether to include or ignore the outlier. If you can think of a reason why this result is so different (e.g., you made a mistake when you took this measurement), you should ignore it and explain why.

Step 3: Calculate the mean of the other nine results.

Add the nine values together and divide by the number of values:

$$\frac{(18.8 + 19.1 + 18.9 + 19.4 + 19.0 + 19.2 + 19.1 + 19.0 + 19.3)\ ppb}{9} = \frac{171.8\ ppb}{9} = 19.1\ ppb$$

19.1 ppb is called the **mean value** of the nine measurements. The mean value is used as the best estimate of the true value.

The best estimate for the concentration of NO_2 is 19.1 ppb

The mean value is 19.1 ppb. This is the best estimate of the concentration of nitrogen dioxide in the sample of air. You cannot be absolutely sure that it is the true value, but you can be sure that:

- the true value is within the **range** 18.8–19.4 ppb
- the best estimate of the true value is 19.1 ppb.

If you had taken only one measurement, you would not have been sure it was accurate. If the range had been narrower, say 19.0–19.3 ppb, you would have been even more confident about your best estimate of the true value.

Synoptic link

You can learn more about working with experimental data in *C8Q Uncertainty in measurements*.

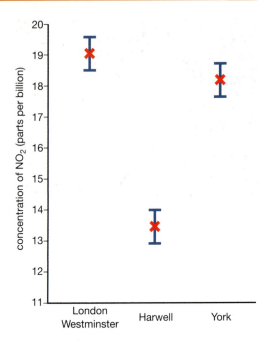

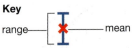

Key

range — mean

MOT exhaust emission analysis. Cars with emissions above the legal limit of carbon monoxide and particulates are not allowed on the road.

Comparing nitrogen dioxide concentrations

The graph on the left shows the mean and range for the concentration of NO_2 in three different places.

Compare London and York. The means are different but the range for London overlaps the range for York. So the true value for London could be the same as the true value for York.

You cannot be confident that London and York's NO_2 concentrations are different.

Compare London and Harwell. The means are different and the ranges do not overlap.

You can be very confident that there is a real difference between the NO_2 concentrations in London and Harwell.

When you compare data, do not just look at the means. To make sure that there is a real difference, check that the ranges do not overlap.

Reducing pollution from cars

The simplest way for you and your family to reduce the emission of air pollutants is to use your car less. Walking and cycling produce no extra air pollutants. Travelling by bus or train produces less air pollution per person than if each person travelled in their own car.

Engineers are continually working on improving the **efficiency** of engines. A more efficient car engine means that a car will burn less fuel to travel the same distance, therefore reducing pollutants. This is good for car owners too because they do not need to buy as much fuel.

Scientists have developed the **catalytic converter**, which changes harmful pollutants into less harmful substances. All new cars have catalytic converters fitted to their exhaust systems.

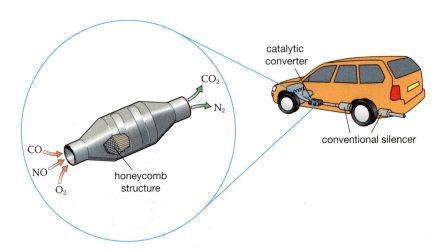

The pollutants carbon monoxide (CO) and nitrogen monoxide (NO) are changed into the less harmful gases of nitrogen and carbon dioxide.

The metal honeycomb structure has a large surface area, which speeds up the chemical reactions.

Key words

➤ efficiency
➤ catalytic converter
➤ gas scrubbing

While carbon dioxide is less harmful than carbon monoxide, levels are still a concern due to the link with climate change. The only way of producing less carbon dioxide is to burn less fossil fuel.

Low-sulfur fuels, such as diesel, reduce the amount of sulfur dioxide in exhaust emissions. This helps reduce acid rain and the problems it causes.

Some vehicles are now powered using electricity stored in batteries. Using electric vehicles in congested cities means the pollutants are not released in the city but at the power station where the electricity is generated (if it burns fossil fuels).

Reducing pollutants from power stations

Individuals, governments, and industries all have a role to play in reducing pollution from power stations. You and your family can help reduce emissions by using less electricity.

Scientists are constantly looking for methods of reducing pollutant levels, either by removing impurities from the fuel before it is burnt or by removing the pollutants from the waste gases before they can reach the atmosphere.

Sulfur dioxide leads to acid rain. Natural gas and fuel oil contain sulfur but can be refined to remove it before they are burnt in a power station, reducing sulfur dioxide emissions.

Sulfur dioxide can also be removed from waste gases (or flue gases) before they escape from the power station chimney using a process called **gas scrubbing**.

Sulfur dioxide is acidic. Acids react with alkaline chemicals. Gas scrubbing is a process that uses an alkali to react with sulfur dioxide and remove it from flue gases.

Charging points for electric cars are now more common.

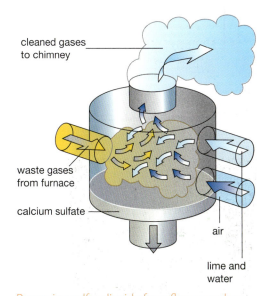

Removing sulfur dioxide from flue gases by scrubbing with an alkaline slurry.

Trees killed by acid rain in the Czech Republic. Sulfur dioxide is a waste gas often produced by power stations. It reacts with water and oxygen to form acid rain.

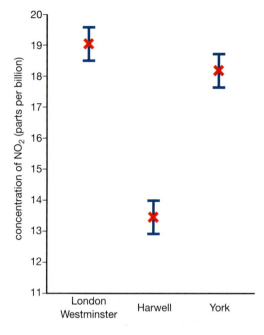

concentration of NO_2 (parts per billion)

London Westminster   Harwell   York

Key

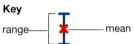

range —⊦×⊦— mean

1 Jess measured the NO_2 concentration in the middle of a town. She took six readings: 22 ppb, 20 ppb, 18 ppb, 24 ppb, 21 ppb, and 23 ppb. Jess used new equipment and was careful taking her measurements.

   a Calculate the mean value of the measurements.

   b Write down the best estimate and the range for the NO_2 concentration in this sample of air.

2 Look at the graph on the left. Does it show that there is a real difference in NO_2 levels between Harwell and York? Explain your answer.

3 Repeat measurements on an air sample produced these results for the NO_2 concentration:

   reading 1 – 39.4 ppb   reading 2 – 45.8 ppb
   reading 3 – 42.3 ppb   reading 4 – 38.7 ppb
   reading 5 – 39.7 ppb   reading 6 – 32.7 ppb

   There had been some problems with the equipment that day.

   a Plot these six readings on a number line.

   b Work out the mean NO_2 concentration and range for this sample.

   c Another sample was taken from a second place in the same town. The mean NO_2 concentration for this sample was found to be 44.1 ppb. Can you say with confidence that the second location had a higher NO_2 concentration than the first? Explain your answer.

4 A scientist took one measurement of NO_2 in an air sample. Explain why you would not have much confidence in the accuracy of the result.

5 All new cars are fitted with a catalytic converter.

   a Which two pollutants are removed by catalytic converters?

   b For each pollutant, state which less harmful gas it is changed into.

   c Why is this not a perfect solution?

6 Hybrid cars use electricity but also have a normal fuel engine for when it is needed.

   a Explain why using a hybrid car could reduce traffic pollution in a town.

   b Explain why using a hybrid car in electric mode can still result in pollutants reaching the atmosphere.

F: Representations of chemical reactions

Describing reactions

You can use pictures to describe the chemical change that happens when charcoal (carbon) burns.

The chemicals before the arrow are the ones that react together. Chemists call them **reactants**.

The chemicals after the arrow are the new chemicals that are made. Chemists call them **products**.

The combustion of charcoal can be summarised in this **word equation**:

carbon + oxygen $\longrightarrow$ carbon dioxide

If you want more detail you can write a **chemical equation**. In the chemical equation for the combustion of charcoal, the names of substances are replaced by their symbol or **chemical formula**.

$$C + O_2 \longrightarrow CO_2$$

The chemical equation is a more useful description of the reaction than the word equation. It tells you how many atoms and molecules are involved and what happens to each atom.

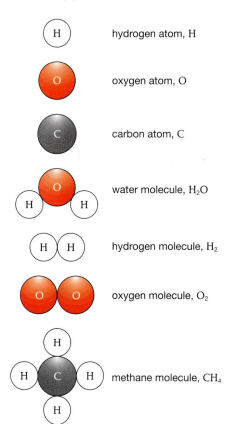

hydrogen atom, H

oxygen atom, O

carbon atom, C

water molecule, H_2O

hydrogen molecule, H_2

oxygen molecule, O_2

methane molecule, CH_4

Find out about

- different ways of representing chemical reactions
- what happens to atoms during chemical reactions
- how to balance a chemical equation

Key words

➤ reactants
➤ products
➤ word equation
➤ chemical equation
➤ chemical formula

When you have had a bonfire, some of the atoms that made up the twigs and leaves are in the ashes left on the ground. The others are in the products released into the air.

Key words
➤ conservation of atoms
➤ conservation of mass
➤ balanced equation

Conservation of atoms

All the atoms present at the beginning of a chemical reaction are still there at the end. No atoms are destroyed and no new atoms are formed. The atoms are conserved. They rearrange to form new chemicals but they are still there. This is called **conservation of atoms**.

For example, when a car engine burns fuel, the atoms in the petrol or diesel are not destroyed. They rearrange to form the new chemicals found in the exhaust gases.

When hydrogen reacts with oxygen to form water, two molecules of hydrogen react with one molecule of oxygen. This produces two molecules of water. We can represent this change by:

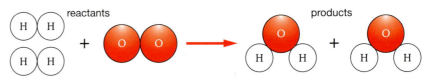

Notice that there are the same numbers of each kind of atom on each side of the equation. All the atoms that are in the reactants end up in the products. The atoms are conserved.

Conservation of mass

All atoms have mass. Because the atoms are conserved, the mass of the reactants is the same as the mass of the products. This is called **conservation of mass**.

Understanding chemical equations

When you write a chemical equation, it is important to remember that the number of each type of atom must be conserved.

The reactants shown in the equation must have the same number of atoms as the products. The equation must be **balanced**.

Before attempting to balance an equation, it is important to understand what the numbers in an equation mean.

The large numbers in front of each formula tell you the ratio of one type of substance to another. In the equation above, the ratio of oxygen molecules to water molecules is 1:2. There are twice as many water molecules as there are oxygen molecules. You will need to change these large numbers to make an equation balanced.

The small, subscript, numbers tell you the number of atoms of each element that are in each substance. You must not change the small numbers otherwise you will change the substance.

H_2O is water that we drink.

H_2O_2 is hydrogen peroxide, which is a totally different substance from H_2O. It can make hair turn blond, but it is not healthy for the hair.

Worked example: Balancing equations 1

Write a balanced chemical equation to show the reaction that takes place when methane burns in a power station.

Step 1: Write a word equation for the reaction.

The names of the reactants and products may be in the question or you may need to use the chemistry that you have learnt.

methane + oxygen $\longrightarrow$ carbon dioxide + water

Step 2: Write the basic chemical equation (unbalanced) by replacing the names of the substances with their symbol or formula.

$CH_4 + O_2 \longrightarrow CO_2 + H_2O$

Step 3: Check whether the equation is balanced.

To do this you need to count each type of atom. If it helps, you could draw pictures of the molecules.

To be balanced there must be the same number of atoms on each side of the equation.

Reactants	Products
$C \times 1$	$C \times 1$
$H \times 4$ (in CH_4)	$H \times 2$ (in H_2O)
$O \times 2$ (in O_2)	$O \times 3$ (in CO_2 and H_2O)

Step 4: Change the number of one of the molecules to make one type of atom balance.

This step tends to involve trial and error. Choose an atom that is easy to make balance to start with.

There are four hydrogens in the reactants but only two in the products so try doubling the number of H_2O molecules in the products.

$CH_4 + O_2 \longrightarrow CO_2 + 2H_2O$

Step 5: Repeat steps 3 and 4 until the equation is balanced.

Reactants	Products
$C \times 1$	$C \times 1$
$H \times 4$	$H \times 4$ (from $2 \times H_2O$)
$O \times 2$	$O \times 4$ (from $2 \times H_2O$)

Now there are not enough oxygen atoms in the reactants so try changing another number.

$CH_4 + 2O_2 \longrightarrow CO_2 + 2H_2O$

Reactants	Products
$C \times 1$	$C \times 1$
$H \times 4$	$H \times 4$
$O \times 4$ (from $2 \times O_2$)	$O \times 4$

The equation is balanced.

Answer:

$CH_4 + 2O_2 \longrightarrow CO_2 + 2H_2O$

Worked example: Balancing equations 2

Write a balanced chemical equation to show the reaction that takes place when nitrogen oxide (NO) is oxidised further to nitrogen dioxide (NO_2).

Step 1: Write a word equation for the reaction.

nitrogen oxide + oxygen $\longrightarrow$ nitrogen dioxide

Step 2: Write the basic chemical equation (unbalanced).

$NO + O_2 \longrightarrow NO_2$

Step 3: Check whether the equation is balanced.

Reactants	Products
N × 1	N × 1
O × 3	O × 2

Step 4: Change the number of one of the molecules to make one type of atom balance.

Hint: Sometimes this is not possible the first time, especially if the number of atoms on the left is not a multiple of those on the right (e.g., three and two). In this case, try to make the number of oxygen atoms on the left a multiple of the number on the right (e.g., four and two).

Step 5: Repeat steps 3 and 4 until the equation is balanced.

$2NO + O_2 \longrightarrow NO_2$

Reactants	Products
N × 2 (from 2 × NO)	N × 1
O × 4 (from 2 × NO and O_2)	O × 2

There are now twice as many atoms in the reactants as the products so the number of NO_2 molecules needs doubling.

$2NO + O_2 \longrightarrow 2NO_2$

Reactants	Products
N × 2	N × 2 (from 2 × NO_2)
O × 4	O × 4 (from 2 × NO_2)

Answer:

$$2NO + O_2 \longrightarrow 2NO_2$$

Questions

1 Inside a power station, sulfur impurities in the coal can react with oxygen from the air to form sulfur dioxide.
   a Write a word equation for this reaction.
   b Write the basic chemical equation for this reaction and show that it is balanced.

2 A catalytic converter works by ensuring that nitrogen monoxide breaks down into nitrogen and oxygen. In addition, carbon monoxide reacts with oxygen to form the less harmful carbon dioxide.
   a Use Worked example 1 to write a balanced chemical equation for the nitrogen monoxide (**NO**) reaction in a catalytic converter.
   b Write a balanced chemical equation for the carbon monoxide reaction.

A: Is energy always released during a chemical reaction?

Exothermic and endothermic reactions

When the gas from a cooking hob burns, the energy released heats the food for you to eat.

Reactions that transfer energy to the surroundings, making the surroundings get a little warmer, are called **exothermic**. There are also reactions that absorb energy from the surroundings, making the surroundings get a little cooler. These are called **endothermic** reactions.

It is possible to tell whether a chemical reaction is exothermic or endothermic by measuring the temperature of the reactants before the reaction starts and the temperature of the products after it has finished.

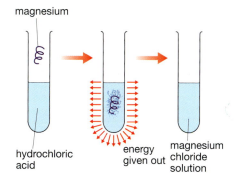

magnesium

hydrochloric acid | energy given out | magnesium chloride solution

When magnesium reacts with hydrochloric acid, energy is given out. The temperature of the products is higher than the temperature of the reactants. The reaction is exothermic.

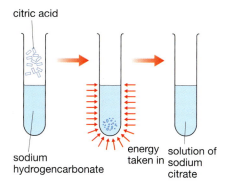

citric acid

sodium hydrogencarbonate | energy taken in | solution of sodium citrate

When citric acid reacts with sodium hydrogencarbonate, the temperature of the products is less than the temperature of the reactants. The reaction is endothermic.

Find out about

- energy changes during reactions
- exothermic and endothermic reactions
- why a fuel needs a spark or flame to make it burn

Key words

➤ exothermic
➤ endothermic

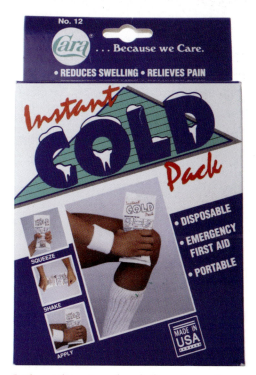

Both exothermic and endothermic reactions have practical uses. A cold pack absorbs energy from an injured muscle, making the muscle a little cooler. The chemical reaction in the cold pack is an endothermic reaction.

Energy-level diagrams

You can keep track of changes in energy during a chemical reaction using an **energy-level diagram**.

An energy-level diagram is constructed like a graph, with energy plotted on the vertical axis and the progress of the reaction along the horizontal axis. The energy of the reactants is therefore plotted on the left and the energy of the products on the right. These are usually labelled using the left-hand-side and right-hand-side of the balanced equation for the reaction.

Energy change

The energy-level diagram for an exothermic reaction is shown below. Energy is given out in this reaction so the products have less energy than the reactants. The energy change of an exothermic reaction is negative. This is shown by a downwards arrow.

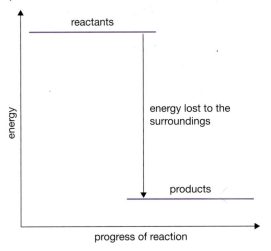

Energy-level diagram for an exothermic reaction.

Energy is taken in during an endothermic reaction so the products have more energy than the reactants. The energy change is positive and shown on the energy-level diagram below by an upwards arrow.

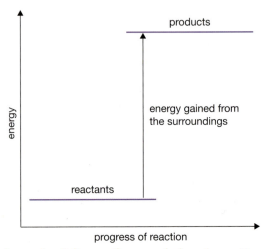

Energy-level diagram for an endothermic reaction.

Activation energy

You may have seen a demonstration where a balloon containing hydrogen and oxygen is lit, making it explode. But why does it not explode before it is lit?

In order for the hydrogen and oxygen to react, the molecules must collide and break apart. It takes energy for molecules to break apart. For every reaction there is a certain minimum energy needed before the reaction can happen. This is called the **activation energy.**

The collisions between molecules have a range of energies. Head-on collisions between fast-moving molecules are the most energetic. If the colliding molecules have enough energy then the collision is successful and a reaction occurs between those molecules.

If the activation energy for a reaction is low, a lot of the collisions will have enough energy to react and the reaction will happen quickly. If the activation energy is high then far fewer collisions will have enough energy to react and the reaction will be very slow.

Heating the reactants gives the molecules more energy, meaning that a greater number of molecules have enough energy to react. The reaction happens more quickly.

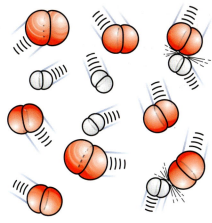

A mixture of hydrogen and oxygen molecules. The molecules that are colliding may react to form new molecules, but only if they have enough energy to start breaking bonds.

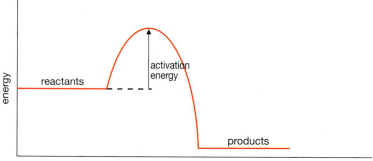

The reaction profile for an exothermic reaction showing the activation energy needed to get the reaction going.

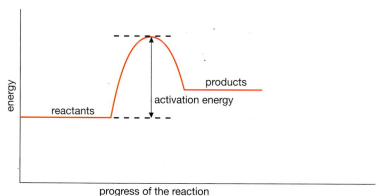

The reaction profile for an endothermic reaction shows that the products have more energy than the reactants.

Catalysts work by lowering the activation energy of a reaction. This means that a greater proportion of molecules can collide successfully and the reaction can happen more quickly.

Questions

1 If you have ever had a plaster cast for a broken bone you may recall the feeling of warmth as the plaster set. If you put a sherbet sweet on your tongue you will feel your mouth getting cold. Both of these are chemical reactions. Which reaction is exothermic and which is endothermic? Explain your answer.

2 The standard test for hydrogen is to hold a lit splint near the test tube and listen for a 'squeaky pop'. Explain why there is no 'squeaky pop' without the lit splint. Use a diagram in your explanation.

3 A catalytic converter speeds up the reactions during which molecules from waste gases react to form less harmful molecules. Use a diagram to help explain how a catalyst can speed up a reaction.

B: How much energy is released when a fuel burns?

Find out about

- energy calculations

Key words

➤ chemical bond
➤ bond strength
➤ relative formula mass

It takes energy to stretch and break an elastic band. Stretching the band stretches the bonds between the atoms in the rubber. Letting go before the band breaks releases energy – perhaps making it fly across the room.

Synoptic link

You can learn more about relative formula mass in C5.3 *How are the amounts of substances in reactions calculated?*

Breaking and making molecules

To calculate how much energy is released when hydrogen burns, it is first necessary to think about how the molecules of the reactants change into the molecules of the product.

The forces that hold atoms together in molecules are called **chemical bonds**. You can think of chemical bonds as tiny springs. In order to get hydrogen to react with oxygen, the tiny springs joining the atoms in the molecules have to be stretched and broken. This takes energy.

When hydrogen reacts with oxygen, the product is water. Water is created as new bonds form between oxygen atoms and hydrogen atoms. Forming bonds releases energy just like relaxing a rubber band.

If more energy is needed to break the bonds than is gained by making new bonds then the reaction is endothermic. However, if more energy is released when the new bonds form than was needed to break the bonds of the reactants then energy is released to the surroundings and the reaction is exothermic.

Hydrogen burning is an exothermic reaction. The energy given out keeps the reactants hot enough for the reaction to continue.

The diagram below shows the energy changes that take place during this process.

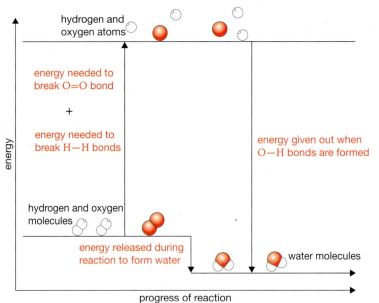

Two H–H bonds and one O=O bond break when hydrogen reacts with oxygen. The atoms recombine to make water as four new O–H bonds form.

Energy-change calculations

The overall energy change that takes place during a chemical reaction can be calculated if the strengths of all the chemical bonds in the reactants and the products are known. The unit for measuring **bond strength** is the kilojoule (kJ). Bond strengths are given for a set quantity of bonds equal to the number of bonds in the **relative formula mass** of that substance.

H

Worked example: Calculating the energy change in a reaction

Calculate the energy change that takes place during the formation of steam from hydrogen and oxygen, and state whether the reaction is exothermic or endothermic.

Step 1: Write a balanced chemical equation for the reaction.

$2H_2 (g) + O_2 (g) \rightarrow 2H_2O (g)$

Step 2: Look up the bond strength for each bond that breaks and each bond that forms.

Bond	Energy change for the formula masses (kJ)
H–H	434
O=O	498
O–H	464

Step 3: Work out which bonds must break, and how many of each type.

2 × H–H must break plus 1 × O=O

Step 4: Calculate the total energy needed to break these bonds.

(2 × 434 kJ) + 498 kJ = 1366 kJ

Step 5: Repeat steps 3 and 4 for the bonds formed.

4 × O–H bonds must form (2 in each H_2O molecule)

4 × 464 kJ = 1856 kJ

Step 6: Calculate the overall energy change.

Energy needed to break bonds – energy given out as new bonds are formed

= 1366 kJ – 1856 kJ = –490 kJ

Step 7: Decide whether the reaction is exothermic or endothermic.

If the answer is negative then this shows that forming the bonds in the products released more energy than was needed to break the bonds in the reactants. Overall, energy was released to the surroundings so the reaction is exothermic.

If the answer is positive then forming the bonds in the products released less energy than was needed to break the bonds in the reactants. Overall, energy was absorbed from the surroundings and so the reaction must be endothermic.

490 kJ were released during the reaction – it was exothermic

Questions

1 Hydrogen reacts with chlorine to form hydrogen chloride.

$H_2 (g) + Cl_2 (g) \rightarrow 2HCl (g)$

  a Which bonds are broken during the reaction?
  b Which bonds are formed during the reaction?

H **c** Use the equation and the data below to calculate the overall energy change for the reaction.

Bond	Energy change for the formula mass (kJ)
H–H	434
Cl–Cl	242
H–Cl	431

  d Is this reaction exothermic or endothermic?
  e Draw an energy-level diagram for the reaction.

C: How can you use hydrogen as an energy supply for transport?

Find out about

- how hydrogen can be used to make a fuel cell
- the advantages and disadvantages of using fuel cells in cars

Key words

➤ electrode
➤ potential difference
➤ fuel cell

When hydrogen is mixed with oxygen at room temperature, the two gases will not react. It takes a hot flame or a spark to heat up the mixture enough for the reaction to start.

Hydrogen can be used to fuel rockets but is too dangerous for day-to-day use in transport.

The Hindenberg airship was filled with hydrogen gas. Sadly it burst into flames shortly before landing.

A simple electric cell

A simple electric cell is formed when two different types of metal **electrode** are placed in a solution. Even a potato can be used as a cell. A **potential difference** is created between the two electrodes.

If you use a device powered by a standard type of cell, you will know that after a while the cell runs out. This happens when the chemicals inside the cell have all been changed into new chemicals.

Many devices, such as phones and cameras, use rechargeable cells.

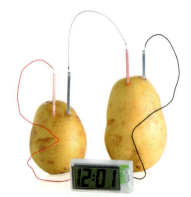

A potential difference is created between the two electrodes, made of different metals, when they are inserted into the potato.

Hydrogen fuel cells

In a **fuel cell** a chemical reaction produces a potential difference, just like in a simple electric cell. The difference is that the chemicals are continuously topped up so that they don't run out. All fuel cells use hydrogen as the energy source. Some fuel cell systems use hydrogen extracted from hydrocarbons, others systems use pure hydrogen. Scientists have developed fuel cells that use pure hydrogen together with oxygen extracted from the air.

When the potential difference is created, hydrogen and oxygen in the cell are converted into water so this is the only waste product to leave the car. A single cell does not produce enough electricity to drive the motor in a car. The cells must be stacked to provide sufficient electricity.

Advantages of hydrogen fuel cell cars

In the short term, the major advantage of using hydrogen fuel cells is that they contribute towards a reduction in carbon dioxide emissions and also reduce air pollution in the area where the car is being used.

Some methods of producing hydrogen will release carbon dioxide and pollutants into the atmosphere elsewhere, but less than the quantities that would have been produced by petrol or diesel cars.

In the long term, hydrogen-fuelled cars could help to reduce our reliance on fossil fuels by using methods of hydrogen production that do not require coal, oil, or gas.

The chassis of a hydrogen fuel cell car at a motor exhibition.

Disadvantages of hydrogen fuel cell cars

There are some disadvantages to hydrogen-fuelled cars that need to be overcome before they become a common sight on our roads:

- Hydrogen storage may not be sufficient for as long a distance as a conventional car fuel tank.

- The fuel cells and motors are not yet as durable (long-lasting) as petrol or diesel engines.

- The technology is still very expensive.

- There is not yet a countrywide network of places to fill up with hydrogen.

- Customers have safety concerns about the use of hydrogen fuel.

Questions

1 Use a chemical equation to explain why hydrogen-fuelled cars can help to improve air quality in a city.

2 The production of hydrogen may require the use of electricity. Explain why the use of hydrogen-fuelled cars would not cut carbon dioxide emissions completely.

3 In 10 years time would you consider buying a hydrogen-fuelled car? Explain your answer. If your answer was 'no', what would have to happen to change your mind?

A hydrogen fuel pump.

A: How do scientists find out whether our climate is changing?

Find out about

- the measurement of the Earth's climate
- uncertainty in measurements

Key words

➤ weather
➤ climate
➤ uncertainty

Climate

The **weather** changes every day or even every hour. You may walk to school in the rain and spend lunch hour in the sun. Weather is localised. It could be rainy in one part of a city and dry in another. **Climate** describes the average conditions over a long time period and a wide area.

Think of some questions about the Earth's climate, for example, 'Will summers in the UK be hotter in 20 years' time?' and 'Is it true that there will be more flooding in future?'.

An understanding of data can help you to find out whether scientists can, or cannot, answer your questions and how certain they can be about their answers.

Monitoring global climate

Monitoring changes in global climate requires the collection of data from tens of thousands of monitoring stations located around the world, both on land and at sea.

Scientists compare the actual temperature recorded at the monitoring station with the long-term average for that region. A positive difference shows that a region is warmer than the long-term average. Scientists then analyse these difference measurements from all across the world. If overall the differences are getting increasingly positive, this shows that the global climate is warming.

Many people think that Manchester is the wettest city in the UK, but data collected by the Meteorological Office shows that Cardiff is the wettest city. This data is more reliable than any individual's experience.

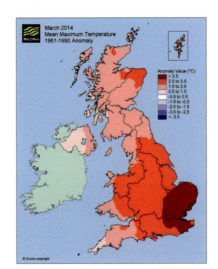
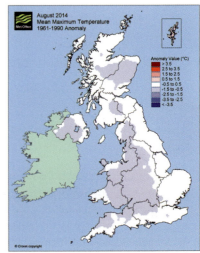

These maps show how the temperatures in the UK in March 2014 and August 2014 were different from the long-term average. A single set of data does not show whether temperatures are increasing over time. For this, data must be collected of a long period so that a trend can be identified. (Contains public sector information licensed under the Open Government Licence v1.0.)

Uncertainty

As you will have discovered when collecting your own data, there are often errors in the measurements made. For this reason, scientists present their data showing the **uncertainty** in the results. This is shown by the grey area on the graph below. The best estimate is shown as a black line.

Uncertainties have decreased over time as more monitoring stations have been built and the equipment in the stations has improved.

It can be clearly seen that even with some level of uncertainty, there is a real difference. The trend shows an increase in global temperatures.

Importantly, data collected by two other organisations that monitor global temperatures also show this increase. This boosts confidence in the conclusion.

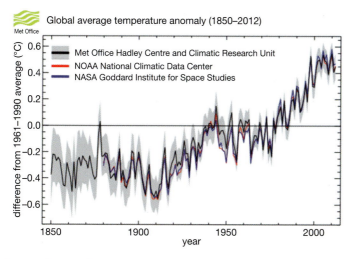

Looking at the graph you can see that the global average surface temperature data from all three organisations agrees. (Contains public sector information licensed under the Open Government Licence v1.0.)

Interpreting new data

At the start of this century the rate of increase in surface temperature seemed to be slowing. This information was initially interpreted in different ways by different people. These were some interpretations:

- Climate change was not happening.
- The predicted rate of warming was wrong.
- It was to be expected as similar pauses have been observed in the past.

Over the same time period, deep ocean temperatures and changes in Arctic ice show an unchanged trend in warming. Scientists are investigating why the trend in this data is unchanged and yet average surface temperature warming has slowed.

Scientists do not necessarily reject an accepted explanation until new data has been fully analysed and explained. During this period there may be a number of views held about the reasons for the unexpected results. The eventual explanation may identify an error in the results or method, or it could reveal previously unknown scientific mechanisms.

Questions

1 Scientists have produced updated datasets of global temperature measurements. For a few monitoring stations they noticed that some readings were very different from the others. *outlier*

  a What name is given to a reading that is very different from other measurements (see C1.1E)?

  b Explain how scientists would take account of these differences.

  c Describe how the uncertainty in the Met Office results has changed since 1850. Suggest a reason for this and explain why it is important.

2 One year the summer is much cooler than normal. Why does this not show that climate change is not happening?

B: What evidence are scientists using to establish whether human activities are causing climate change?

Find out about

- the greenhouse effect
- the contribution of human activity to climate change

Key words

➤ climate change
➤ greenhouse effect
➤ greenhouse gas
➤ correlation
➤ causal link

What controls the Earth's climate?

The Earth absorbs energy from the Sun and radiates energy back into space as it cools. If the Earth radiates less energy than it absorbs then it will become gradually warmer.

Climate change could be caused by an increase in energy from the Sun and/or a decrease in the amount of energy radiated by the Earth.

Scientific research has shown that an increase in solar radiation is not the cause for the current climate warming trend although it may have been responsible for separate warmer periods in the past.

Climate change must therefore be caused by a decrease in the amount of energy radiated by Earth as it cools.

Greenhouse effect

The amount of energy that the Earth radiates back into space is reduced by the atmosphere.

The Earth's atmosphere has always contained a small quantity of carbon dioxide. Without it the Earth's surface temperature would be much colder and more similar to that on the Moon.

The diagram below shows how radiation from the Sun passes through the atmosphere, warming the Earth's surface.

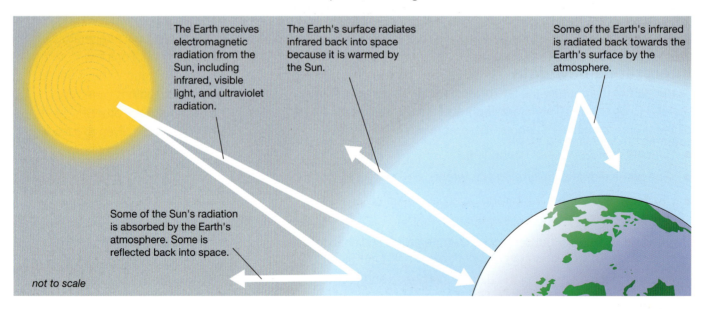

The Earth receives electromagnetic radiation from the Sun, including infrared, visible light, and ultraviolet radiation.

The Earth's surface radiates infrared back into space because it is warmed by the Sun.

Some of the Earth's infrared is radiated back towards the Earth's surface by the atmosphere.

Some of the Sun's radiation is absorbed by the Earth's atmosphere. Some is reflected back into space.

not to scale

The atmosphere lets in infrared radiation from the Sun, but prevents the infrared radiation emitted by the Earth from escaping. This is because the Earth's radiation has lower frequencies than the Sun's, and these frequencies are absorbed by the atmosphere.

Less energy is radiated than is absorbed, which results in a warming effect known as the **greenhouse effect**. The main gases that absorb energy radiating from the Earth are carbon dioxide, water vapour, and methane. These gases are known as **greenhouse gases**. Without the greenhouse effect, all the water on Earth would be frozen. The planet would have a climate unsuitable for life.

Human activity

Since the Industrial Revolution at the end of the 19th century, humans have burnt increasing quantities of fossil fuels.

The data in the two graphs show that at the same time as carbon dioxide levels have increased, the Earth's temperature has also increased. This shows a **correlation**, but did the rising carbon dioxide levels cause the increase in temperature?

The greenhouse effect provides an explanation for how carbon dioxide can increase the Earth's temperature, so there is a **causal link**.

Was this temperature increase caused by the human addition of carbon dioxide to the atmosphere or could it have resulted from other factors, such as natural causes?

All living things respire and release carbon dioxide into the atmosphere. However, for millions of years the amount of carbon dioxide released into the atmosphere was equal to the amount used in photosynthesis and dissolved in the oceans.

Extra carbon dioxide can be added to the atmosphere by erupting volcanoes, but scientists have found that this would not have produced enough to explain the size of the increase in atmospheric carbon dioxide observed.

Scientists have measured oxygen levels in the atmosphere and have found that levels have decreased. This adds further evidence that combustion of fossil fuels by humans is affecting the composition of the atmosphere.

As a result, scientists from around the world have concluded that the observed warming of the Earth, or climate change, is almost certainly caused by the activities of humans.

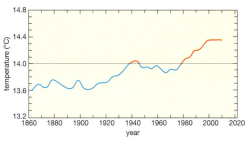

This graph of data from weather stations shows that the Earth's surface temperature has risen over the past 150 years.

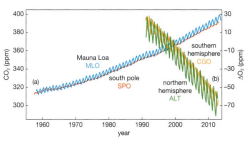

This graph shows that while carbon dioxide levels are increasing in the atmosphere, oxygen levels are decreasing.

Questions

1   'When sales of ice cream increase, so does the number of people experiencing symptoms of hay fever.'
    Explain why this statement shows a correlation but not a causal link.

2   Since the Industrial Revolution, humans have burnt increasing quantities of fossil fuels.
    a Describe the correlation between this statement and the graph to show CO_2 levels over time.
    b Use the science you have learnt in this chapter to explain the causal link between the two measurements.

C: What are the potential impacts of climate change and what can we do about them?

Find out about

- how scientists use computers to model climate change
- the impacts of climate change
- ways to mitigate carbon dioxide and methane emissions

Key words

➤ climate model
➤ mitigate
➤ carbon sink

Climate modelling

Climate scientists use scientific theories and climate data to create highly complex **climate models** using super computers. They can use these to investigate how the different parts of our climate system interact and how changing part of the system affects the rest.

Scientists use historical data to test their models. Does the model predict temperature changes that match the observations? If so, the model increases confidence in the link between the production of carbon dioxide by human activities and climate change. If not, then scientists need to programme other factors into their model.

Comparing climate models

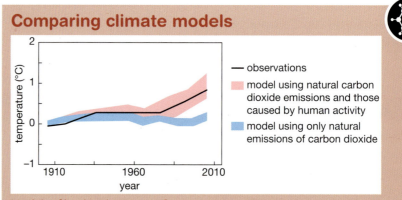

Models of land and ocean surface temperature over time.

An investigation used historical data (black) to compare two models. One model includes data about carbon dioxide emissions from natural causes only (blue). The second model includes carbon dioxide emissions from both human activity and natural sources (pink). The output of the second model better matches the observations.

Predicting the impacts of climate change

Climate models are used to make predictions about the impacts of a particular increase in global temperature.

As more data becomes available and more is learnt about how the Earth's climate system operates, the better the models and predictions become.

Current climate models predict several changes that will result from increases in global temperature. Climate systems are interlinked so one impact, such as the melting of sea ice, can contribute to another impact, such as rises in sea levels.

It is important to understand that there is always a lot of natural variability in weather from year to year. The predictions relate to patterns in the weather and not individual weather events.

Ice

The Arctic ice sheet is predicted to continue to shrink and thin. Snow coverage in the Northern Hemisphere is predicted to decrease and glacier melting increase.

Ocean temperatures

Models show that ocean warming will continue, including warming at greater depths. An alteration in deep-ocean temperatures could change the behaviour of existing ocean currents, including those that currently keep the British Isles and Northern Europe warm. This could lead to a drop in temperatures in the UK.

Sea levels

A combination of ocean warming and melting ice is predicted to raise sea levels, resulting in the flooding of low-lying land areas.

Extreme weather patterns

Greater differences in weather between places is predicted. Climate models predict more extreme temperatures, an increase in length and intensity of droughts, and more intense rainfall. However, current models are not detailed enough to make predictions about some complex systems. For example, it is not yet possible to predict how tropical cyclones will be affected.

Impact on crops

The impacts of climate change will significantly affect where crops can be grown. Models predict that some farmland will be flooded as sea levels rise. There is an increased risk of storm surges and other land could be contaminated by salt water. In areas where rainfall decreases, droughts will make it difficult to grow some crops. In contrast, where rainfall becomes extreme, valuable crops may be washed away due to flooding.

What can we do to reduce the risks of climate change?

Climate models show a range of impacts that have to be considered by governments when planning action. What actually happens will depend upon how the emissions of greenhouse gases are changed. Given the potential impacts predicted by climate models, scientists are working hard to devise ways to **mitigate** (lessen) the impacts of climate change.

Greenhouse gases build up and mix in the atmosphere so mitigation will only be effective if countries around the world collaborate to tackle the problem. Action must take place on a global scale if it is to be effective.

Mitigating the effect of carbon dioxide emissions

Carbon dioxide levels are increasing in the atmosphere because carbon dioxide is being added to the atmosphere faster than natural processes can remove it.

Processes that remove carbon dioxide from the atmosphere are known as **carbon sinks**. For instance, a large amount of carbon dioxide dissolves in our oceans.

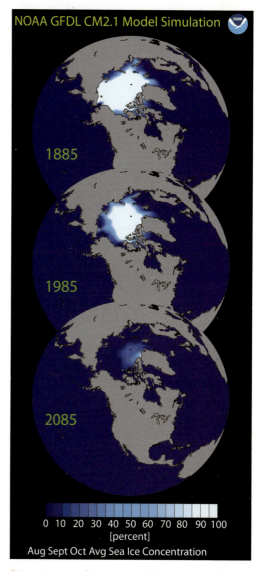

NOAA GFDL CM2.1 Model Simulation

1885

1985

2085

0 10 20 30 40 50 60 70 80 90 100
[percent]
Aug Sept Oct Avg Sea Ice Concentration

Scientists use climate models to predict the loss of sea ice in the Arctic.

The glacier was here in
Le glacier était ici en
1908

The Athabasca glacier in the Canadian Rockies has lost over 60% of its ice mass in less than 150 years.

The Amazon rainforest is a significant carbon sink for atmospheric carbon dioxide.

Congestion charging has been introduced in some city centres to attempt to encourage people to use public transport.

Carbon capture technology aims to capture carbon dioxide before it enters the atmosphere, and store it underground.

The vast areas of rainforest that cover the Earth are another important carbon sink. Carbon dioxide is removed from the atmosphere by plants that use it for photosynthesis. Preserving large areas of forest that currently exist and planting new trees is one way to mitigate carbon dioxide emissions.

Vehicles

Reducing emissions from vehicles powered by fossil fuel engines would not only improve air quality locally, it would also help to mitigate carbon dioxide emissions.

Governments can introduce regulations to either encourage or enforce reductions in emissions from vehicles:

- Increasing taxes on diesel and petrol may encourage people to use their cars less.
- Introducing legislation on the efficiency of engines in new cars could lower carbon dioxide emissions as cars would need to burn less fuel for every mile driven.
- Influencing the pricing of public transport to make it cheaper could encourage more people to use it.

Power stations

The simplest way to reduce carbon dioxide emissions from fossil fuel power stations is for people to use less electricity. Emissions of carbon dioxide could be reduced further by increasing the proportion of electricity generated by nuclear power stations and renewable technologies, such as solar and wind power, and tidal and hydroelectric generation.

Scientists and engineers are developing **carbon capture** systems to directly reduce emissions from fossil-fuel-burning power stations. These would reduce carbon dioxide emissions by removing carbon dioxide from the waste gases and storing it underground.

Making decisions

Decisions on strategies to mitigate emissions must be evaluated by consideration of the risks and benefits of any actions.

When a decision is made it can be controversial because many people may still object to the risks being taken, or to the effect the decision will have on their lives.

Nuclear power would reduce carbon dioxide emissions but many people are worried about the risks.

Tidal power stations are built in river estuaries. They have benefits in terms of reducing carbon dioxide emissions but some people have concerns about the risk to estuary wildlife.

Mitigating methane emissions

One molecule of methane has a greater greenhouse effect than one molecule of carbon dioxide so it is very important that emissions of methane are reduced.

In the UK, methane is released into the atmosphere from landfill, farming, the oil and gas industry, and coal mining.

While action is needed on a global scale to reduce methane emissions, local community decisions can also make a contribution. To make these decisions people need to know how and why methane is formed and released.

Tidal power station in France.

Landfill gas power stations use the methane gas produced by decomposing matter.

Source	Why methane is produced	Mitigation methods
Landfill	Biodegradable waste in a landfill site that is not exposed to oxygen will produce methane gas as it decomposes.	Reduce the amount of biodegradable material going to landfill by using alternative disposal methods, such as composting or incineration. Capture the methane that is produced and either burn it off, use it as a fuel, or burn it to generate electricity.
Farming	Methane is emitted from the digestive systems of cattle and sheep. Methane is also emitted from manure that is being stored.	Emissions from animals could be reduced by cutting down on the number of animals (through improved productivity in terms of meat or milk output) or by altering their diet. Manure spread directly onto fields does not produce methane. Methane from manure could be captured and used as fuel for electricity generation. Manure could be added to an anaerobic digester, which is a technology that encourages the production of methane that is then used for fuel.

Different decisions may be made in different circumstances. In one area a local authority may decide to build an incinerator for biodegradable waste or give permission for a farmer to build an anaerobic digester. The decisions made will take account of local opinion, the needs of the community, and the financial implications.

Questions

1 Explain how climate change could lead to the UK getting cooler and not hotter.

2 Explain why climate change could give rise to both more floods and longer droughts.

3 Suggest how the farming in the UK may change as the climate changes.

4 Explain why increasing the use of wind power could help mitigate climate change.

5 Many of the suggested methods of mitigating methane emissions result in the formation of carbon dioxide. Give one example and explain why these are still considered useful in counteracting climate change.

6 Suggest three actions that could be taken in your local community to reduce emissions of:
   a carbon dioxide
   b methane.

A: Increasing the availability of potable water

Find out about

● sources of water
● treatment of water
● how benefits must be balanced with risks
● how to test for chlorine

Potable water

If you want a drink of water all you have to do is turn on the tap. Elsewhere in the world many people still do not have access to safe **potable** (drinkable) water. As the world population increases there is a greater need for potable water. Climate change could make the problem even worse. Even in the UK, demand could outstrip supply in a particular region during periods of drought.

Water sources

Around the world people take their water from a variety of sources. Whatever the source is, it is essential that the water is made safe to drink.

Water treatment varies depending upon the source of the water. Water sources must be checked to ensure that the water does not contain any toxic substances arising either from the natural environment or human pollution, such as industrial waste.

Surface water

Surface water is collected from rivers, lakes, streams, ponds, and reservoirs. This water can be treated to make it clean and safe to drink.

Groundwater

In the south of England, where population and demand is highest, about 70% of public water is obtained from deep under the ground.

Water found deep underground fell as rain hundreds or thousands of years ago. Groundwater may be obtained by digging wells or by boring into deep underground aquifers (underground layers of water-bearing rock). This water is already of high quality because it has been filtered naturally by passing through layers of sand and rock. To make it safe to drink it needs to be disinfected to kill any harmful microorganisms.

If climate change increases temperatures in the UK, this would lead to a decrease in the availability of ground water. On the other hand, extreme rainfall could cause groundwater flooding.

Groundwater can help provide good quality water around the world.

Seawater

Where other sources of water are scarce, seawater may be used. First the salt must be removed. This process of **desalination** could become a key method for producing potable water in coastal regions that lack sufficient freshwater.

Distillation can be used to evaporate the water and separate it from the dissolved salt before cooling to condense the water vapour.

Filtering does not usually remove dissolved substances from a solute; however, some desalination plants now use a super-fine **membrane**. The membrane is a very thin material that will only let through specific substances. In this case, salty water is forced through the membrane at high pressure. Water molecules pass through but the dissolved salts do not.

Lanzarote is a small island in the Atlantic Ocean with a growing population and many visitors. This desalination plant is the latest to be built to meet an increased demand for water.

Water treatment

In the UK most of the water supply comes from groundwater and surface water. However, the population is ever increasing and demand for water is getting higher. Wastewater from the sewage system is treated and made safe to return to the river system, where it may be extracted further downstream and used again.

Treatment of wastewater

Step 1 Screening
This removes large solids.

Step 2 Settlement
Solid material settles to the bottom of large tanks.

Step 3 Aeration and biological treatment
Bacteria break down remaining organic matter. Aeration provides oxygen for the bacteria.

Step 4 Final settlement
Any remaining fine particles are allowed to settle to the bottom of the tank.
Water is then released back into the river system. A final filtration stage can make the water even cleaner.

Treatment of surface water

Step 1 Screening
This removes large solid items such as leaves or plastic.

Step 2 Clarification
Water is left in large tanks which allows solid matter to settle to the bottom. Chemicals may be added to encourage solid particles to clump together, making it easier for them to sink. This makes the water clearer.

Step 3 **Filtration**
Any very small particles remaining suspended in the water are left behind in the filter as the water passes through.

Step 4 **Disinfection**
Chlorine is added to kill any microorganisms in the water. This is called **chlorination**.

Chlorination

The last time you were at a swimming pool you will have smelled the chlorine from compounds added to water to kill any microorganisms that it may contain.

Chlorination of water is critical as untreated water can contain microorganisms that can cause deadly diseases, such as cholera, dysentery, and typhoid. These diseases can rapidly sweep through a community if water is not properly treated.

Chlorine

Chlorine gas is toxic to breathe in. It can irritate the eyes, respiratory system, and skin. It can make symptoms worse if you have breathing difficulties, such as asthma. It is also very toxic to aquatic organisms if it dissolves in water and reaches the river system.

All containers of chemicals must be labelled with symbols to warn users of the **hazards**. This could be in the laboratory, at work, or in the home. The symbols are used because they are clear and can be understood by everyone.

Chlorine tablets are used to treat swimming pool water. Pool workers must be trained in handling the chemical and wear protective clothing, including goggles.

Reducing health hazards

Chlorination processes in the UK ensure that low levels are used that are not dangerous to health.

Chlorine can react with organic compounds found naturally in water and produce trichloromethane, which can cause health hazards. The **risk** of this can be reduced by removing organic waste from the water before the chlorination stage of the treatment process. Trichloromethane is a member of the trihalomethane group of chemicals (THMs).

Research into health risks of chlorination

One potential risk of chlorination that scientists have researched is whether the THMs produced could cause birth defects in babies whose mothers drank the water.

Scientists looked for patterns between the number and types of birth defects and the level of THMs in the water of the area of Taiwan where mothers lived. They concluded that drinking water with higher levels of THMs did not increase the risk of birth defects in general but that there was an increased risk of a baby being born with a specific defect of the heart. Although the scientists found a correlation, they could not find a cause. They could not explain how THMs could cause this heart defect. They published their work in a peer-reviewed journal. Other scientists are now looking to see whether the same effects are found in other places.

You might find an individual case of a mother who drank water with high levels of THMs and yet sadly had a baby with a different birth defect. Does this immediately provide convincing evidence against the findings? No, it does not, because one case cannot provide evidence for a correlation. A correlation is identified from a collection of data that shows overall trends and patterns.

The National Health Service (NHS) in the UK reports that mothers should not be concerned by this research and identifies a number of problems with how the survey was carried out. They also note that it is unclear whether levels of THMs from chlorination of water in Taiwan are similar to the UK so the findings may not be relevant here.

In general, countries consider that the benefits of chlorination in terms of reducing waterborne disease outweigh the risks. The risks can be reduced by regulating the way in which chlorination is carried out.

Questions

1   The treatment of wastewater has many steps. Explain why:
    a   screening is used at the start of the process rather than a filter
    b   aeration is needed
    c   filtration does not remove harmful bacteria.

2   There are some similarities and differences between the treatment of surface water and wastewater.
    a   Give two similarities between the treatment of surface water and wastewater.
    b   Give one way in which the treatments are different.
    c   Explain why wastewater is not necessarily disinfected.

3   Why did the publication of the research in a peer-reviewed journal make journalists more likely to trust the source of information?

4   One mother in the survey drank water that was unfortunately very high in THMs levels and yet her baby was born fit and healthy. Explain whether this individual case provides evidence for or against the conclusion of the report that there was a correlation between THM levels and some birth defects.

Science explanations

C1 Air and water

In this chapter you have learnt how the pollution of the air and water can affect people's health and the environment. In order to improve air and water quality it is important to understand where air pollutants come from and how they are made.

You should know:

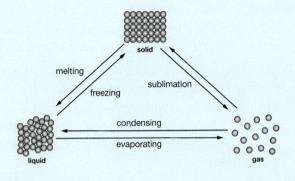

- the main features of the particle model and be able to use it to explain changes of state
- how the Earth's early atmosphere was formed and how scientists think an oxygen-rich atmosphere developed
- why the incomplete burning of fuels produces particulate carbon and a poisonous gas, carbon monoxide
- why some fuels produce sulfur dioxide gas when they burn, and why this gas gives rise to acid rain if released into the air
- why the waste gases from fuels burning inside a furnace or engine produce nitrogen oxide gas
- that, when released into the air, nitrogen oxide combines with more oxygen to make nitrogen dioxide gas, which can also contribute to acid rain
- that technological developments, such as catalytic converters and gas scrubbing, can reduce the amounts of pollutants released into the atmosphere
- the names and symbols of common elements and compounds, and how to write balanced chemical equations for the reactions described in this chapter
- the tests to identify oxygen, hydrogen, carbon dioxide, and chlorine gas
- the difference between endothermic and exothermic reactions, and how they can be described by a reaction profile
- how to identify the activation energy in a reaction profile and how to calculate the energy changes during a chemical reaction
- that a chemical cell produces a potential difference due to a chemical reaction S
- the advantages and disadvantages of using fuel cells
- that burning fossil fuels changes the atmosphere by adding extra carbon dioxide (contributing to global warming)
- that all bodies emit and absorb radiation, and how this links to the greenhouse effect
- about the causes of climate change and its possible effects on agriculture, weather, and sea levels
- how the effects of increasing levels of greenhouse gases might be mitigated, including consideration of the scale, risk, and possible environmental implications
- the principal methods for increasing the availability of drinking water, including testing wastewater, groundwater, and sea water.

Ideas about Science

Scientists use data, rather than opinions, to justify their explanations. They collect large amounts of data when they investigate the causes and effects of air pollutants. They can never be sure that a measurement tells them the true value of the quantity being measured.

If you take several measurements of the same quantity, the results are likely to vary. This may be because:

- you have to measure several individual examples, for example, exhaust gases from different cars of the same make
- the quantity you are measuring is varying, for example, the level of nitrogen oxides in exhaust gases
- of the limitations of the measuring equipment or because of the way you use the equipment.

The best estimate of the true value of a quantity is the mean of several measurements. The true value lies in the spread of values in a set of repeat measurements.

A measurement may be an outlier if it lies outside the range of the other values in a set of repeat measurements.

A correlation shows a link between a factor and an outcome, for example, as the level of carbon dioxide in the atmosphere increases, the average global temperature rises.

A correlation does not always mean that the factor causes the outcome.

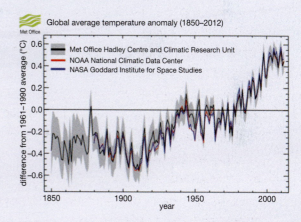

Scientists publish their results so that their data and claims can be checked by others. Scientific claims are only accepted once they have been evaluated critically by other scientists.

Reviewers check claims to make sure the scientists who did the work have checked their findings by repeating them.

Scientists may come to different conclusions about the same data, but as more evidence is produced most scientists will come to an agreement about the best explanation for all the observations.

Some applications of science, such as using fuels, can have a negative impact on quality of life or the environment. However, science can also provide solutions to problems left by earlier technologies.

C1 Review questions

1   The particle model may be used to explain different properties of matter.

  a   Which feature(s) of the particle model can be used to explain why:

   i   molten silver fills the shape of a mould

   ii   an iron nail has a fixed shape.

  H **b**   The low density of aluminium, compared with other metals, cannot be explained by the particle model. Explain why. You should refer to a limitation of the particle model in your answer.

2   Explain whether each of these measures would reduce the amount of nitrogen dioxide pollution in the air.

  ● using low-sulfur fuels

  ● adjusting the balance between public and private transport

  ● encouraging people to use less electricity

3   Some barbecues use propane from a gas cylinder as a fuel.

  a   Write down a word equation for the reaction that occurs when propane (C_3H_8) burns.

  b   Write a balanced chemical equation for the reaction.

4   For each process in the table below, complete the columns to show which gas was added to the atmosphere and/or which was removed.

Process	Gas added to atmosphere	Gas removed from the atmosphere
volcanic emissions	water vapour and…	
cooling (condensation)		
photosynthesis by bacteria		
photosynthesis by plants		
oxidation of rocks		

5 A student mixed sodium hydroxide solution with hydrochloric acid in a polystyrene beaker. They reacted to form sodium chloride and water.

The temperature at the start was 20 °C. The maximum temperature reached during the reaction was 26 °C.

a State whether the reaction was exothermic or endothermic.

b Draw an energy-level diagram for the reaction. Remember to label the reactants and products.

6 Hydrogen reacts with fluorine. The equation for the reaction is:

$H_2(g) + F_2(g) \rightarrow 2HF(g)$

a Use the data in the table to calculate the energy change for the reacting masses in the equation above.

b State whether the reaction is exothermic or endothermic.

Process	Energy change for the formula masses (kJ)
breaking one H–H bond	434 needed
breaking one F–F bond	158 needed
forming one H–F bond	562 given out

7 a Describe what is meant by climate.

b January and February 2014 were warmer than the average for the UK. Does this provide evidence that climate change isn't happening? Explain your answer.

8 The most important users of desalination plants are countries in the Middle East where the climate is very hot and dry. Desalination allows countries to create potable water from sea water.

a i Give one example of a water source that these countries are lacking.

ii Explain why desalination is not a suitable process for all countries with a similar climate.

b i State which separation technique is used during desalination.

ii Name the two physical changes which take place during the desalination process.

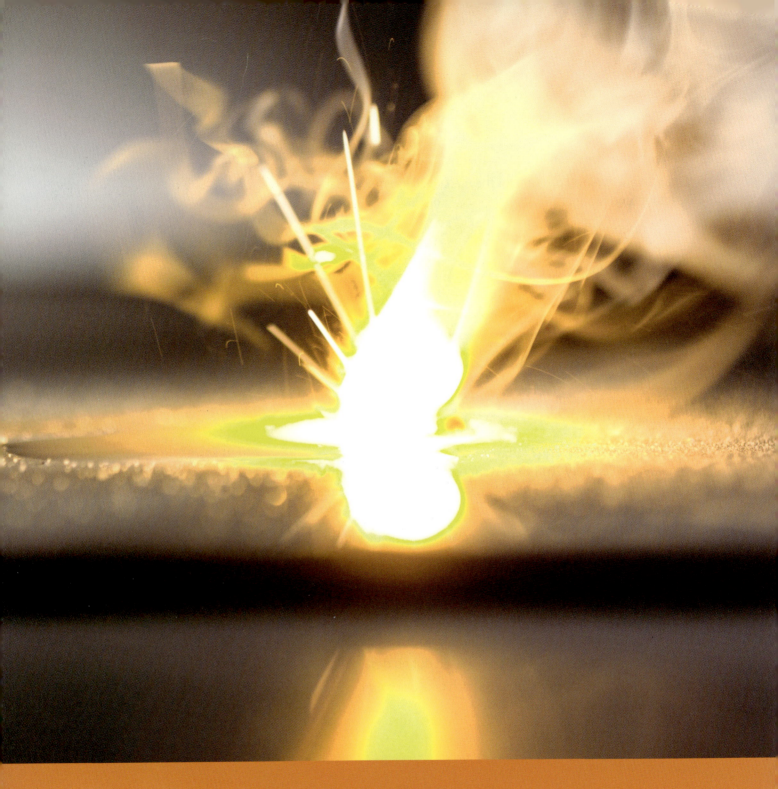

C2 Chemical patterns

Why study chemical patterns?

Copper makes ideal water pipes but another metal, potassium, reacts violently with water. Why do these elements behave so differently? How can we predict their properties? Scientists use what they know about elements and compounds to choose substances with the best properties for particular purposes. They use their knowledge of chemical reactions to work out how to make these substances.

What you already know

- Substances are made of tiny particles, called atoms. Every element has its own type of atom and its own chemical symbol.

- Compounds are made up of atoms of two or more elements that are joined together. You can use chemical formulae to represent compounds.

- During a chemical reaction, atoms are rearranged. The atoms in the reactants are joined together differently to the atoms in the products.

- In a chemical reaction, the total mass of the reactants is the same as the total mass of the products. This is the law of conservation of mass.

- Equations are used to represent chemical reactions. The formulae of the reactants are on the left of the arrow, and the formulae of the products are on the right of the arrow.

- In metal displacement reactions, more reactive metals displace less reactive metals from their compounds.

- Catalysts speed up some chemical reactions. The catalyst is not used up in the reaction.

- The Periodic Table lists all the elements, roughly in order of relative atomic mass.

- In the Periodic Table, each vertical column is called a group of elements. Elements in the same group have similar properties. The horizontal rows are called periods. There is a gradual change in properties from left to right of a period. You can use the position of an element in the Periodic Table to predict its properties.

- In the Periodic Table, the elements on the left of the stepped line are metals, and the elements on the right of the stepped line are non-metals.

- Most metals are shiny, strong, and hard. They are good conductors of heat and electricity. Most have high melting and boiling points.

- Most non-metals are in the gas state at room temperature. Those that are in the solid state are dull, brittle, and soft. They are poor conductors of heat and are electrical insulators.

The Science

In this chapter you will discover how the tiny particles that make up atoms give elements their properties. You will use the Periodic Table to find patterns in the physical and chemical properties of elements, and to make predictions about their reactions.

You will also explain some strange observations. Why does salt solution conduct electricity, when solid salt is an insulator? And why do some elements react violently while others do not react at all?

Ideas about Science

This chapter traces the development of two vital ideas in chemistry – atoms and the Periodic Table.

You will start by finding out about atoms from the earliest thoughts of Greek philosophers to the discoveries of even tinier subatomic particles that continue today.

Next you will learn about organising elements. How did a Russian chemist's leap of imagination help to make sense of other chemists' data and lead to the development of the Periodic Table?

C2.1 How have our ideas about atoms developed over time?

A: Changing models of the atom

Find out about

- Dalton's atomic model
- evidence for electrons
- how and why ideas about atoms developed over time
- the sizes of atoms, molecules, and nuclei
- protons, neutrons, and isotopes
- how to calculate atomic number and mass number

Key word

➤ atom

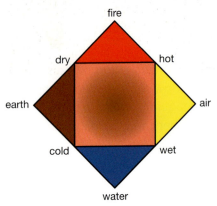

The four elements of the ancient Greeks, and their qualities.

Early ideas

What is the Universe made of? Why is the world as it is? People have always wondered about these questions. They developed models to explain what they saw around them.

Night sky with stars and nebula.

More than two thousand years ago, early Greek philosophers believed that matter was made up of four elements – earth, water, air, and fire. These elements, they said, combine together to make all the materials on Earth.

The Greek thinkers had little evidence for their model. But it did help to explain some observations. For example, heating cold, wet water makes hot, wet air – what we now call steam.

Other early philosophers developed a different model to explain matter. Leucippus and Democritus believed that matter was made up of tiny particles. They called these particles **atoms**, from the Greek word *atomon*, which means indivisible. Atoms, they said, are surrounded by empty space.

The philosophers did not do experiments. They developed their models using creative thought.

Dalton's atomic model

Before 1800, there was little evidence for atoms. Then along came John Dalton, who taught Mathematics and Natural Philosophy in Manchester. He experimented with gases and read about the investigations of other chemists.

Dalton knew that water existed in three states – as a solid, as a liquid, and as a gas. He pointed out that this is evidence for the idea that everything is made up of...

...a vast number of extremely small particles, or atoms of matter, bound together by a force of attraction that is more or less powerful according to circumstances.

Dalton used this model to explain his observations. Water vapour can occupy the same space as air, he said, because their particles mix together.

Dalton knew that water was made up of atoms of two elements, hydrogen and oxygen. He wondered if all hydrogen atoms are alike. Do they have the same shape and the same mass?

Dalton used evidence from experiments to reason that water particles are identical. If water particles are identical, then their hydrogen atoms must all have the same mass and shape as each other. The same must also be true for oxygen atoms.

Dalton's atomic model

Dalton published his atomic theory in 1805. It still forms part of the basic ideas of modern chemistry.

- Everything is made up of atoms.
- Atoms cannot be split up, created, or destroyed.
- The atoms of an element have the same mass and size.
- Atoms of different elements have different masses and sizes.
- Atoms of different elements join together to make compounds.
- Atoms are rearranged in chemical reactions.

Dalton created symbols for the elements. He also listed their atomic masses. Scientists now have better methods for measuring atomic masses, so we now know that some of Dalton's values are incorrect.

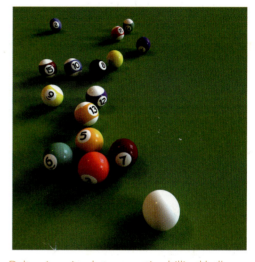

Dalton collected gases from a marsh for some of his experiments. One of the gases he collected was methane. (This picture is in Manchester Town Hall.)

Dalton imagined atoms as tiny billiard balls.

The plum-pudding model of the atom

Scientists were investigating gases. They put a tiny amount of a gas in a sealed tube with a fluorescent screen. They passed a high-voltage electric current through the gas. The screen glowed green and the scientists suggested that rays from the negative electrode travelled through the gas and hit the screen. They called them cathode rays.

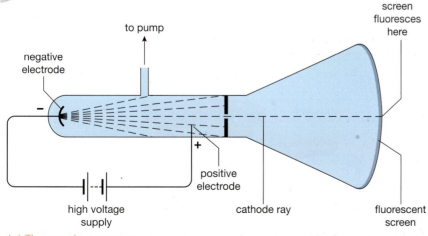

J. J. Thomson's apparatus.

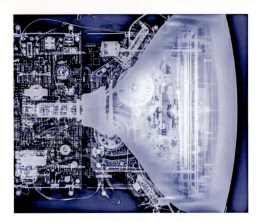

Until flat screens were developed, cathode ray tubes were at the heart of every television.

A plum pudding.

Some scientists thought that cathode rays were made up of electrically charged particles. British scientist J. J. Thomson tested this idea by passing the rays between magnets and between electrically charged pieces of metal. The rays changed direction. This led Thomson to conclude that: 'Cathode rays are charges of negative electricity carried by particles of matter.'

Then Thomson asked another question: 'What are these particles? Are they atoms, or molecules, or matter in a still finer state of subdivision?'

Thomson repeated his experiments with many different gases. The cathode rays from every gas had the same properties. He concluded that cathode rays are made up of particles with a mass that is much less than the mass of an atom. Each particle has the same negative charge. These particles are called **electrons**.

Thomson wondered where the negative particles came from. The only possible place would be from inside the atoms of the gases that the electricity travelled through. Thomson had discovered the first **subatomic particle**.

Thomson's discovery suggested that Dalton's model was not quite right – it *is* possible to split up atoms. Thomson had found a limitation of Dalton's model.

Thomson knew that atoms have no overall electrical charge. He also knew that they contain negatively charged particles. How could he explain both of these observations?

Thomson needed a new model. He said that an atom consists of tiny, negatively charged electrons dotted about in a positively charged sphere. He thought the electrons moved around in rings within the sphere. Other people called this model the **plum-pudding model**. The electrons reminded people of plums in a pudding.

The nuclear model of the atom

Ernest Rutherford knew about Thomson's plum-pudding model. Rutherford used the model to make a prediction:

If you fire positive particles at gold foil almost all of them will travel straight through the foil, because the electrical charge in an atom is evenly distributed. A few will change direction slightly because they pass close to a negative electron.

Rutherford asked some other scientists, Hans Geiger and Ernest Marsden, to do an experiment to test this prediction.

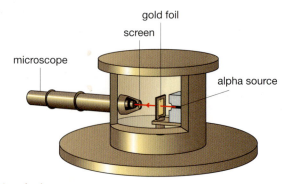

Geiger and Marsden's apparatus.

The scientists fired positive particles at a thin piece of gold foil. Most of the positive particles travelled straight through, as predicted. But a few changed direction dramatically. And about one positive particle in every 10 000 bounced back off the foil. The scientists were amazed. Rutherford said:

It was quite the most incredible event that has ever happened to me in my life. It was almost as incredible as if you fired a 15-inch artillery shell at a piece of tissue paper and it came back and hit you.

Rutherford realised that Thomson's plum-pudding model had limitations. The electrical charges are not evenly distributed throughout the atom. By 1911 he had suggested a new model of the atom:

- Atoms have a central **nucleus**. Most of the mass of an atom is in its nucleus. The nucleus is positively charged.

- The nucleus is surrounded by a big empty space in which negatively charged electrons are orbiting, like planets in a solar system.

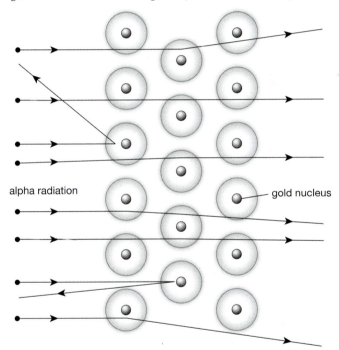

alpha radiation   gold nucleus

Most of Geiger and Marsden's positively charged particles travelled straight through the gold foil. Those that went near the nucleus changed direction.

Bohr's model

Niels Bohr noticed a limitation of Rutherford's model. He said that, if Rutherford's model was correct, electrons would spiral in towards the positively charged nucleus. But this clearly does not happen.

Bohr used mathematics to suggest an improvement to Rutherford's model. He said that electrons move around in certain fixed orbits. These are called **energy levels**, or **electron shells**.

Synoptic link

You can learn more about the nuclear model of the atom in P5.1B *What is an atom?*

Key words

➤ order-of-magnitude
➤ proton
➤ neutron
➤ atomic number
➤ mass number

How big is an atom?

The diameter of a typical atom is about 0.000 000 01 cm. This is the same as 0.000 000 000 1 m, which you can write as 1×10^{-10} m, or 10^{-10} m. This is an **order-of-magnitude** estimate.

If you could place one hundred million atoms side by side, they would cover 1 cm.

A molecule is a group of atoms, strongly joined together. Petrol includes a compound called decane. A molecule of decane includes 10 carbon atoms, joined in a line. You can estimate the length of a decane molecule like this:

estimate for length of decane molecule = diameter of one carbon atom × number of atoms
$$= 1 \times 10^{-10} \text{ m} \times 10$$
$$= 1 \times 10^{-9} \text{ m}$$

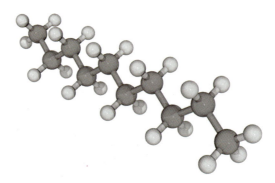

Petrol includes a compound called decane. The length of a decane molecule is approximately 1×10^{-9} m.

The size of a nucleus

The nucleus of an atom is tiny compared to its atom. An order-of-magnitude estimate for the diameter of a typical nucleus is 2×10^{-15} m. This means that the diameter of an atom is 50 000 times greater than the diameter of its nucleus.

Worked example: Comparing the size of an atom to objects you can see

In a model, if the Earth represents an atom, estimate the diameter of the nucleus of the atom.
The diameter of the Earth = 13×10^6 m

Step 1: Write down what you know, with the units.

diameter of an atom is 50 000 times greater than the diameter of its nucleus

diameter of Earth = 13×10^6 m

Step 2: Write down the relationship you will use.

$$\frac{\text{diameter of nucleus of atom}}{\text{diameter of atom}} = \frac{1}{50\,000} = \frac{\text{diameter of nucleus in model}}{\text{diameter of Earth}}$$

Step 3: Substitute the quantities into the equation and calculate the diameter of the nucleus in model.

$$\frac{1}{50\,000} = \frac{\text{diameter of nucleus in model}}{13 \times 10^6 \text{ m}}$$

$$\text{diameter of nucleus in model} = \frac{13 \times 10^6 \text{ m}}{5 \times 10^4 \text{ m}}$$

Answer:

diameter of nucleus in model = 260 m

Inside the nucleus

By 1932 scientists had discovered that the nucleus is made up of two kinds of subatomic particles, called **protons** and **neutrons**. This is the model of the atom that we will use in this book.

As you know, the mass of an atom is concentrated in a tiny, central nucleus. The nucleus consists of protons and neutrons. Protons have a positive charge. Neutrons have no charge. The mass of a proton is roughly the same as the mass of a neutron.

Outside the nucleus electrons are orbiting. An electron has a negative charge. The mass of an electron is so small that you can often ignore it.

An atom has the same number of protons and electrons. This means that the total positive charge is equal to the total negative charge. Overall, an atom has no electrical charge. It is neutral.

If an atom were the size of the Earth, its nucleus would be about the size of a cruise liner.

cloud of electrons

nucleus

Particle	Mass	Charge
electron	0.0005	−1
proton	1	+1
neutron	1	0

All atoms consist of protons, neutrons, and electrons. This diagram is not to scale.

Atomic number and mass number

What does the image on the right show?

The image shows the clearest ever view of atoms in a silicon crystal. The image is magnified approximately 10 million times, and was produced by a transmission electron microscope. Silicon is used to make microchips. These are a vital component of electronic devices, such as, mobile phones, computers, and games consoles.

What is inside a silicon atom? Every atom of an element has the same number of protons. This is the **atomic number** of the element. Silicon has 14 protons, so its atomic number is 14. The number of protons in an atom is the same as the number of electrons. This means that the atomic number also gives the number of electrons in an atom. Silicon has 14 electrons.

Every element has its own atomic number. These are shown in the Periodic Table.

But the nucleus of silicon also contains neutrons. Protons and neutrons give an atom almost all its mass. The total number of protons and neutrons in an atom is its **mass number**.

mass number = number of protons + number of neutrons

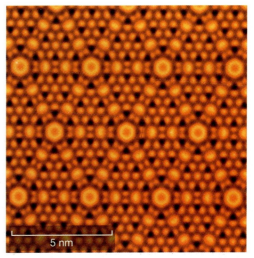

Atoms in a silicon crystal.

5 nm

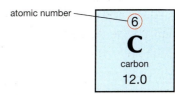

atomic number

⑥

C

carbon

12.0

Box of Periodic Table with atomic number circled.

Worked example: Mass number and atomic number

... of an element has 15 protons and 16 neutrons. Work out its atomic number and mass ... Then identify the element.

... **Write down** ... **know.**	proton number = 15 neutron number = 16
... **work out the** ... **number.**	atomic number = number of protons **atomic number = 15** mass number = number of protons + number of neutrons
Step 3: Work out the mass number.	mass number = 15 + 16 **mass number = 31**
Step 4: Use the Periodic Table and the atomic number to identify the element.	The element with atomic number 15 is **phosphorus**.

Worked example: Proton number and neutron number

The mass number of an arsenic atom is 75. Use this piece of data and the Periodic Table to find out the number of protons, neutrons, and electrons in an arsenic atom.

Step 1: Write down what you know.	mass number = 75
Step 2: Use the Periodic Table to find the atomic number of arsenic.	atomic number = 33
Step 3: Use the atomic number to give the number of protons and electrons.	number of protons = atomic number. **number of protons = 33** number of electrons = number of protons **number of electrons = 33**
Step 4: Write down the equation for mass number. Re-arrange it and substitute numbers to find the number of neutrons.	mass number = number of protons + number of neutrons
Step 5: Substitute in values you know. Work out the number of neutrons.	75 = 33 + number of neutrons number of neutrons = 75 − 33 = 42
Answer:	**number of neutrons = 42**

Isotopes

Dalton thought that every atom of an element had the same mass. But we now know that this is not quite true. For example, there are different types of hydrogen atom:

- Most (99.98%) hydrogen atoms have one proton and no neutrons.
- A few hydrogen atoms have one proton and one neutron.
- A few hydrogen atoms have one proton and two neutrons.

Atoms of the same element with the same number of protons and different numbers of neutrons are called **isotopes**. The diagrams in the margin show three isotopes of hydrogen.

Many elements have several isotopes. For example, 75% of chlorine atoms have nuclei with 17 protons and 18 neutrons. The mass number of this isotope is 35 (17 + 18 = 35).

But 25% of chlorine atoms have nuclei with 17 protons and 20 neutrons. The mass number of this isotope is 37 (17 + 20 = 37).

You can show the isotopes of chlorine like this:

$$^{35}_{17}\text{Cl} \text{ (chlorine-35)}$$

$$^{37}_{17}\text{Cl} \text{ (chlorine-37)}$$

The mass number is at the top, and the atomic number is at the bottom.

In the Periodic Table, the lower number in each square gives the relative atomic mass of the element. For most elements, where almost all the atoms are of one isotope, the relative atomic mass is the same as the mass number. This means that, for most elements, you can use data from the Periodic Table to calculate the numbers of protons, neutrons, and electrons in atoms.

Scientists use information related to isotopes to find out about past climates and bird migrations. The fraction of water molecules in rain that includes a $^{2}_{1}\text{H}$ atom depends on the climate. This means that scientists can use information from ice cores to learn about past climates.

Fabiola Gianotti is a scientist at the Large Hadron Collider.

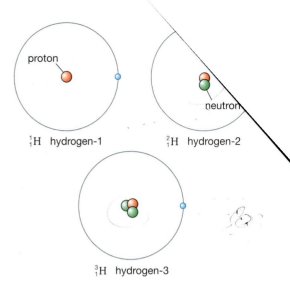

$^{1}_{1}\text{H}$ hydrogen-1 $^{2}_{1}\text{H}$ hydrogen-2

$^{3}_{1}\text{H}$ hydrogen-3

Hydrogen has three isotopes. Each has a different number of neutrons and the same number of protons.

Scientists analysed isotopes of hydrogen, carbon, and nitrogen in feathers of Burrowing Owls. They use this information to trace the birds from their wintering grounds in Texas and Mexico, back to their breeding grounds in Canada.

Synoptic link

You can learn more about climate change in C1 *Air and water*.

Key word

➤ isotope

Still searching

Today, scientists are searching for even smaller subatomic particles. They use particle accelerators such as the Large Hadron Collider to smash atoms up. They examine the debris to learn more about subatomic particles.

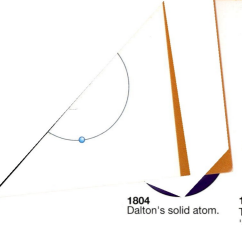

1804
Dalton's solid atom.

1913
The Bohr–Rutherford 'Solar System' atom, in which electrons orbit around a very small nucleus.

1924
A model of the atom in which the electrons are no longer treated as particles but pictured as occupying energy levels, which give rise to regions of negative charge around the nucleus (charge clouds).

1932
The atom in which the nucleus is built up from neutrons as well as protons.

2000+
The present-day atom in which the protons and neutrons in the nucleus are built up from many kinds of particles.

Atomic models from 1800 to the present. The diameter of an atom is about ten million times smaller than a millimetre. These diagrams are distorted. For atoms at this scale, the nuclei would be invisibly small.

Questions

1 Describe the early Greek model of the four elements.

2 **a** Describe two pieces of evidence that Dalton used to help him develop his atomic model.
   b State one observation that Dalton used his model to explain.

3 Draw Dalton's symbols for the two elements that make up water.

4 Identify the main difference between the methods that the early Greeks used to develop their four-element model and the method that Dalton used to develop his atomic model.

5 Describe Thomson's plum-pudding model of the atom.

6 Geiger and Marsden did an experiment in which they fired positively charged particles at a thin sheet of gold foil.
   a Describe the evidence they collected.
   b Describe the explanation that Rutherford developed to explain Geiger and Marsden's evidence.

7 Compare Thomson's model of the atom to Rutherford's model of the atom.

8 Estimate the number of atoms that would fit, side by side, along a line that is 1 m long.

9 Give an order-of-magnitude estimate for the radius of a typical atom.

10 In a model, if a car represents the nucleus of an atom, estimate the diameter of an atom.
The length of a typical car is 4 m.

11 **a** An atom has nine protons and 10 neutrons. Give its atomic number and mass number.
   b An element has an atomic number of 31 and a mass number of 70. Give the number of protons, neutrons, and electrons in an atom of the element, and identify the element from the Periodic Table.

12 The symbols below represent some isotopes of oxygen.

$$^{16}_{8}\text{O} \qquad ^{17}_{8}\text{O} \qquad ^{18}_{8}\text{O}$$

   a State how the isotopes are similar and how they are different, in terms of protons and neutrons.
   b Give the number of neutrons in one atom of each isotope.

A: Developing the Periodic Table

A leap of imagination

By 1860 chemists had found about 60 elements. They asked questions about them:

- What are the patterns in the properties of the elements?
- How many elements are there?
- Can you use patterns to help find more elements?

An Italian chemist worked out the atomic masses of the known elements. He gave this data to other scientists, including Dmitri Mendeleev, from Russia.

Mendeleev was looking for patterns in the properties of elements. One day, in 1869, he made some small cards. On each card he wrote the name of an element, its atomic mass, and some of its properties.

Mendeleev tried arranging the cards in different ways. Then he had a brilliant idea. He placed the elements in atomic mass order, grouping together elements with similar properties. Through thinking creatively about the data, he had come up with the **Periodic Table**.

Missing elements

Mendeleev was confident that his Periodic Table was correct. He reversed the positions of iodine and tellurium because, when placed in atomic mass order, their properties did not fit the pattern.

Mendeleev used the Periodic Table to predict the existence of elements that had not yet been discovered. He left gaps for these elements.

One of the gaps was between silicon and tin. Mendeleev predicted the properties of this element. Later, Clemens Winkler discovered the missing element. He called it germanium. Its properties were just like those Mendeleev had predicted. Scientists soon found other missing elements, including gallium and scandium.

The fact that Mendeleev's predictions were correct increased other scientists' confidence in his great explanation. His Periodic Table is considered by many scientists to be the foundation of modern chemistry.

Electrons in atoms

When Mendeleev developed the Periodic Table scientists did not know about electrons. We now know that the position of an element in the Periodic Table is related to the arrangement of electrons in its atoms.

Find out about

- how the elements are arranged in the Periodic Table
- how Mendeleev developed the Periodic Table
- the link between electronic structure and the position of an element in the Periodic Table

Key word

➤ Periodic Table

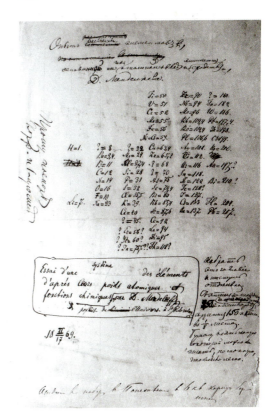

Mendeleev's Periodic Table.

Electrons in atoms whizz around outside the nucleus. They occupy different energy levels, or shells. Electrons fill the shells nearest to the nucleus first. Each shell holds a maximum number of electrons:

- The first shell holds up to two electrons.
- The second shell holds up to eight electrons.

You can also use numbers to show the **electronic structure** (arrangement of electrons) of an atom. The first number gives the number of electrons in the first shell, the second number gives the number of electrons in the second shell, and so on.

Element	Electronic structure
hydrogen	1
lithium	2.1
magnesium	2.8.2

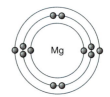

Hydrogen has one electron. It occupies the first shell.

Lithium has three electrons. There are two electrons in the first shell, and one in the second shell.

Magnesium has 12 electrons. There are two in the first shell, and eight in the second shell. These shells are full. The third shell has two electrons.

Electronic structure and the modern Periodic Table

The diagram below gives the symbols and electronic structures of the first 20 elements of the Periodic Table. The elements are arranged as they are in the modern Periodic Table.

H 1									He 2
Li 2.1	Be 2.2			B 2.3	C 2.4	N 2.5	O 2.6	F 2.7	Ne 2.8
Na 2.8.1	Mg 2.8.2			Al 2.8.3	Si 2.8.4	P 2.8.5	S 2.8.6	Cl 2.8.7	Ar 2.8.8
K 2.8.8.1	Ca 2.8.8.2	Transition elements							

There are patterns in the electronic structures:

- Across a horizontal **period**, from one element to the next, the number of electrons increases by one. So from left to right across a period, an electron shell is filled up.
- All the elements in the same vertical **group** of the Periodic Table have the same number of electrons in their outermost shell.

The electronic structure of an element gives the element its properties. The repeating pattern of electronic structures explains the repeating patterns of properties in the Periodic Table.

- Elements in the same group of the Periodic Table have similar properties.

- There is a gradual change in the properties of the elements going across a period.

The number of electrons in an atom is the same as the number of protons. This means that, in the modern Periodic Table, the elements are in atomic number order. This order places iodine in Group 7, with other elements with similar properties. Tellurium is also in a group with similar elements. Mendeleev was right to swap the positions of these elements.

(1)	(2)											(3)	(4)	(5)	(6)	(7)	(0)
1			Key														**18**
1 **H** Hydrogen 1.0	**2**		atomic number **symbol** name relative atomic mass									13	14	15	16	17	2 **He** Helium 4.0
3 **Li** Lithium 6.9	4 **Be** Beryllium 9.0											5 **B** Boron 10.8	6 **C** Carbon 12.0	7 **N** Nitrogen 14.0	8 **O** Oxygen 16.0	9 **F** Fluorine 19.0	10 **Ne** Neon 20.2
11 **Na** Sodium 23.0	12 **Mg** Magnesium 24.3	**3**	**4**	**5**	**6**	**7**	**8**	**9**	**10**	**11**	**12**	13 **Al** Aluminium 27.0	14 **Si** Silicon 28.1	15 **P** Phosphorus 31.0	16 **S** Sulfur 32.1	17 **Cl** Chlorine 35.5	18 **Ar** Argon 39.9
19 **K** Potassium 39.1	20 **Ca** Calcium 40.1	21 **Sc** Scandium 45.0	22 **Ti** Titanium 47.9	23 **V** Vanadium 50.9	24 **Cr** Chromium 52.0	25 **Mn** Manganese 54.9	26 **Fe** Iron 55.8	27 **Co** Cobalt 58.9	28 **Ni** Nickel 58.7	29 **Cu** Copper 63.5	30 **Zn** Zinc 65.4	31 **Ga** Gallium 69.7	32 **Ge** Germanium 72.6	33 **As** Arsenic 74.9	34 **Se** Selenium 79.0	35 **Br** Bromine 79.9	36 **Kr** Krypton 83.8
37 **Rb** Rubidium 85.5	38 **Sr** Strontium 87.6	39 **Y** Yttrium 88.9	40 **Zr** Zirconium 91.2	41 **Nb** Niobium 92.9	42 **Mo** Molybdenum 95.9	43 **Tc** Technetium	44 **Ru** Ruthenium 101.1	45 **Rh** Rhodium 102.9	46 **Pd** Palladium 106.4	47 **Ag** Silver 107.9	48 **Cd** Cadmium 112.4	49 **In** Indium 114.8	50 **Sn** Tin 118.7	51 **Sb** Antimony 121.8	52 **Te** Tellurium 127.6	53 **I** Iodine 126.9	54 **Xe** Xenon 131.3
55 **Cs** Caesium 132.9	56 **Ba** Barium 137.3	57–71 Lanthanides	72 **Hf** Hafnium 178.5	73 **Ta** Tantalum 180.9	74 **W** Tungsten 183.8	75 **Re** Rhenium 186.2	76 **Os** Osmium 190.2	77 **Ir** Iridium 192.2	78 **Pt** Platinum 195.1	79 **Au** Gold 197.0	80 **Hg** Mercury 200.6	81 **Tl** Thallium 204.4	82 **Pb** Lead 207.2	83 **Bi** Bismuth 209.0	84 **Po** Polonium	85 **At** Astatine	86 **Rn** Radon
87 **Fr** Francium	88 **Ra** Radium	89–103 Actinides	104 **Rf** Rutherfordium	105 **Db** Dubnium	106 **Sg** Seaborgium	107 **Bh** Bohrium	108 **Hs** Hassium	109 **Mt** Meitnerium	110 **Ds** Darmstadtium	111 **Rg** Roentgenium	112 **Cn** Copernicium 1		114 **Fl** Flerovium		116 **Lv** Livermorium		

* The Lanthanides (atomic numbers 58–71) and the Actinides (atomic numbers 90–103) have been omitted.

The modern Periodic Table.

Questions

1 Describe the pattern in electronic structures for the first 20 elements of the Periodic Table.

2 Explain how the discovery of germanium supported Mendeleev's decision to leave gaps in his Periodic Table.

3 Predict the number of electrons in the outer shell of a tellurium atom. Then explain why iodine and tellurium are not placed in the Periodic Table in mass number order.

B: Metals and non-metals in the Periodic Table

Find out about

- properties of metals and non-metals
- the location of metals and non-metals in the Periodic Table
- how the properties of elements determine their uses
- the discovery of technetium

Key words

➤ metal
➤ non-metal

Metal and non-metal properties

The pictures show uses of the element palladium.

An electrical connector.

Palladium jewellery.

The uses of palladium show that it is a typical **metal** element. It is shiny and has a high density. Its melting point and boiling point are high, so it is in the solid state at room temperature. It is a good electrical conductor.

Nitrogen makes up 78% of the air. It is a typical **non-metal** element. It has low melting and boiling points, and is an electrical insulator. Very few non-metal elements are in the solid state at room temperature, but those that are have low densities and are not shiny.

Element	Melting point (°C)	Boiling point (°C)	Does it conduct electricity?
palladium	1550	3980	yes
nitrogen	−210	−196	no

You can look at the properties of an element to decide whether it is a metal or a non-metal. You can also look at its position in the Periodic Table.

Metals are on the left of the stepped line and non-metals are on the right.

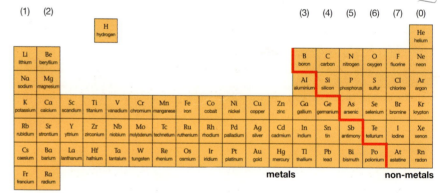

Note: This Periodic Table does not include all the elements.

Metals are on the left of the stepped line and non-metals are on the right.

Using metals and non-metals

The properties of metals and non-metals make them suitable for different uses.

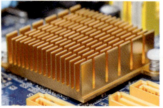

Metals are good thermal conductors. This aluminium and copper heat sink conducts heat away from the central processing unit in a computer.

Sulfur is one of the few non-metals that is solid at room temperature. It is brittle, so it breaks easily if you drop it. It is also soft, so you can scratch it easily. Sulfur is used to make sulfuric acid.

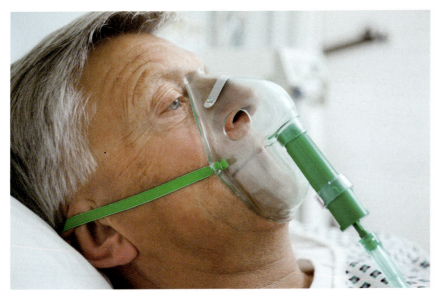

Oxygen makes up 21% of the air. Your body – and other living organisms – uses this non-metal element for respiration. This hospital patient is getting oxygen from an oxygen cylinder.

Missing metal

By 1925 most of the gaps in the Periodic Table had been filled. But there were spaces beneath manganese for two undiscovered elements.

A group of German scientists – Ida Tacke, Walter Noddack, and Otto Berg – looked for these metal elements by bombarding a mineral with a beam of electrons. They claimed they had detected the element with atomic number 43.

Other scientists tried to reproduce this work. They could not detect the new element. They said that Tacke and her colleagues were wrong to claim that they had found the missing element.

In 1937, Italian scientists used a different method to detect the missing element. They called it technetium. Other scientists confirmed their claim, so it was widely accepted.

Many years later, in 1999, David Curtis reproduced the work of Tacke, Noddack, and Berg. He detected the element with atomic number 43. Perhaps Tacke, Noddack, and Berg had, in fact, discovered the missing element.

Ida Tacke was one of the team of scientists who searched for the undiscovered elements below manganese.

Questions

1 State three properties that are typical of metals. For each property, describe an example of a use of a metal for which that property is important.

2 An element has a melting point of 1675 °C and a boiling point of 3260 °C. It is a good conductor of electricity. Predict whether the element is a metal or a non-metal.

3 Explain why, in 1936, scientists did not believe Tacke, Noddack, and Berg's claim that they had found the element with atomic number 43. Then explain why scientists now think that they might, in fact, have found this element.

C: The Group 1 elements

Find out about

- the physical properties of the Group 1 elements
- the reactions of the Group 1 elements with moist air, water, and chlorine

Key words

➤ alkali metals

Lithium batteries last for a long time. They are used in electronic devices, including laptops, tablets, and heart pacemakers.

Sodium is a thermal conductor. It is used as a coolant in this nuclear reactor.

How many battery-powered devices do you use? Many of these include lithium batteries.

Lithium is in Group 1, along with sodium, potassium, rubidium, caesium, and francium. The elements in the group have similar properties.

(1)	(2)											(3)	(4)	(5)	(6)	(7)	(0)
							H										He
Li	Be											B	C	N	O	F	Ne
Na	Mg											Al	Si	P	S	Cl	Ar
K	Ca	Sc	Ti	V	Cr	Mn	Fe	Co	Ni	Cu	Zn	Ga	Ge	As	Se	Br	Kr
Rb	Sr	Y	Zr	Nb	Mo	Tc	Ru	Rh	Pd	Ag	Cd	In	Sn	Sb	Te	I	Xe
Cs	Ba	La	Hf	Ta	W	Re	Os	Ir	Pt	Au	Hg	Tl	Pb	Bi	Po	At	Rn
Fr	Ra	Ac	Rf	Db	Sg	Bh	Hs	Mt	Ds	Rg							

Group 1 the alkali metals

The location of Group 1 in the Periodic Table.

Group 1 elements are corrosive and flammable. Your teacher will use gloves when handling them, and you must wear eye protection when observing their reactions.

Physical properties

Some properties of the Group 1 elements are similar to the properties of other metals. For example, they are good electrical and thermal conductors.

Some physical properties of the Group 1 elements are not like those of other metals. For example, they have low melting and boiling points, they are soft, and they have low densities (so they float on water).

Reactions with water

A teacher adds a small piece of lithium to some water. There is a fizzing sound as bubbles of hydrogen gas are formed.

The other product of this chemical reaction is lithium hydroxide. This dissolves in water as it forms, making an alkaline solution.

$$\text{lithium} + \text{water} \rightarrow \text{lithium hydroxide} + \text{hydrogen}$$

The reaction with sodium is more exciting. During the reaction, enough energy is transferred by heating to melt the sodium. A sphere of liquid sodium skates around the surface of the water, propelled by bubbles of hydrogen. An alkaline solution of sodium hydroxide also forms.

$$\text{sodium} + \text{water} \rightarrow \text{sodium hydroxide} + \text{hydrogen}$$

Potassium reacts violently with water. The metal moves around quickly on the surface of the water. Hydrogen gas forms and immediately catches fire.

$$\text{potassium} + \text{water} \rightarrow \text{potassium hydroxide} + \text{hydrogen}$$

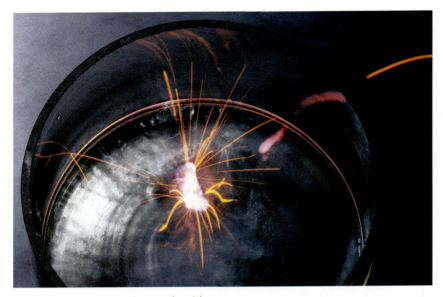

Potassium reacts very vigorously with water.

Sodium reacts vigorously with chlorine.

All the Group 1 elements react with water to make hydrogen gas and an alkaline solution. This is why the Group 1 elements are called the **alkali metals**. The reactions of the alkali metals with water show a trend – from top to bottom of the group, the reactions get more vigorous.

Reactions with chlorine

Hot sodium burns in chlorine gas with a bright yellow flame. The reaction produces white crystals of sodium chloride (table salt).

$$\text{sodium} + \text{chlorine} \rightarrow \text{sodium chloride}$$

The other Group 1 elements react in a similar way when added to chlorine. Lithium produces lithium chloride. Potassium produces potassium chloride.

All three chlorides are white solids at room temperature. They dissolve in water to make colourless solutions. The solutions are not acidic or alkaline – they are neutral.

The reactions of the Group 1 elements with chlorine show a trend. The reaction with lithium is least vigorous. Going down the group, the reactions get more vigorous.

Reactions with moist air

Group 1 elements are stored in oil. This stops oxygen and water vapour from coming into contact with them. For example, moist air reacts with sodium to make sodium oxide, which coats the surface of the sodium.

Questions

1 A scientist is planning an experiment to drop a small piece of rubidium into water.
   a Predict what he would observe if he were to carry out this experiment.
   b Write a word equation for the reaction.
   c Suggest why this experiment will not be carried out by your teacher.

2 For rubidium chloride, predict:
   a its colour in the solid state
   b the colour of its solution
   c whether its solution is acidic, alkaline, or neutral.

D: The Group 7 elements

Find out about

- the physical properties of the Group 7 elements
- reactions of the Group 7 elements with the Group 1 elements
- displacement reactions of Group 7 elements

Key words

➤ halogen
➤ displacement reaction

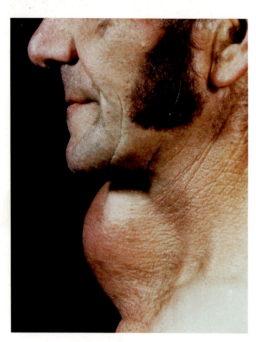

This person has a goitre. Scientists discovered that adding an iodine-containing compound to salt prevents the disease.

Adding tiny amounts of chlorine (or some of its compounds) to water makes it safe to drink and swim in.

Some compounds that include Group 7 elements, or **halogens**, make important contributions to health.

The elements themselves are toxic or harmful. Bromine is also corrosive. Your teacher will use them in a fume cupboard, and will wear gloves when using bromine.

The location of Group 7 in the Periodic Table.

Physical properties

The Group 7 elements are non-metals. Like other non-metals, they do not conduct electricity. They have low melting and boiling points.

Element	Melting point (°C)	Boiling point (°C)	State at room temperature
chlorine	−101	−35	gas
bromine	−7	59	liquid
iodine	114	184	solid

- dense, yellow-green gas
- smelly and poisonous
- melting point −101 °C
- boiling point −35 °C

- deep-red liquid with red-brown vapour
- smelly and poisonous
- melting point −7 °C
- boiling point 59 °C

- grey solid with purple vapour
- smelly and harmful
- melting point 114 °C
- boiling point 184 °C

There are trends in the properties of the Group 7 elements:

- Their melting and boiling points increase from top to bottom of the group.

- Their colours get darker from top to bottom of the group.

The elements exist as diatomic molecules. Each molecule consists of a pair of atoms, strongly joined together. Their chemical formulae are Cl_2, Br_2, and I_2.

In the gas state, the forces of attraction between a molecule and its neighbours are weak. This explains the low melting points.

Chemical reactions

The halogens react with the Group 1 elements. For example, sodium reacts with bromine to make sodium bromide:

$$sodium + bromine \rightarrow sodium\ bromide$$

The other halogens also react with sodium. The reactions get less vigorous from top to bottom of the group. This is different from Group 1, in which the reactions get more vigorous going down the group.

The halogens also take part in **displacement reactions**. If you add chlorine solution to potassium bromide solution, this is what you will see:

	Appearance
chlorine solution (before reaction)	pale green
potassium bromide solution (before reaction)	colourless
mixture after reaction	orange

The orange substance is bromine. It is a product of the reaction. In the reaction, chlorine displaces bromine from potassium bromide:

$$chlorine + potassium\ bromide \rightarrow potassium\ chloride + bromine$$

In a displacement reaction, a more reactive element displaces a less reactive element from a compound. In Group 7, the elements at the top of the group are more reactive than the elements at the bottom of the group. This means that these displacement reactions can take place:

$$chlorine + potassium\ iodide \rightarrow potassium\ chloride + iodine$$

$$bromine + potassium\ iodide \rightarrow potassium\ bromide + iodine$$

Adding chlorine solution to potassium bromide solution makes bromine, which here appears yellow/orange.

Questions

1 Write word equations for the reactions of:
   a potassium with chlorine
   b lithium with bromine.

2 For each pair of reactants, predict whether or not there will be a reaction. Give a reason for each of your decisions.
   a fluorine and potassium chloride
   b bromine and sodium fluoride
   c bromine and sodium iodide

3 Predict which of the following reactions will be most vigorous, and give a reason for your decision.
   • potassium and chlorine
   • lithium and iodine
   • potassium and fluorine

C2.3 How do metals and non-metals combine to form compounds?

A: The Group 0 elements

Key words

➤ inert
➤ noble gases

Food packed in a modified atmosphere.

✱ Electronic configurations
2,8,8,16,16, 32,32,48,48.

What keeps packaged food fresh?

The containers do not contain air. Instead, they are filled up with argon gas. Argon makes up 1% of air. It is in Group 0 of the Periodic Table. The reason for using argon gas to fill in food containers is due to the unique features of Group 0 elements such as argon.

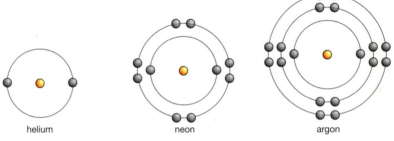

The location of Group 0 in the Periodic Table.

Electronic structure

The Group 0 elements each have a full outer shell of electrons.

The electronic structure of helium.

The electronic structure of neon.

The electronic structure of argon.

Their electronic structures mean that the Group 0 elements are stable. They do not take part in chemical reactions – they are **inert**. That is why they are also called the **noble gases**. Chemists have used the word 'noble' for many years to describe elements that do not react easily.

The stable electronic structures of noble gas atoms mean that they exist as single atoms. They do not join together to form molecules.

The forces of attraction between the single atoms of the Group 0 elements are weak. This means that they have low melting and boiling points. They are all in the gas state at room temperature.

Questions

1 Give the meaning of the word *inert*.

2 Explain why the Group 0 elements do not take part in chemical reactions.

3 Describe the trend in melting points for the Group 0 elements.

4 Give the state of xenon at −110 °C.

Element	Melting point (°C)	Boiling point (°C)
helium	−270	−269
neon	−249	−246
argon	−189	−186
krypton	−157	−152
xenon	−112	−108

B: Ions and reactions

Forming ions

A sodium atom has just one electron in its outer shell. This electron arrangement is not very stable. In chemical reactions, a sodium atom gives away this outer electron. The atom then has eight electrons in its outer shell, just like neon in Group 0. This electronic structure is stable.

When a sodium atom gives away its outer electron, it still has 11 positively charged protons in its nucleus. But it has only 10 negatively charged electrons. Overall, the atom now has a single positive charge. An electrically charged atom is called an **ion**. The formula of this ion is Na^+.

What happens to the electron that a sodium atom gives away? It is added to the outer shell of an atom of another element. This gives the other atom a stable electronic structure.

For example, a chlorine atom has seven electrons in its outer shell. This electron arrangement is not very stable. In chemical reactions, a chlorine atom gains an extra electron. The atom then has eight electrons in its outer shell, just like argon in Group 0. This electronic structure is stable.

The chlorine atom still has 17 positively charged protons in its nucleus. But it now has 18 negatively charged electrons. Overall, the atom now has a single negative charge. The formula of this ion is Cl^-.

Find out about

- how ions are formed during reactions
- how the position of an element in the Periodic Table links to its electron and atomic structure

Key words

➤ ion

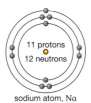
The subatomic particles in a sodium atom.

The subatomic particles in a sodium ion.

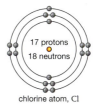
The subatomic particles in a chlorine atom.

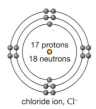
The subatomic particles in a chloride ion.

Models of ions

You can represent atoms and ions using simple dot-and-cross diagrams.

The diagrams below show a potassium ion and a fluoride ion.

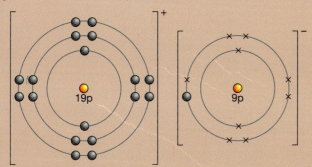

This type of model is useful because it helps to explain why metals join with non-metals in the way that they do. However, the model has limitations. For example, it does not show how the subatomic particles are moving.

Electrons and reactions

The electronic structure of the atoms of an element gives the element its properties. All Group 1 elements have one electron in their outer shell. They give away this electron in chemical reactions. This is why Group 1 elements have vigorous reactions.

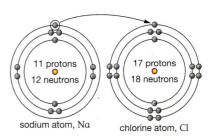
When sodium and chlorine react together, each sodium atom transfers one electron to a chlorine atom.

You saw in the previous section that the Group 1 elements get more reactive going down the group. This is because atom size increases down the group. The outer electrons in bigger atoms are less strongly attracted to the positive nucleus. It is easier for these atoms to lose their outer electrons.

Caesium is near the bottom of Group 1. It is the most reactive element of the group. Caesium reacts violently with water, producing caesium hydroxide and hydrogen gas. Hydrogen bubbles form and the alkaline caesium hydroxide turns phenolphthalein indicator bright pink.

It is not just Group 1 metals that lose electrons in reactions. All metals lose electrons to form positively charged ions. Magnesium reacts with oxygen to make magnesium oxide. In this reaction, magnesium atoms give their two outer electrons to oxygen atoms. Magnesium and oxide ions form, with full outer shells. The formulae of the ions are Mg^{2+} and O^{2-}.

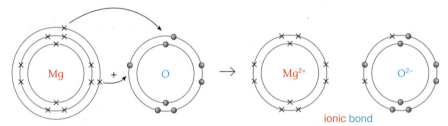

When magnesium burns, magnesium atoms give two electrons to oxygen atoms.

Atoms of all Group 7 elements have seven outer electrons. They need to gain one electron to form a stable ion. This explains the vigorous reactions of Group 7 elements. The reactions of the Group 7 elements get less vigorous from top to bottom.

Some of the other non-metal elements, for example, those in Group 6, also gain electrons in reactions. Ions of non-metal elements are negatively charged.

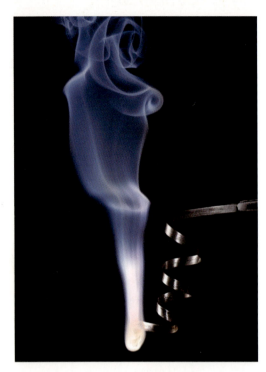

Magnesium burns with a bright white flame. The white 'smoke' is small particles of magnesium oxide.

Questions

1 What is an ion?

2 Explain why each potassium atom gives away one electron when potassium reacts with bromine.

3 Calculate the number of protons, neutrons, and electrons in a sodium ion, Na^+. Its atomic number is 11 and its mass number is 23.

4 Explain why the reaction of potassium with chlorine is more vigorous than the reaction of lithium with chlorine.

C: Properties of ionic compounds

Ionic structure

What gives these crystals their shapes?

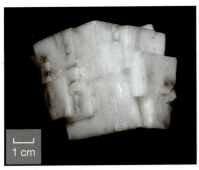

Calcium carbonate.

Sodium chloride.

Vanadinite crystals.

Water evaporates from sodium chloride solution, leaving sodium chloride crystals.

Find out about

- properties of ionic compounds
- how the properties of ionic compounds are related to ionic bonding

Key words

➤ ionic compound
➤ ionic bonding
➤ electrostatic force
➤ giant ionic lattice

The crystals in the pictures are made up of ions. They are **ionic compounds**. In an ionic compound, millions of ions are packed together in a regular pattern. This gives the crystals their shapes.

Sodium chloride is an ionic compound. As a sodium chloride crystal forms, millions of sodium ions, Na^+, and millions of chloride ions, Cl^-, pack closely together. The ions are held together strongly by the attraction between their opposite charges. This is **ionic bonding**. The **electrostatic forces** of attraction between ions act in all directions.

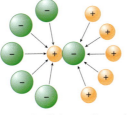

The Na^+ attracts other Cl^- ions that are close to it. The Cl^- attracts other Na^+ ions that are close to it.

Another five Cl^- ions can fit around the Na^+, making six in total, and another five Na^+ ions can fit around the Cl^- ion, making six in total.

Each of these ions then attracts other ions of the opposite charge, and the process continues until millions and millions of oppositely charged ions are all packed closely together.

How a sodium chloride crystal forms.

The structure of sodium chloride is an example of a **giant ionic lattice**. Unlike compounds such as water, which is made up of individual water molecules (H_2O), there are no individual sodium chloride molecules.

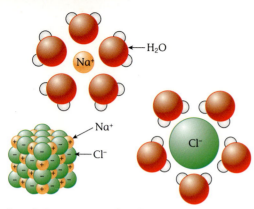

In solution, water molecules surround the positive and negative ions.

Solubility

You can dissolve 90 g of sodium nitrate in a small beaker of water at room temperature. Why is sodium nitrate so soluble?

You can dissolve all this sodium nitrate in 100 g of water.

Sodium nitrate is an ionic compound. When you add any ionic compound to water, its ions separate from each other. Water molecules surround the ions.

Why do water molecules surround positive and negative ions? A water molecule is **polar**. This means that its charges are not evenly distributed. The oxygen part of the molecule has a small negative charge, and the hydrogen parts of the molecule have a small positive charge.

In sodium chloride solution, water molecules surround the positive and negative ions.

Melting points

The table shows the melting points of some chlorides. Can you spot a pattern?

Compound	Is it ionic?	Melting point (°C)
hydrogen chloride	no	−114
lithium chloride	yes	610
phosphorus trichloride	no	−91
potassium chloride	yes	772
sodium chloride	yes	808
sulfur dichloride	no	−122

Ionic compounds have high melting points compared to compounds that contain other types of bonds. When you heat up an ionic compound, energy is transferred to the giant ionic lattice. This energy disrupts the electrostatic forces of attraction, breaking up the regular pattern of ions. The compound melts and becomes liquid.

All ionic compounds have high melting points because their oppositely charged ions are strongly attracted to each other. Large amounts of energy must be transferred to them in order to disrupt their many strong bonds.

Modelling ionic compounds

Chemists use models to show how ions are arranged in an ionic compound. Knowing how the ions are arranged helps them to predict the properties of the compound.

Different coloured spheres are used to make a three-dimensional (3D) physical model.

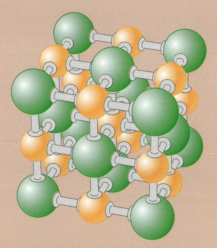

Physical model of sodium chloride.

A 3D diagram can help visualise the structure. A computer generated 3D diagram can be looked at from different angles.

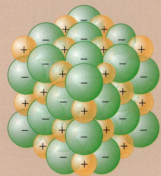

3D diagram of sodium chloride.

A two-dimensional (2D) diagram is easier to draw.

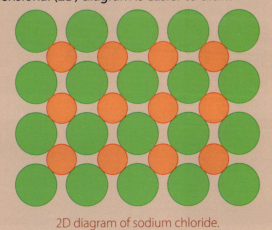

2D diagram of sodium chloride.

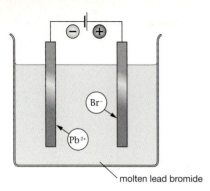

Lead bromide is an ionic compound. Liquid lead bromide conducts electricity because its ions are free to move towards the electrodes.

However, all the models have limitations. For example:

- None of the models shows the forces of attraction between the ions.
- None of the models gives a scale.
- The 2D drawing does not show the 3D arrangement of the ions in space.

Another problem is that none of these models shows how the ions were formed. You can use dot-and-cross diagrams to show how electron transfers form ions. However, these do not show how the ions are arranged in space.

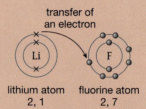

When lithium fluoride forms from its elements, one electron is transferred from a lithium atom to a fluorine atom. A lithium ion, **Li⁺**, and a fluoride ion, **F⁻**, are formed.

Conducting electricity

Which of the substances in the pictures can conduct electricity?

Metals, such as copper, are good conductors of electricity. Sulfur is a non-metal element. It does not conduct electricity.

Solid sodium chloride is a non-conductor. Its charged particles are held in fixed positions, so they cannot move around. This means that they cannot carry electrical charges from one place to another. Sodium chloride solution is different. Its ions are free to move. They can carry electrical charge from place to place. Sodium chloride in the liquid state conducts electricity. In the liquid state, as in the solution, its ions are free to move.

Ionic compounds cannot conduct electricity in the solid state. They do conduct electricity in the liquid state, and when dissolved in water.

Copper.

Sulfur.

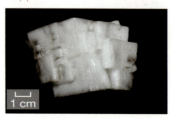

Sodium chloride in the solid state.

Sodium chloride solution.

Questions

1 Describe the bonding in sodium chloride.

2 Explain why potassium chloride has a high boiling point.

3 Explain why potassium chloride conducts electricity in the liquid state but not in the solid state.

4 Describe in detail what happens when sodium bromide dissolves in water.

5 State one purpose of using models to show the arrangement of ions in an ionic compound.

6 Outline three disadvantages of the models shown in the modelling box.

7 Give one advantage of a 3D diagram compared to a 2D diagram for showing the arrangement of ions in an ionic compound.

8 Draw a diagram to show the electron transfer that occurs when lithium fluoride forms from its elements.

A: Writing formulae

Every element and compound has its own **chemical formula**. Chemists work out chemical formulae by experiments, and you can look them up in data tables.

The formula of sodium chloride is $NaCl$ because there is one sodium ion, Na^+, for every chloride ion, Cl^-. There are no molecules in sodium chloride, only ions.

Not all ions have single positive or negative charges. The formula of lead bromide is $PbBr_2$. In this compound there are two bromide ions for every one lead ion. All compounds are overall electrically neutral, so the charge on a lead ion must be twice that on a bromide ion. A bromide ion has a single negative charge, Br^-, and a lead ion has a double positive charge, Pb^{2+}.

The diagram shows the charges on some simple ions. If you know the charges of the ions in a compound, you can work out its formula.

(1)	(2)				(3)	(4)	(5)	(6)	(7)	(0)
			H^+							
Li^+							N^{3-}	O^{2-}	F^-	
Na^+	Mg^{2+}				Al^{3+}	no simple ions		S^{2-}	Cl^-	no ions formed
K^+	Ca^{2+}	transition metals form more than one ion, (e.g., Fe^{2+}, Fe^{3+})							Br^-	
Rb^+	Sr^{2+}								I^-	
Cs^+	Ba^{2+}									
1+	2+				3+		3−	2−	1−	
	metals positive ions							non-metals negative ions		

The charges on simple ions are linked to their positions in the Periodic Table.

Worked example: Writing a formula

Work out the formula of barium bromide.

Step 1: Use the table of ions to write down the formulae of the ions.
Ba^{2+} and Br^-

Step 2: Work out how many ions of each type are needed to make a neutral compound.
A neutral compound needs two Br^- ions for every one Ba^{2+} ion.

Step 3: Write down the formula of the compound. Give the metal ion first.
Barium bromide = $BaBr_2$

Find out about

- how to write formulae of ionic compounds

Key words

➤ chemical formula

Questions

1 Work out the formulae of these ionic compounds. Use the table of ions to help you.
   a potassium iodide
   b sodium bromide
   c calcium bromide
   d aluminium chloride
   e magnesium nitride
   f aluminium sulfide

2 Work out the charge on the following ions. Use the table of ions to help you.
   a copper (Cu) in $CuCl_2$
   b zinc (Zn) in ZnO
   c iron (Fe) in Fe_2O_3
   d hydroxide (OH) in KOH

B: Writing equations

Find out about
- how to write balanced chemical equations

Key words
➤ chemical equation

Chemists write balanced **chemical equations** to summarise reactions.

A balanced chemical equation shows:
- the reactants and products
- the states of the reactants and products
- the relative amounts of the reactants and products.

A chemical equation is not the same as a maths equation. A maths equation includes an equal sign. This means that the two sides of the equation are equal. A chemistry equation includes an arrow. This means *reacts to make*. A balanced equation must have equal numbers of each type of atom on either sides of the arrow.

Worked example: Balancing an equation

Balance this equation: $Li + F_2 \longrightarrow LiF$

Step 1: Count the number of atoms (or ions) of each element on each side of the arrow.

Lithium – one on left, one on right
Fluorine – two on left, one on right

Step 2: Write a number to the left of LiF so that there are same number of fluorine atoms on each side.

$Li + F_2 \longrightarrow 2LiF$

Step 3: Find out if the equation is balanced by counting the number of atoms of each element on each side.

Lithium – one on left, two on right
Fluorine – two on left, two on right
The equation is not balanced.

Step 4: Write a number to the left of Li so that there are the same number of lithium atoms on each side.

$2Li + F_2 \longrightarrow 2LiF$

Step 5: Check that the equation is balanced by counting the number of atoms of each element on each side.

Lithium – two on left, two on right
Fluorine – two on left, two on right
The equation is balanced.

Answer:

$2Li + F_2 \longrightarrow 2LiF$

To write a chemical equation, follow the rules below:

RULES FOR WRITING BALANCED EQUATIONS

Step 1: Write down the word equation.

Step 2: Write the correct formula for each reactant and product under its name.

Step 3: Balance the equation by putting numbers in front of the formulae, if necessary.

NEVER change the formula of a compound or element to balance the equation.

Worked example: Writing a balanced equation

Write a balanced equation to show the reaction of sodium with chlorine to make sodium chloride.

Step 1: Write a word equation.

sodium + chlorine $\longrightarrow$ sodium chloride

Step 2: Write down the formulae for the reactants and products.

$Na + Cl_2 \longrightarrow NaCl$

Step 3: Balance the equation.

$2Na + Cl_2 \longrightarrow 2NaCl$

You must not change any of the formulae. You balance the equation by writing big numbers to the left of the formulae. These numbers then refer to the whole formula.

Step 4: Add state symbols. State symbols usually show the states of the elements and compounds at room temperature and pressure. Substances may be in the solid (s), liquid (l), or gas (g) state, or they may be dissolved in water (aq).

$2Na(s) + Cl_2(g) \longrightarrow 2NaCl(s)$

Answer:

$2Na(s) + Cl_2(g) \longrightarrow 2NaCl(s)$

Here are some word equations and balanced symbol equations for some reactions you might have observed. Study them carefully to check that you know how to work them out for yourself.

hydrogen + chlorine $\longrightarrow$ hydrogen chloride
$H_2(g) + Cl_2(g) \longrightarrow 2HCl(g)$

sodium + water $\longrightarrow$ sodium hydroxide + hydrogen
$2Na(s) + 2H_2O(l) \longrightarrow 2NaOH(aq) + H_2(g)$

zinc + bromine $\longrightarrow$ zinc bromide
$Zn(s) + Br_2(l) \longrightarrow ZnBr_2(s)$

Zinc and bromine react vigorously to make zinc bromide.

Questions

1 Balance the following equations. Include state symbols.
   a $H_2 + Br_2 \longrightarrow HBr$
   b $Fe + Br_2 \longrightarrow FeBr_2$
   c $K + H_2O \longrightarrow KOH + H_2$
   d $NaI + Br_2 \longrightarrow NaBr + I_2$

2 Write balanced symbol equations for these reactions:
   a the reaction of lithium with water
   b the reaction of potassium with chlorine
   c the reaction of potassium bromide with chlorine solution.

A: Properties and uses of transition metals

Find out about

● the properties of transition metals
● how the properties of transition metals give them their uses

Key words

➤ transition metal
➤ catalyst

A car.

Breakfast cereal high in iron.

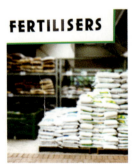
Fertiliser.

Catalysts, coins, and batteries

Look at the three pictures in the margin. How are they linked?

The answer is iron. The car is made from steel, which is mainly iron. The cereal contains iron, which enables your blood to carry oxygen around your body. Iron also speeds up the reaction of nitrogen with hydrogen to produce ammonia, a vital raw material for making fertilisers.

Iron is in the central block of the Periodic Table. All the elements in this block are **transition metals**. The transition metals have similar properties to each other. Their properties give them a huge number of important uses.

Physical properties

The transition metals have some properties that are similar to those of the Group 1 metals. For example, they conduct electricity and have shiny surfaces when freshly cut. But there are differences, too. Compared with the Group 1 metals, the transition metals:

● are stronger and harder

● have higher densities

● have higher melting points (except for mercury, which is liquid at room temperature).

Name	Type of metal	Melting point (°C)	Density (g/cm³)
lithium	Group 1 metal	180	0.53
sodium	Group 1 metal	98	0.97
iron	transition metal	1538	7.9
copper	transition metal	1085	8.9
chromium	transition metal	1890	7.1
gold	transition metal	1063	19
silver	transition metal	961	10

Chemical reactions

The transition metals are less reactive than the alkali metals. For example, at room temperature, the alkali metals react quickly with water and oxygen. The transition metals react slowly, if at all.

Copper, silver, and gold do not react with water or oxygen at room temperature. This explains why they make beautiful and long-lasting jewellery. This property also explains why gold electrical connectors never tarnish. Copper is used for water pipes. This is partly because copper does not react with water at room temperature and partly because it is cheaper than other unreactive metals, such as gold.

Copper water pipes.   Silver jewellery.   Gold electrical connectors.

Iron reacts with water and oxygen at room temperature, but slowly. The product is hydrated iron oxide, or rust.

Colours, catalysts, and ions

Many transition metals form ions with different charges. For example, iron forms Fe^{2+} ions and Fe^{3+} ions. This means that iron forms two oxides, FeO and Fe_2O_3. The oxide ion has a charge of −2 in both of these oxides.

Transition metals form compounds of many colours.

Chromium(III) chloride $CrCl_3$ is violet and chromium(III) nitrate $Cr(NO_3)_3$ is green. Potassium(VI) chromate K_2CrO_4 is yellow and potassium(VI) dichromate $K_2Cr_2O_7$ is orange.

Transition elements are also important **catalysts**. This means they speed up reactions. In catalytic converters, platinum, palladium, and rhodium catalyse reactions in which dangerous exhaust gases are converted to less harmful substances. Nickel catalyses the reaction of hydrogen with vegetable oils to make margarine.

Questions

1 List five uses of transition metals.

2 Draw two bar charts for the metals in the table on the opposite page – one to display the melting-point data and one to display the density data.

3 Describe two similarities and two differences between the physical properties of alkali metals and transition metals.

4 Describe two ways in which the chemical properties of a Group 1 metal and a transition metal are different.

5 Work out the charge on the chromium ion in the compound Cr_2O_3.

Science explanations

C2 Chemical patterns

Chemists have identified patterns and come up with theories to make sense of the world, and to explain how roughly 100 elements can give rise to such a huge variety of chemical compounds.

You should know:

- that the chemists' model of the atom has a tiny central nucleus, containing protons and neutrons, surrounded by electrons
- that the number of electrons is equal to the number of protons and how these numbers relate to atomic number and mass number
- the relative charges and masses of protons, neutrons, and electrons
- the typical size of atoms and small molecules, and how this relates to other objects
- how Mendeleev organised the elements in the first Periodic Table and how he left gaps for elements that were later discovered
- that the position of elements in the Periodic Table is related to the arrangement of their electrons and, therefore, to their atomic number
- how the chemistry of an element is largely determined by the number and arrangement of the electrons in its atoms and, therefore, to its atomic number
- the difference between metals and non-metals on the basis of their physical and chemical properties
- that Group 1 elements are the alkali metals, which react with moist air, water and chlorine, becoming more reactive down the group
- how the Group 7 halogens react with Group 1 elements and their displacement reactions with other metal halides
- how the observed properties of Groups 1, 7, and 0 depend on the outer shell of electrons of the atoms and how to predict properties from given trends down the groups
- how possible reactions and probable reactivity of elements can be predicted from their positions in the Periodic Table
- how ionic compounds form, in terms of electrostatic forces and transfer of electrons
- how to represent ionic substances using dot-and-cross diagrams
- how to write formulae of ionic compounds
- that the properties of an ionic compound are related to the types of bonds between the ions
- how to use word equations and symbol equations to describe reactions, including the physical state symbols
- the properties of transition metals, using copper, iron, chromium, silver, and gold as examples ⓢ
- how to describe the limitations of particular representations and models of ions and ionically bonded compounds, including dot-and-cross diagrams, and 3D representations.

1804
Dalton's solid atom.

1913
The Bohr–Rutherford 'Solar System' atom, in which electrons orbit around a very small nucleus.

1924
A model of the atom in which the electrons are no longer treated as particles but pictured as occupying energy levels, which give rise to regions of negative charge around the nucleus (charge clouds).

1932
The atom in which the nucleus is built up from neutrons as well as protons.

2000+
The present-day atom in which the protons and neutrons in the nucleus are built up from many kinds of particles.

Atomic models from 1800 to the present. The diameter of an atom is about ten million times smaller than a millimetre. These diagrams are distorted. For atoms at this scale, the nuclei would be invisibly small.

Ideas about Science

Scientific explanations are based on data but they go beyond the data and are distinct from them. An explanation has to be thought up creatively to account for the data. A new explanation may explain a range of phenomena not previously thought to be linked. The explanation should also allow predictions to be made about new situations or examples.

In the context of the discovery of new elements and the development of the Periodic Table, you should be able to:

- give an account of scientific work and distinguish between statements that report data and statements of explanatory ideas (hypotheses, explanations, and theories)

- identify where creative thinking is involved in the development of an explanation, such as Mendeleev's insight that he had to leave gaps for undiscovered elements

- recognise data or observations that are accounted for by (or conflict with) an explanation, such as the data from spectra that could be explained by the existence of new elements

- suggest reasons for accepting or rejecting a proposed scientific explanation, such as the way that the shell model of atomic structure accounts for the arrangement of elements in the Periodic Table

- identify the best of given scientific explanations for a phenomenon, such as Mendeleev's use of his Periodic Table to predict the existence of unknown elements

- understand that when a prediction agrees with an observation, this increases confidence in the explanation on which the prediction is based but does not prove it is correct. For example, Mendeleev correctly predicted the properties of missing elements in his Periodic Table.

Mendeleev's Periodic Table.

Models are used in science to help explain ideas and to test explanations. A model identifies features of a system and rules by which the features interact. It can be used to predict possible outcomes. The scientific model of a nuclear atom can be used to explain a range of observations and to predict the outcomes of experiments. Dot-and-cross diagrams are another kind of model – they help represent chemical interactions. For a variety of models used in chemistry you should be able to:

- describe the main features of the model

- suggest how the model can be used to explain or investigate a phenomenon

- use the model to make a prediction

- identify limitations of the model.

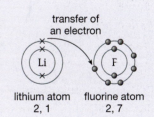

transfer of
an electron

lithium atom
2, 1

fluorine atom
2, 7

C2 Review questions

1 Ernest Rutherford used Thomson's 'plum-pudding' model of the atom to predict observations in an experiment.
Other scientists carried out the experiment. They fired positive particles at gold foil and observed what happened to them.

  a What did Rutherford predict would happen when the positive particles were fired at gold foil?

  b What observations were made during the experiment?

  Rutherford used these observations to suggest a new model of the atom.

  c Describe Rutherford's new model of the atom.

  d Suggest how this model accounts for the observations in the experiment.

2 The mass number of an iron atom is 56.

  a Use the Periodic Table to find the atomic number of iron.

  b Now use the mass number and atomic number to work out how many of the following subatomic particles are in this atom of iron:

    i protons

    ii neutrons

    iii electrons

  c One isotope of iron has a mass number of 56. Another isotope has a mass number of 59.

    i Describe one other way that these isotopes are different.

    ii Describe two ways that these isotopes are similar.

3 The table shows part of the Periodic Table, with only a few elements included.

  a Using only the elements in the table, write down the symbols for:

    i a metal that floats on water

    ii an element with similar properties to silicon (**Si**)

    iii three elements that have molecules made up of two atoms

    iv the most reactive metal.

  b Predict the state of fluorine, **F**, at room temperature. Give a reason for your prediction.

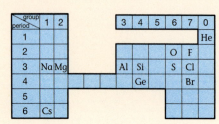

4 The table shows the properties of three halogens.

Halogen	Boiling point (°C)	Formula of compound with iron
chlorine	−34.7	$FeCl_3$
bromine	58.8	$FeBr_3$
iodine	184	

  a **i** Describe the trend in boiling points.

    ii Predict the formula of iron iodide.

    iii The formula of a bromide ion is Br^-. Work out the charge on the iron ion.

  b **i** Describe the pattern of reactivity of the halogens going down the group.

    ii Give examples of reactions that show this pattern.

    iii Explain the pattern using ideas about electron arrangement and the formation of ions.

  c How does the model of atomic structure with electrons in shells explain the similarities between the halogens?

5 The table shows data about some elements in Group 7.

Element	Formula of molecule	State at 20°C	Melting point in °C
fluorine	F_2	gas	−220
chlorine	Cl_2	gas	−101
bromine	Br_2	liquid	59
iodine	I_2	solid	114

  a One of the melting points is wrong. Which melting point in the table is wrong? Explain how you made your choice.

  b Estimate the correct value for this melting point.

  c Astatine is another element in Group 7. What is the formula for a molecule of astatine?

6 The diagrams show the electrons in the outer shell of different elements from the Periodic Table.

  a **i** Which diagram could be for an element in Group 0?

    ii Explain why elements in Group 0 are unreactive.

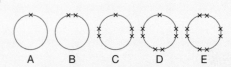

  b Which diagram could be for an element with properties similar to sodium?

  c Element **D** will form an ion with a single negative charge.

    Draw a diagram to show the electron arrangement for the outer shell of the ion of element **D**.

7 Chlorine reacts with metals to make metal chlorides.

The table gives information about metals and metal chlorides.

Metal	Number of electrons in outer shell of atom	Formula of metal ion	Formula of metal chloride
lithium	1	Li^+	LiCl
sodium	1	Na^+	NaCl
beryllium	2	Be^{2+}	$BeCl_2$
magnesium	2	Mg^{2+}	$MgCl_2$
aluminium	3	Al^{3+}	$AlCl_3$

  a Describe two links between the information in the columns in the table.

  b Write the word and the symbol equation for the reaction of lithium and chlorine to make lithium chloride.

  c Potassium has one electron in the outer shell of its atom and gallium has three electrons. Write down the formula of:

    i potassium chloride

    ii gallium chloride.

  d Copper reacts with chlorine to make copper chloride, $CuCl_2$.

    What are the symbols for the two ions in this compound?

C3 Chemicals of the natural environment

Why study chemicals of the natural environment?

Metals and crude oil are both extracted from the natural environment. Metals and compounds from crude oil are essential to our technological society and bring significant benefits to people's lives. For example, they are used in transport, construction, and manufacturing. However, the extraction of metals and crude oil is using up finite resources and can have an impact on the quality of the environment. Scientists are working in these areas to ensure that we have sustainable quality of life in the future.

What you already know

- Some elements are metals and some are non-metals.
- The properties of metals and other materials can be measured in experiments.
- Scientists and engineers decide how to use materials by evaluating data and information about their properties and how they react.
- There are different states of matter (solid, liquid, and gas) and these can be explained using the particle model.
- Symbols for elements and compounds show the numbers and types of atoms and molecules that they contain.
- How to interpret and write equations for reactions using words and formulae.
- Properties of materials can be explained by understanding how the atoms are held together by chemical bonds (e.g., in ionic compounds).
- Models can be used to represent atoms and molecules, and how they bond together.
- Acids have low pH values and react with metals and other compounds to make salts.

The Science

The method used to extract metals is linked directly to their reactivity. Models can be used to represent the structure of metals and carbon compounds from crude oil. They support our understanding of the properties and reactions of these substances. Data and information enable chemists to identify trends and patterns, and to explain why metals and carbon compounds react as they do. An understanding of these patterns helps to identify ways of using metals and carbon compounds more effectively.

Ideas about Science

Metals and carbon compounds enhance the quality of our lives because they provide the materials to make many of the manufactured goods that we use. Electrolysis as a means of extracting reactive metals was developed in the late 19th century by several scientists who thought creatively and built on each other's work.

Data from experiments is used to test hypotheses and to make and test predictions about the behaviour of metals and carbon compounds. Chemists also work to develop new methods to counteract and reduce the impact of extraction of materials from the natural environment and to increase sustainability.

C3.1 How can the properties of metals be explained?

A: Structure and properties of metals

Find out about
- properties of metals
- a model for metallic bonding

Key words
➤ alloy
➤ giant structure
➤ malleable
➤ metallic bonding
➤ ion

Most metals have high melting points.

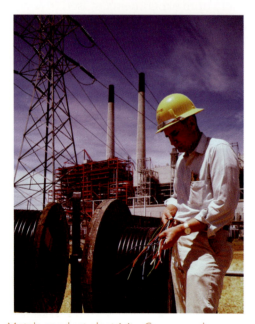
Metals conduct electricity. Copper and aluminium are used as conductors in power lines and household electricity cables.

Metal properties

Metals have been part of human history for thousands of years. Even though new technology has made many new materials available, people still depend on metals for construction, for vehicles, and to carry electricity. This is due to the unique properties of metals.

Engineers continue to develop new uses for metals and **alloys**. Alloys are composed of metals or a metal and a non-metal. For instance, brass is an alloy made of copper and zinc, whereas steel is made of iron and carbon. Many other metals and alloys, including aluminium, magnesium, titanium, bronze and Inconel, are now used in engineering.

Models for the structure and bonding in metals

Most metals have high melting points and are strong, and all conduct electricity. To explain why they have these properties, materials scientists need to know something about the structure of metals.

Giant structure of metals

Scientists use models to describe what they have discovered about the structures of metals. A simple diagram shows how metal atoms are arranged in a **giant structure**. This model shows metal atoms:
- as tiny spheres
- arranged in a regular pattern
- packed closely together in a giant structure.

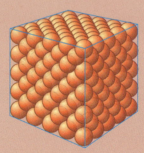
The arrangement of atoms in copper.

This model explains some of the properties of metals.

You can see how closely together the atoms of copper are packed. The atoms must be held to each other by strong chemical bonds, because copper is strong and has a high melting point. Copper is flexible and **malleable** because the atoms are arranged in layers that can slide over each other to allow copper to stretch and change shape.

The model that shows atoms in a metal as spheres is limited:
- It does not show how the bonds are formed.
- It does not explain why metals conduct electricity.

The metallic bonding model

A more complex model is needed to represent **metallic bonding.**

'sea' of freely
moving electrons

lattice of
positive ions

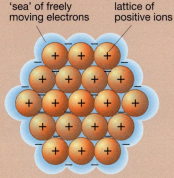

A representation of metallic bonding. This model shows the atoms sharing outer electrons, which hold the charged atoms (metal ions) together.

Metal atoms tend to share the electrons in their outer shell easily. In the solid metal, the atoms lose these shared electrons and become positively charged **ions.** The electrons are no longer held by any one particular atom; they drift freely between the metal atoms. The strong attraction between the 'sea' of negative electrons and the positively charged metal ions holds the structure together.

The electrons move through the giant structure. This explains why metals conduct electricity well. When an electric current flows through a metal wire, the free electrons drift from one end of the wire towards the other. Although the electrons are free, the metal ions are packed closely together in a regular lattice.

Metallic bonding explains why metals are strong and why they conduct electricity well. However, the model does not explain why different metals have different properties. Although it appears to show metals containing ions, metals behave like a collection of atoms and do not have typical ionic properties.

Many metals are strong. The titanium hull of this research submarine is strong enough to withstand the pressure at a depth of 6 km. Titanium is also used to make hip joints and racing cars.

Metals are malleable so can be bent or pressed into shape without breaking. Aluminium sheet can be moulded under pressure to make cans.

Questions

1 Use the diagrams of the two models for the structure and bonding in metals to explain why:
   a metals are strong and have high melting points
   b metals can be stretched or hammered into different shapes
   c metals conduct electricity.

2 Write down some uses of metals that depend on each of the properties of metals from question **1**.

A: Extracting metals from ores

Find out about

- extracting metals from their ores
- metal-extraction issues
- how carbon reacts with metal oxides
- **H** greener ways of extracting metals

Metal ores

Minerals are naturally occurring inorganic materials that have a chemical formula. They should not be confused with rocks, which do not have a specific chemical composition. Rocks that contain metal compounds are called **ores**. Over hundreds of millions of years, rich deposits of ores have built up in certain parts of the Earth's crust. People mine for ores and extract metals from the metal compounds. There is a huge demand for all types of metals to support technological lifestyles.

Metal ores

Metals have different reactivities. A few unreactive metals, such as gold, are found uncombined. Gold has been used by humans for more than 5000 years. More **reactive metals**, such as copper and iron, occur as compounds. This means that they could not be used until people found out how to **extract** them from the metal compounds in their ores. The metal compounds are often **oxides** or sulfides of the metals. Extracting metals from minerals involves **reduction** reactions to remove the oxygen from the metal.

Worldwide there is a huge demand for copper metal. It is the main metal used to make electricity cables and electrical circuitry in appliances. Water pipes are often made from copper.

Gold is so unreactive that it occurs uncombined. Most metals occur as compounds.

Our lifestyle depends on using large amounts of metals such as copper.

Metal	Name of the ore	Chemical in the mineral
aluminium	bauxite	aluminium oxide, Al_2O_3
copper	copper pyrite	copper iron sulfide, $CuFeS_2$
gold	gold	gold, Au
iron	haematite	iron oxide, Fe_2O_3
sodium	rock salt	sodium chloride, $NaCl$

Extracting metals: the issues

Mining extracts both metal ores and waste rock. The metal compounds are usually only present in tiny amounts with a large amount of waste rock. Over time, as people mine more and more metals, the deposits left in the Earth have smaller and smaller percentages of useful compounds. Scientists now mine copper from rock that contains less than 1% copper. This means that mining enough copper to meet demand creates large holes and produces a very large amount of waste rock.

There is a range of factors to consider when deciding how a metal should be extracted.

	Metal	Method
MORE REACTIVE ↑	potassium sodium calcium magnesium aluminium	electrolysis of molten ores
	zinc iron tin lead copper	reduction of ores using carbon
LESS REACTIVE	silver gold	metals occur uncombined

To meet demand for copper, we quarry millions of tonnes of copper ore. Mining and extracting copper has an impact on the environment.

How can the ore be reduced?

The more reactive the metal, the harder it is to reduce its ore. The table above compares the methods used to reduce different ores.

Is there a good supply of ore?

Metal ores are mined in different parts of the world. If ore is not very pure, it may not be worth using. The cost of concentrating the ore may be too great. The more valuable the metal is, the lower the quality of ore that can be used.

What are the energy costs?

It takes energy to extract metals. This is especially true if the metal is extracted using electricity. For example, a quarter of the cost of producing aluminium is the cost of electricity.

What is the impact on the environment?

Metals such as iron and aluminium are produced on a huge scale. Millions of tonnes of ore are needed. Mining this ore can have a big environmental impact. This is why it is important to recycle metals. It takes about 250 kg of copper ore to make 1 kg of copper. So 1 kg of recycled copper means that 250 kg of ore need not be dug up. The negative impact of mining ore is reduced if the waste material can be used. Some hard core for roads and the foundations of buildings comes from waste mining rock.

Even if all waste copper is recycled, there is not enough to meet demand. In the next 25–50 years the known world reserves of copper and some other metals may run out.

Key words

➤ mineral
➤ ore
➤ reactive metals
➤ extract
➤ oxide
➤ reduction

Using carbon to extract metals

Many metals can be extracted from their ores by heating with carbon. This works for some metals but not for very reactive metals. Reactive metals, such as aluminium, are extracted by **electrolysis**.

Carbon works well to extract metals from ores that contain metal oxides, but some metal ores contain metal sulfides. Zinc ores include zinc blende (**ZnS**), zinc spar (**ZnCO₃**), and zincite (**ZnO**). Zinc blende, which is zinc sulfide (**ZnS**), is the main ore for zinc extraction. Sulfide ores can be heated in air so that the sulfide compounds become oxides, for example, turning zinc sulfide into zinc oxide.

To extract zinc from zinc oxide, the oxygen is removed so that ZnO becomes **Zn**. This process of removing the oxygen is called reduction. Carbon is used as a **reducing agent** to remove oxygen.

$$\text{zinc oxide} + \text{carbon} \longrightarrow \text{zinc} + \text{carbon monoxide}$$

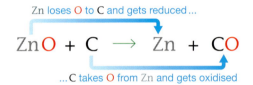

Zn loses O to C and gets reduced ...

$$ZnO + C \longrightarrow Zn + CO$$

... C takes O from Zn and gets oxidised

Further **oxidation** of the carbon monoxide forms carbon dioxide, which is less harmful.

Carbon is often used as a reducing agent to extract metals because:

- Carbon, in the form of coke, can be made cheaply from coal.
- At high temperatures, carbon has a strong tendency to react with oxygen, so it is a good reducing agent.
- The carbon monoxide formed is a gas, so it is not left behind to make the zinc impure.

Carbon can also be used to extract iron and copper. These reactions can be summarised as:

$$\text{iron oxide} + \text{carbon} \longrightarrow \text{iron} + \text{carbon dioxide}$$
$$\text{copper oxide} + \text{carbon} \longrightarrow \text{copper} + \text{carbon dioxide}$$

H Green methods of extracting metals

Scientists are developing new methods to extract metals from old mining waste heaps, which contain very low percentages of useful metals. These methods are 'green' because they make use of waste material and have much lower energy costs than traditional mining. Their negative impact on the environment is less than traditional mining. Also, some metals are toxic so these methods remove toxic waste and make it into a useful resource. At the moment the amount of metal that can be produced is far too small to make much difference, but these methods may be more important as our supplies of metals run out.

Synoptic link

You can learn more about electrolysis in C3.3 *What are electrolytes and what happens during electrolysis?*

Synoptic link

You can learn more about oxidation in C1.1 *How has the Earth's atmosphere changed over time, and why?*

H

Using bacteria to extract metals

Bioleaching uses bacteria present in old spoil heaps to oxidise sulfide minerals. There are several different types of process. For one process, the optimum conditions for enzymes in the bacteria to work include an acidic pH, so the spoil heaps are sprayed with sulfuric acid. The run off from the heaps is carefully collected. It contains dilute copper sulfate ($CuSO_4$). Copper can be extracted from this in a **displacement reaction** with iron or by electrolysis.

Phytoextraction

Phytoextraction uses plants to take metal ions from soils or mining waste. Phytoextraction is not yet used to extract useful metals but is very useful for taking toxic metals out of the soil to 'clean up' the environment around old mines. Although the metals are toxic even to the plants, some types of plant have a very high tolerance for toxic metals and can remove large amounts of these metals from the soil. They store the metals in their leaves.

In the UK, there are now strict government regulations that prevent companies from dumping waste that contains toxic metals. Phytoextraction is useful to treat old mine workings or to help to clean up waste heaps caused by cheap metal extraction in other countries where these regulations are not in place.

Waste from old mines contains tiny amounts of valuable minerals. Some waste also contains toxic heavy metals, which can cause problems in the environment.

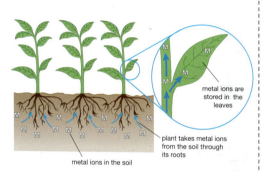

metal ions are stored in the leaves

plant takes metal ions from the soil through its roots

metal ions in the soil

Phytoextraction removes toxic metals from soil. The plant stores the metals in its leaves.

Questions

1 Suggest explanations for these facts:
 a The Romans used copper, iron, and gold but not aluminium.
 b Iron is cheap compared to many other metals.
 c Gold is expensive, even though it is found uncombined in nature.
 d About half the iron we use is recycled, but nearly all the gold is recycled.
 e The tin mines in Cornwall have closed, even though there is still some tin ore left in the ground.

2 Copper can be extracted from copper sulfide in a two-step process.
 a In step 1, copper sulfide (CuS) is heated in air to make copper oxide (CuO) and sulfur dioxide. Write a word equation for this reaction.
 b What problems might be caused by making sulfur dioxide on a large scale?
 c In step 2, copper oxide is heated with carbon. Write a word and formula equation to show how this reaction makes copper and carbon dioxide.

3 Why do oxidation and reduction always go together when carbon extracts a metal from a metal oxide?

4 Think about how mining copper affects people and the environment. What are the main benefits and costs of mining copper? Make a table to summarise your ideas.

H 5 A headline in a scientific magazine says: 'Bacteria and Plants: Hopes for a green mining future'.
 Write a short article to go with this headline. Make it clear what the benefits and limitations of the new approaches are likely to be.

B: Reactivity of metals

Key words

➤ acid
➤ formula
➤ salt

Potassium is a very reactive metal. It gives out so much energy when it reacts with water that the hydrogen gas catches fire as it forms.

Magnesium reacts too slowly with water to follow the reaction easily in an experiment. The metal sometimes floats because tiny bubbles of hydrogen form on the surface and keep it afloat.

Patterns in reactivity

The method of extraction for a metal depends on its reactivity. There are patterns in the reactivity of metals. These are shown by how they react with water and with **acids**.

Only very reactive metals, such as those in Group 1 and Group 2 of the Periodic Table, react quickly enough with water to see a fast change in an experiment.

- Very reactive metals react explosively with acids.
- Less reactive metals react steadily with acids.
- Unreactive metals do not react with either water or acids.
- Reactions of metals with water and acids make hydrogen gas so that you can see fizzing or bubbling. This helps you to work out the order of reactivity of the metals.

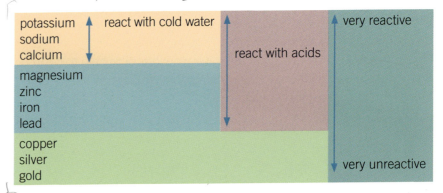

Patterns in reactivity: reaction with water

Group 1 and Group 2 metals are reactive metals. Metals from both groups react with water.

When metals react with water they make a metal hydroxide and hydrogen.

$$\text{metal} + \text{water} \longrightarrow \text{metal hydroxide} + \text{hydrogen}$$

When Group 1 and Group 2 metals react with water, the metals are more reactive lower down the group.

- Group 2 metals are less reactive than Group 1 metals in the same period.
- The word equations are the same.
- The **formulae** of the metal hydroxides are different for Group 1 and Group 2 metals. This is because Group 1 metals form ions with a 1+ charge, Group 2 ions have a 2+ charge, for example, Na^+ and Ca^{2+}.
- The formula for a hydroxide ion is always OH^-.

Remember that in a formula the charges on the ions must balance.

Worked example: Writing formulae

1. *Write the formula for sodium hydroxide.*

Step 1: Write down the symbols for the ions, including their charge.

Sodium hydroxide contains the ions:

Na^+ (in Group 1 so one positive charge)

OH^- (with one negative charge)

Step 2: How many OH^- are needed to balance the + charge of the metal ion?

one OH^- will balance one Na^+

Step 3: Write down the formula.

formula of sodium hydroxide: **NaOH**

2. *Write the formula for calcium hydroxide.*

Step 1: Write down the symbols for the ions, including their charge.

Calcium hydroxide contains the ions:

Ca^{2+} (in Group 2 so two positive charges)

OH^- (with one negative charge)

Step 2: How many OH^- are needed to balance the + charge of the metal ion?

two OH^- will balance one Ca^{2+}

Step 3: Write down the formula.

formula of calcium hydroxide: **Ca(OH)$_2$**

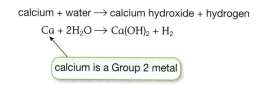

sodium is a Group 1 metal

$$\text{sodium + water} \longrightarrow \text{sodium hydroxide + hydrogen}$$
$$2Na + 2H_2O \longrightarrow 2NaOH + H_2$$

$$\text{calcium + water} \longrightarrow \text{calcium hydroxide + hydrogen}$$
$$Ca + 2H_2O \longrightarrow Ca(OH)_2 + H_2$$

calcium is a Group 2 metal

Patterns in reactivity: reaction with acid

The reactivity of less reactive metals can be investigated by looking at their reactions with acids.

Metals react with acids to make a metal **salt** and hydrogen.

$$\text{metal + acid} \longrightarrow \text{metal salt + hydrogen}$$

$$\text{magnesium + sulfuric acid} \longrightarrow \text{magnesium sulfate + hydrogen}$$
$$Mg + H_2SO_4 \longrightarrow MgSO_4 + H_2$$

Magnesium is in Group 2. Its ion has a 2+ charge.

$MgSO_4$ contains these ions... Mg^{2+} SO_4^{2-}

$$\text{magnesium + hydrochloric acid} \longrightarrow \text{magnesium chloride + hydrogen}$$
$$Mg + 2HCl \longrightarrow MgCl_2 + H_2$$

$MgCl_2$ contains two Cl^- ions to balance the 2+ charge on the Mg^{2+}

Mg^{2+} Cl^- Cl^-

- All metals make a salt. The name and formula of the salt depends on the metal and the acid in the reaction.

- The formula of the salt can be worked out using the charge on the metal ion and the formula of the negative ion from the acid.

- When writing an equation, work out the correct formula for the salt before you start balancing.

Magnesium reacts quickly with acids. The bubbles are hydrogen gas.

Acid	Formula of acid	Name of salt	Negative ion in salt
sulfuric acid	H_2SO_4	metal sulfate	SO_4^{2-}
hydrochloric acid	HCl	metal chloride	Cl^-
nitric acid	HNO_3	metal nitrate	NO_3^-

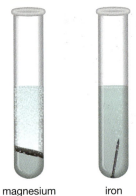

magnesium   iron

More reactive metals react faster with acid. Sometimes the difference between the reactivity of metals can be easily seen by comparing how quickly the bubbles of hydrogen are made.

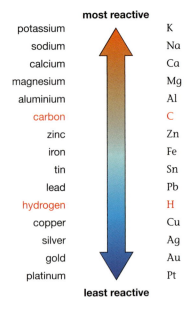

most reactive		
potassium	K	↑
sodium	Na	
calcium	Ca	
magnesium	Mg	
aluminium	Al	
carbon	C	
zinc	Zn	
iron	Fe	
tin	Sn	
lead	Pb	
hydrogen	H	
copper	Cu	
silver	Ag	
gold	Au	↓
platinum	Pt	
least reactive		

A coating of copper forms on any iron object if it is put into copper sulfate solution. Iron displaces copper.

Investigating reactivity

The reactivity of metals can be investigated by measuring how quickly the metal reacts with water or with an acid. These experiments provide data to test hypotheses and predictions about the reactivity of metals.

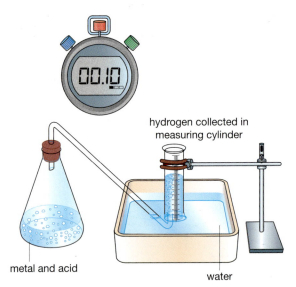

hydrogen collected in measuring cylinder

metal and acid   water

Measuring the volume of gas made in a certain time is one way to collect data for comparing reactivity.

● Collecting data from experiments allows us to put metals in order of reactivity.

● It is helpful to put carbon and hydrogen in the list, even though they are not metals:

   ■ Metals less reactive than carbon can be extracted by heating their oxides with carbon.

   ■ Metals more reactive than carbon are extracted using electrolysis.

   ■ Metals more reactive than hydrogen react with acids, metal less reactive than hydrogen do not.

Displacement reactions

If iron is put into a solution containing copper ions, copper metal forms on the surface of the iron. This reaction can be used to extract copper or to plate copper onto iron objects. This works because iron is more reactive than copper.

A displacement reaction happens when a more reactive metal reacts with a compound of a less reactive metal. Displacement reactions cannot happen in the opposite direction. Less reactive metals do not react with compounds of more reactive metals.

Data from displacement reactions can be used to work out the order of reactivity of metals. The order of reactivity is the same whether you look at data from metal reactions with water, with acids, or in displacement reactions.

	magnesium chloride	zinc chloride	lead chloride	copper chloride
magnesium		zinc forms	lead forms	copper forms
zinc	no change		lead forms	copper forms
lead	no change	no change		copper forms
copper	no change	no change	no change	

Some results of displacement reactions between metals and solutions of metal chloride compounds.

Evidence that a displacement reaction is happening may be that:

● a new solid forms (this is the displaced metal)

● the solution changes colour

● the temperature changes.

Explaining displacement reactions

Iron reacts with copper sulfate because iron is more reactive than copper. A more reactive metal can displace a less reactive metal from its compounds.

In a displacement reaction:

● the more reactive metal always forms an ion

● the less reactive metal always turns into the metal

● you can use the reactivity list to work out whether a displacement reaction will happen

● reactions do not happen 'backwards'. Copper cannot react with iron sulfate because copper is less reactive than iron.

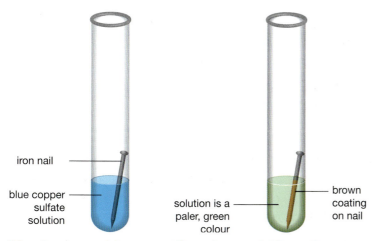

iron nail

blue copper sulfate solution

solution is a paler, green colour

brown coating on nail

When iron reacts with copper sulfate, a brown solid forms. This is copper. The blue solution changes colour. It becomes green as the copper ions in the solution are replaced with iron ions.

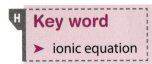

Key word

➤ ionic equation

Metals and metal ions

When metals react they form compounds. The compounds contain positive metal ions. This means that when a metal reacts it always forms an ion.

More reactive metals react more easily because they form ions more easily. It takes more energy to change ions of reactive metals back into metal atoms.

Ions of less reactive metals can be changed into metals more easily. This is why it is easier to extract less reactive metals from metal compounds in their ores.

Ideas about changing metals into metal ions and back again can be used to explain displacement reactions. The more reactive metal forms metal ions, while the less reactive metal turns from its ion into a metal.

iron + copper sulfate ⟶ copper + iron sulfate

> Iron atoms become iron ions. $FeSO_4$ contains Fe^{2+} ions

$Fe + CuSO_4 \rightarrow Cu + FeSO_4$

> Cu^{2+} ions in $CuSO_4$ become copper atoms

The sulfate ions are not involved in the reaction between copper sulfate and iron. You can leave them out when you write an **ionic equation.**

$$\text{iron} + \text{copper ions} \longrightarrow \text{copper} + \text{iron ions}$$

$$Fe(s) + Cu^{2+}(aq) \longrightarrow Cu(s) + Fe^{2+}(aq)$$

- (s) shows that **Fe** (iron) and **Cu** (copper) are solids.
- (aq) shows that Fe^{2+} ions and Cu^{2+} ions are dissolved in water.

In an ionic equation you only need to show the ions that take part in the reaction. This makes it very clear to see that iron has formed an ion (because it is more reactive) and that a copper ion has changed into solid copper metal.

In ionic equations:

- the numbers of each type of atom on each side must balance
- the total charges shown on each side must balance
- ions that do not take part in the reaction can be left out.

Results from displacement reactions can be used to work out which metal is the most reactive.

The thermite reaction used to be used to join railway rails together. Aluminium powder displaces iron from iron oxide. The reaction gives off so much heat that the iron is a liquid and runs into the gaps between the rails.

Questions

1 Write word equations for these reactions:
   a potassium and water
   b magnesium and water

2 Work out the formulae of these metal hydroxides:
   a potassium hydroxide
   b magnesium hydroxide

3 Use your answers to question **2** to help you to write balanced symbol equations for the reactions in question **1**.

4 Write word equations for these reactions:
   a zinc and sulfuric acid
   b calcium and hydrochloric acid

5 Work out the formulae of these metal hydroxides and metal salts:
   a zinc sulfate (the symbol of a zinc ion is Zn^{2+})
   b calcium chloride
   c sodium nitrate

6 Use your answers to question **5** to help you to write balanced symbol equations for the reactions in question **4**.

7 Use the list of metals in order of reactivity to decide whether each of these statements is true or false:
   a Magnesium is more reactive than zinc.
   b The most reactive Group 1 metal in the list is sodium.
   c Heating calcium oxide and carbon gives calcium and carbon dioxide.
   d Tin can be extracted from tin oxide by heating with carbon.
   e Copper reacts with sulfuric acid.
   f Iron reacts with nitric acid.

8 a Which metal in the table on page 93 would you expect to react most slowly with nitric acid?
   b Write a word equation for the reaction with nitric acid.

9 How does the table on page 93 show that magnesium is the most reactive metal and copper is the least reactive metal?

10 Andy tests a metal. He finds that it reacts with lead chloride to make lead. It does not react with zinc chloride. Andy wants to find out the name of the metal.
   a Use the order of reactivity to suggest what the identity of the metal might be.
   b What further experiments could Andy do to make certain he has identified the metal correctly?

11 Use the order of reactivity to decide whether a reaction would happen between the following metals and metal compounds:
   a magnesium and lead chloride
   b calcium and zinc sulfate
   c copper and magnesium sulfate
   d zinc and copper sulfate

12 Write word equations for any reactions that will happen in question **11**.

13 Write ionic equations for any reactions that will happen in question **11**.
   Use these symbols for ions to help you: magnesium (Mg^{2+}), lead (Pb^{2+}), copper (Cu^{2+}), calcium (Ca^{2+}), zinc (Zn^{2+}).

A: What is electrolysis?

Find out about

- the discovery of electrolysis
- electrolysis of molten salts
- reactions at the electrodes
- **H** half equations
- oxidation and reduction
- electrolysis of aqueous solutions

Key words

➤ electrode
➤ cathode
➤ anode

Davy and Faraday both gave lectures to other scientists and to the public to explain what they had found out. Some scientists were very interested in their ideas but they were often challenged and some mocked them because they thought that electricity would never have any practical use.

Discovery of electrolysis

In the late 18th century scientists began to do experiments with electricity. Scientists had recently discovered how to control electric currents and were interested in finding out what would happen if they passed electricity through different materials. They found that:

- metals conduct electricity but are not changed when it passes through
- some salts conduct electricity when they are in liquid form (when they are melted or dissolved in water)
- when salts conduct electricity they break down into new substances.

Electrolysis is the process of breaking down compounds using electricity. The term electrolysis is based on two Greek words: 'electro' meaning 'electricity' and 'lysis' meaning 'splitting'.

The discovery of electrolysis was very important because, for the first time, it was possible to extract very reactive metals from their compounds. In 1807 and 1808, Humphry Davy (an English chemist) used electrolysis to isolate the elements potassium, sodium, barium, strontium, calcium, and magnesium. This was the first time anyone had been able to do this as these elements are too reactive to be extracted by heating their ores with carbon.

Faraday's theory of ions

Michael Faraday worked with Humphry Davy. He began as an assistant, and then established himself as a leading scientist in his own right. In 1833 he began to study electrolysis. He had to be creative to come up with an explanation that would account for all his observations.

Faraday suggested that compounds must contain charged particles. Since opposite charges attract each other, he could imagine the negative **electrode** attracting positively charged particles and the positive electrode attracting negatively charged particles.

The charged particles move towards the electrodes. When they reach the electrodes, they turn into atoms. This explains why chemical changes happen during electrolysis.

Faraday named these moving, charged particles ions, from a Greek word meaning 'to go'. This was the first time the word ion had been thought about or used. Eventually, Faraday's ideas led to the models scientists use today to explain how ionic compounds behave.

The nature of electric current was not known when Faraday studied electrolysis. He named the two electrodes the **cathode** (which we now know attracts positive ions) and the **anode** (which attracts negative ions).

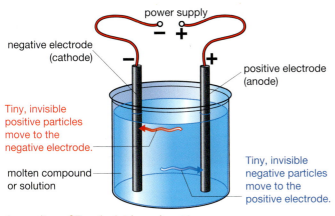

An outline of Faraday's ideas about ions.

negative electrode (cathode)
positive electrode (anode)
power supply

Tiny, invisible positive particles move to the negative electrode.

molten compound or solution

Tiny, invisible negative particles move to the positive electrode.

Electrolysis of molten salts

The apparatus below is used to investigate what happens when electricity passes through a molten salt.

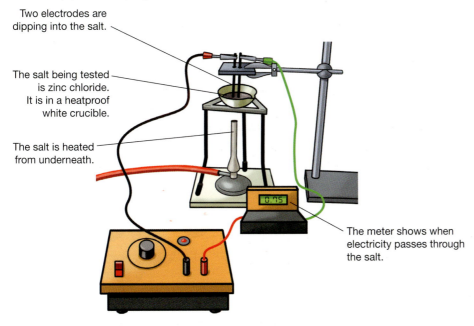

Two electrodes are dipping into the salt.

The salt being tested is zinc chloride. It is in a heatproof white crucible.

The salt is heated from underneath.

The meter shows when electricity passes through the salt.

At first there is no reading on the meter, showing that the solid zinc chloride does not conduct electricity. As it is heated, the zinc chloride melts. As soon as the zinc chloride melts, there is a reading on the meter and the compound begins to decompose chemically. The most obvious change is the bubbling around the positive electrode. The gas made is chlorine.

After a while, the current is switched off. When cool enough, shiny pieces of zinc metal can be seen. The electric current splits the compound into its elements: zinc and chlorine.

$$\text{zinc chloride} \longrightarrow \text{zinc} + \text{chlorine}$$

$$ZnCl_2 \longrightarrow Zn + Cl_2$$

Key words

➤ electrolyte
➤ binary compound
➤ half equation

Explaining electrolysis

Faraday's ionic model is still used today to explain electrolysis. Ionic theory states that all salts are made of ions. For example, sodium chloride is made of sodium ions and chloride ions. Sodium ions, Na^+, are positively charged. The chloride ions, Cl^-, are negatively charged. These oppositely charged ions attract each other.

A crystal of sodium chloride consists of millions and millions of Na^+ and Cl^- ions closely packed together. In the solid, these ions cannot move towards the electrodes and the compound cannot conduct electricity. The ions can move when sodium chloride is hot enough to melt or when it is dissolved in water. Molten ionic compounds and ionic compounds dissolved in water are both **electrolytes** because they conduct electricity due to their freely moving ions.

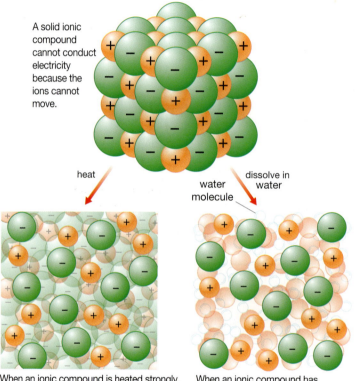

A solid ionic compound cannot conduct electricity because the ions cannot move.

heat

dissolve in water

water molecule

When an ionic compound is heated strongly, the solid melts and the ions start to move around. The moving ions carry charge so that the molten compound conducts electricity.

When an ionic compound has dissolved, it can conduct electricity because its ions can move independently among the water molecules.

Electrolysing molten sodium chloride breaks it down to make sodium and chlorine.

$$sodium\ chloride \longrightarrow sodium + chlorine$$

$$2NaCl \longrightarrow 2Na + Cl_2$$

Molten sodium chloride conducts electricity because its ions can move towards the electrodes.

Sodium ions are positively charged (Na^+). They are attracted to the negative electrode where they gain an electron and turn into sodium atoms.

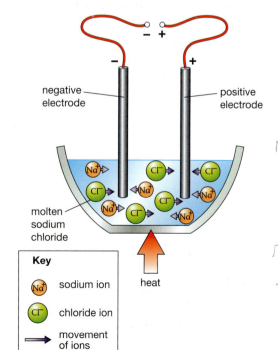

negative electrode

positive electrode

molten sodium chloride

Key

Na⁺ sodium ion

Cl⁻ chloride ion

→ movement of ions

heat

Chloride ions are negatively charged (Cl^-). They are attracted to the positive electrode where they lose an electron and join together to make chlorine molecules.

Explaining the changes at each electrode

Sodium chloride is called a **binary compound** because it only contains two elements.

One element is a metal (sodium) and the other is a non-metal (chlorine).

During electrolysis of molten binary compounds:

- The metal always forms at the negative electrode.
  This is because the metal ion in the compound has a positive charge and so is attracted to the negative electrode.

- The non-metal always forms at the positive electrode.
  This is because the non-metal ion in the compound has a negative charge and so is attracted to the positive electrode.

The metal ion always has a positive charge. When it reaches the negative electrode it gains electrons from the electrode. This cancels out the positive charge and turns it into an atom.

The non-metal ion always has a negative charge. When it reaches the positive electrode it loses electrons and gives them to the electrode. The ions turn into atoms.

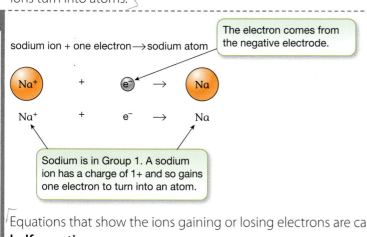

The electron comes from the negative electrode.

sodium ion + one electron → sodium atom

$$Na^+ + e^- \rightarrow Na$$

Sodium is in Group 1. A sodium ion has a charge of 1+ and so gains one electron to turn into an atom.

two chloride ions → chlorine molecule + two electrons

$$2Cl^- \rightarrow Cl_2 + 2e^-$$

Each chlorine ion loses an electron and then they join together to make a chlorine molecule.

Equations that show the ions gaining or losing electrons are called **half equations.**

In a half equation:

- the numbers of each type of atom must balance
- the total charge on each side of the equation must be the same.

So for zinc chloride:

At the negative electrode:

$$Zn^{2+} + 2e^- \rightarrow Zn$$
zinc ions   zinc

A zinc ion has a 2+ charge and so gains two electrons to turn into an atom.

At the positive electrode:

$$2Cl^- \rightarrow Cl_2 + 2e^-$$
chloride ions   chlorine

Mnemonic

OILRIG

Oxidation Is Loss (of electrons)
Reduction Is Gain (of electrons)

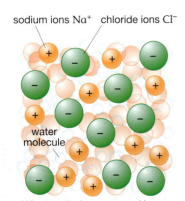

sodium ions Na⁺ $\quad$ chloride ions Cl⁻

When an ionic compound has dissolved, it can conduct electricity because its ions can move independently among the water molecules.

An aqueous solution contains ions mixed with water molecules.

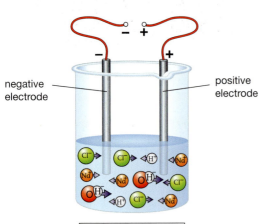

negative electrode

positive electrode

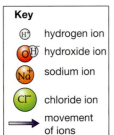

Key

⊕	hydrogen ion
🔴	hydroxide ion
🟠	sodium ion
🟢	chloride ion
➡	movement of ions

Ions from the water (H⁺ ions and OH⁻ ions) and ions from the dissolved salt (Na⁺ ions and Cl⁻ ions) are all attracted to the electrodes.

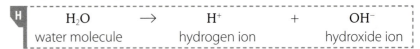

Oxidation and reduction

Ideas about oxidation and reduction can be used to explain the changes at the electrodes.

- Oxidation happens when electrons are lost.
- Reduction happens when electrons are gained.

This means that positive metal ions are always *reduced* during electrolysis because they always gain electrons. Negative non-metal ions are always *oxidised* during electrolysis because they always lose electrons.

Electrolysis of aqueous solutions

An **aqueous solution** of an ionic salt contains ions mixed with water molecules. For example, an aqueous solution of sodium chloride contains sodium ions (Na^+) and chloride ions (Cl^-) mixed in with the water molecules.

A very tiny proportion of the water molecules split into hydrogen ions and hydroxide ions.

$$H_2O \rightarrow H^+ + OH^-$$
water molecule $\quad\quad\quad$ hydrogen ion $\quad\quad$ hydroxide ion

The hydrogen ions and hydroxide ions are attracted to the electrodes with the other ions and may also form products.

- Hydrogen ions, H^+, are positively charged and move to the negative electrode.
- Hydroxide ions, OH^-, are negatively charged and move to the positive electrode.

Predicting products at the negative electrode

Salts are ionic compounds. They always contain positive metal ions and negative non-metal ions.

- The positive metal ions are attracted to the negative electrode.
- The hydrogen ions, H^+, from the water are also attracted to the negative electrode.

The product at the negative electrode depends on the reactivity of the metal compared with hydrogen.

- If the metal is *less* reactive than hydrogen, the metal forms at the negative electrode.
- If the metal is *more* reactive than hydrogen, hydrogen gas forms at the negative electrode.

Sodium is a very reactive metal. During electrolysis of sodium chloride solution, hydrogen gas forms at the negative electrode. The Na^+ ions stay in the solution.

This is the half equation to show what happens.

$$2H^+(aq) + 2e^- \rightarrow H_2(g)$$
hydrogen ions $\quad\quad\quad\quad\quad\quad\quad\quad$ hydrogen gas

Notice the state symbols:

- (aq) means that the H^+ ions are in aqueous solution
- (g) shows that hydrogen is a gas

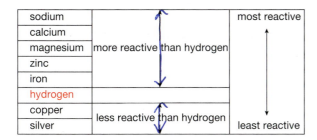

sodium			most reactive
calcium			
magnesium	more reactive than hydrogen		
zinc			
iron			
hydrogen			
copper	less reactive than hydrogen		least reactive
silver			

Notice that most metals are more reactive than hydrogen. Hydrogen is almost always made at the negative electrode during the electrolysis of aqueous solutions unless the solution contains ions of a very unreactive metal, such as, copper or silver.

If the salt contains ions of an unreactive metal, the metal forms at the negative electrode instead of hydrogen. The product is the same as the electrolysis of molten compounds.

$$Cu^{2+}(aq) \quad + \quad 2e^- \quad \rightarrow \quad Cu(s)$$
$$\text{copper ions} \qquad\qquad\qquad \text{solid copper}$$

The positive hydrogen ions or metal ions are always reduced at the negative electrode because they always gain electrons.

Predicting products at the positive electrode

Negative ions in the solution are attracted to the positive electrode. Negative ions come from the dissolved ionic compound and the water (hydroxide ions, OH^-).

During electrolysis of solutions of most ionic compounds, the OH^- from the water lose electrons to form oxygen and water.

$$4OH^-(aq) \quad \rightarrow \quad O_2(g) \quad + \quad 2H_2O(g) \quad + \quad 4e^-$$
$$\text{hydroxide ions} \qquad \text{oxygen gas} \qquad\qquad \text{water}$$

Some ionic compounds give a different product at the positive electrode. If the ionic compound is a chloride, chlorine gas forms because the chloride ions lose electrons instead.

$$2Cl^-(aq) \quad \rightarrow \quad Cl_2(g) \quad + \quad 2e^-$$
$$\text{chloride ions} \qquad \text{chlorine gas}$$

The negative hydroxide ions or chloride ions are always oxidised at the negative electrode because they always lose electrons.

Electrolysis of solutions of most salts makes hydrogen gas at the negative electrode and oxygen gas at the positive electrode. You can see bubbles of gas forming at the electrodes. The gases can be collected and tested.

Negative ion in aqueous solution		Product at positive electrode	Equation
sulfate	SO_4^{2-}	oxygen	$4OH^-(aq) \rightarrow O_2(g) + 2H_2O(g) + 4e^-$
nitrate	NO_3^-		
hydroxide	OH^-		
chloride	Cl^-	chlorine	$2Cl^-(aq) \rightarrow Cl_2(g) + 2e^-$

Questions

1 Why do you think Davy's work was so important to other chemists?

2 What big idea occurred to Faraday that enabled him to explain how solutions of salts and molten salts conduct electricity?

3 Why was it important for Davy and Faraday to communicate their ideas to other scientists?

4 Why do solid compounds made of ions not conduct electricity?

5 Look back at C3.1, which explains how metals conduct electricity. Make a list of similarities and differences between what happens when a metal and an ionic compound conduct electricity.

6 Predict what products are made when these molten compounds are electrolysed:
   a lead bromide
   b copper chloride
   c potassium oxide

7 Electrolysis of molten salts is always done in a fume cupboard. Explain why this is necessary.

8 When molten zinc chloride is electrolysed, the zinc formed is a liquid. Copy the symbol equation for the electrolysis of zinc chloride and add state symbols to show the states of each substance.

H 9 Write down the half equations for the electrolysis of sodium chloride and zinc chloride.

H Which elements are oxidised in each electrolysis? Which are reduced? Explain your reasoning.

10 Write half equations to show what happens to these ions during electrolysis.
Li^+ Ca^{2+} Fe^{3+} Br^- O^{2-}

11 Look back at the ionic equations in C3.2. What are the similarities and differences between writing ionic equations and half equations?

12 Predict the products of electrolysis of solutions of these salts:
   a sodium nitrate
   b calcium sulfate
   c copper chloride
   d silver nitrate

H 13 a Write half equations for the reactions at the electrodes in question 12.
   Include state symbols. (The symbol for a silver ion is Ag^+).
   b Make a table to show which element has been reduced and which has been oxidised for each salt.

14 Very reactive metals are extracted from their ores by electrolysis. The metal compounds are always heated to very high temperatures until they are molten. This takes very large amounts of energy. Explain why it is not possible to extract these metals by electrolysis of the metal compounds dissolved in water.

B: Using electrolysis

Using electrolysis to extract aluminium

Worldwide we use hundreds of millions of tonnes of aluminium every year. To meet this enormous demand, aluminium is extracted on a very large scale.

Aluminium is a reactive metal. Reactive metals such as aluminium are extracted from their ores using electrolysis. The main ore of aluminium is bauxite. This mainly consists of aluminium oxide, Al_2O_3. Aluminium cannot be extracted by heating aluminium oxide with carbon because aluminium is more reactive than carbon, and the Al^{3+} ions in the oxide need a large amount of energy to turn them into aluminium atoms.

The ore is processed before electrolysis to remove impurities.

The diagram below shows the equipment used to extract aluminium by electrolysis.

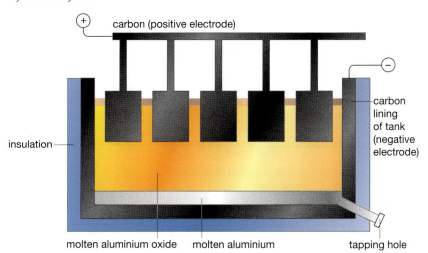

Equipment for extracting aluminium from its oxide by electrolysis.

Aluminium oxide, Al_2O_3, is an **ionic compound**.

The aluminium oxide is heated until it melts. Molten Al_2O_3 contains freely moving Al^{3+} and O^{2-} ions.

The steel tank is lined with carbon. This is the negative electrode.

The Al^{3+} ions are attracted to the negative electrode. The ions gain electrons and form aluminium. Because it is very hot in the tank, the aluminium is a liquid and forms a pool at the bottom of the tank, which runs out of the 'tapping hole'.

The positive electrodes are blocks of carbon. O^{2-} ions are attracted to the positive electrodes where they lose electrons to form oxygen gas.

Oxygen gas reacts with the carbon electrodes to make carbon dioxide. The carbon is used up in this reaction so the carbon electrodes have to be continuously replaced.

Reactions at the electrodes

Electrolysis turns ions into atoms. Aluminium ions are positively charged, so they are attracted to the negative electrode. It is a flow of electrons from the power supply into this electrode that makes it negative.

Find out about

- extracting aluminium
- H reduction and oxidation
- uses of electrolysis
- environmental impacts

Synoptic link

You can learn more about metals and their ores in C3.2 *How are metals with different reactivities extracted?*

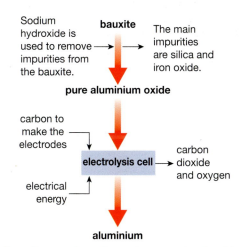

Lightweight objects are often made from aluminium. It is used to make technological tools, vehicles, aeroplanes, and drinks cans.

Flowchart to show the main stages involved in making aluminium from bauxite.

Key word

➤ ionic compound

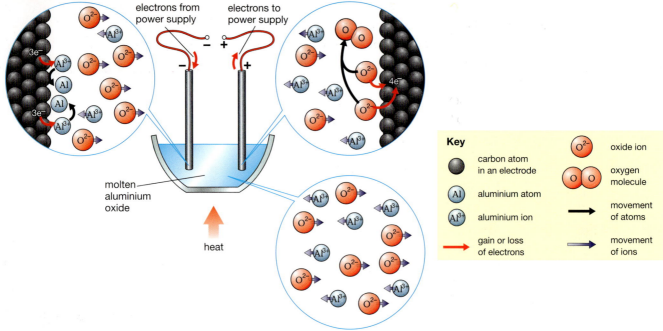

Changes to ions during the electrolysis of molten aluminium oxide.

Positive aluminium ions gain electrons from the negative electrode and turn into atoms. This is shown in a half equation:

$$Al^{3+} \quad + \quad 3e^- \quad \rightarrow \quad Al$$

| aluminium ion | electrons from the negative electrode | aluminium atom |

The aluminium ions are *reduced* because they gain electrons.

Metal extraction is often referred to as 'reduction' because the positive metal ion always gains electrons to form the metal, but reduction and oxidation always happen at the same time.

Oxygen ions are negatively charged, so they are attracted to the positive electrode. They give up electrons to the positive electrode and turn into atoms. Oxide ions turn into oxygen atoms, which pair up to make oxygen molecules. This is the half equation for the reaction:

$$O^{2-} \quad \rightarrow \quad O \quad + \quad 2e^-$$

| oxide ion | oxygen atom | electrons lost to the positive electrode |

$$O \quad + \quad O \quad \rightarrow \quad O_2$$

| oxygen atom | oxygen atom | oxygen molecule |

Oxygen ions lose electrons. They are *oxidised*.

Finally, the hot oxygen gas reacts with the carbon electrode to make carbon dioxide.

$$C \quad + \quad O_2 \quad \rightarrow \quad CO_2$$

| carbon from the positive electrode | oxygen gas | carbon dioxide gas |

Essential electrolysis: some benefits

Electrolysis is currently the only viable method of extracting reactive metals in large enough quantities to meet demand. Our modern lives are dependent on having access to very large amounts of reactive metals. Electrolysis is also used to purify metals. When the metal forms on the negative electrode, it is very pure. Metals used in electrical appliances and smart technical applications need to be of very high purity.

Key word
➤ electroplate

Metals for use in electrical wiring at home and in industry must be very pure. Impurities increase resistance, which can cause malfunction or overheating. Copper wiring needs to be close to 100% purity.

Electrolysis is also used to **electroplate** very thin layers of metals onto the surface of objects. The object is used as the negative electrode. The metal forms on the negative electrode as a thin coating. This very thin coating means that only very tiny amounts of valuable metals need to be used.

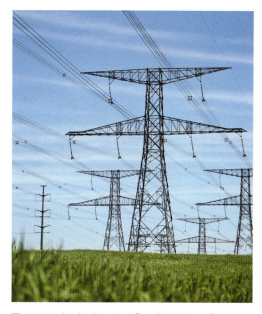

There is a high demand for aluminium for overhead electricity cables.

High-performance applications such as the electrical contacts in air bags need very high purity metals. Air bags use tiny amounts of gold to make the electrical contacts.

Recovery of valuable metals from scrap and from old mining waste heaps relies on electrolysis. For example, bioleaching of old gold or copper mines makes acidic solutions of ions of gold or copper at very dilute concentrations. The metals can be recovered by passing an electric current through the solution. The valuable metals collect on the negative electrode. A similar process can be used to recover valuable metals, such as gold and other precious metals, from old electronic equipment.

There is a large demand for high-purity metals for technical applications. The manufacture of circuit boards uses electrolysis as part of the process to 'print' very thin layers of metal on the surface of the board.

Essential electrolysis: some costs

Very large amounts of metal ores are processed to meet our demand for metals such as aluminium. Aluminium ore (bauxite) is mined by 'strip mining'. It is dug out of the ground in huge quarries, spoiling the landscape and destroying habitats and ecosystems. In addition, the quarrying produces 'red mud', which contains corrosive and toxic chemicals. In many countries there are strict laws about how mines are managed and the safe storage of the waste. However, these countries may import aluminium from countries where the laws are not as strict. The imported aluminium may be cheaper but the mining is not as closely controlled. The waste may not be handled properly and may contaminate water supplies and the surrounding land.

Purifying aluminium ore before electrolysis uses chemicals such as concentrated sodium hydroxide and produces harmful waste that needs safe disposal. Electrolysis uses huge amounts of energy, both to heat and melt the ores and for the electrolysis itself. In most countries, this energy comes from fossil fuels. Burning fossil fuels may reduce air quality and the carbon dioxide produced is a greenhouse gas, which leads to climate change.

Chemists are working to reduce the impact of the extraction processes. For example, they look for safe methods to treat and dispose of harmful waste. For aluminium extraction the usual chemical process uses a chemical called cryolite, which acts as a solvent for the aluminium ore and means that the electrolysis can run at a lower temperature, reducing the energy demand.

They have also found many new ways of recovering metals from mining waste and scrapped metals or appliances. However, the very large amounts of metals that are needed to meet demand cannot be met by recycling alone. This means that it is impossible to remove all of the environmental impacts from the extraction of these metals.

Bauxite mines are huge and produce waste 'red mud'.

Electricity for electrolysis to extract reactive metals is mostly generated from fossil fuels.

Questions

1 a Draw a diagram to show the electron arrangements in:
   i an aluminium atom
   ii an aluminium ion.
  b Use your diagrams to explain why the extraction of aluminium is a reduction reaction.

2 Look at the flow chart for making aluminium from bauxite.
  a Make a list of everything that is used up in the process.
  b Make a list of all the waste products of the process.

c Aluminium oxide is one of the most abundant compounds in the Earth's crust yet aluminium is an expensive metal. Use information about the extraction of aluminium to explain why this statement is true.

3 Joe says 'The environmental impact of extracting aluminium is too high. We need to stop all extraction now'. What would you say to answer Joe's point?

4 Valuable metals used to make electrical appliances and technical gadgets can be recovered using electrolysis. In practice, only a tiny percentage of the metals used to make these appliances is ever recovered. Make a list of reasons why.

A: Separating and using crude oil

Crude oil

Crude oil is a thick, sticky, dark-coloured liquid that formed over millions of years from the remains of tiny plants and animals called plankton. It is pumped out of the Earth's crust from wells under the ground or sea.

Crude oil is a mixture of **hydrocarbon** compounds. These compounds are very important as modern life is crucially dependent upon hydrocarbons. Almost all of the fuel used for transporting people and products (including petrol and diesel) comes from crude oil. Hydrocarbons are also used to make many different every-day products, including cosmetics, washing products, colourings and dyes, polymers, packaging, paints, and fibres. Most manufactured products use crude oil at some stage in their production.

Fractional distillation

Because it is a mixture, crude oil is not very useful as it is. It needs to be separated into groups of molecules of similar size, called **fractions**. This is done by **fractional distillation**. When crude oil has been refined in this way it becomes very useful indeed, which is why it is so valuable.

Find out about

- how crude oil is separated into fractions
- why crude oil fractions have so many different uses
- issues about how we choose to use crude oil

Key words

➤ crude oil
➤ hydrocarbon
➤ fraction
➤ fractional distillation

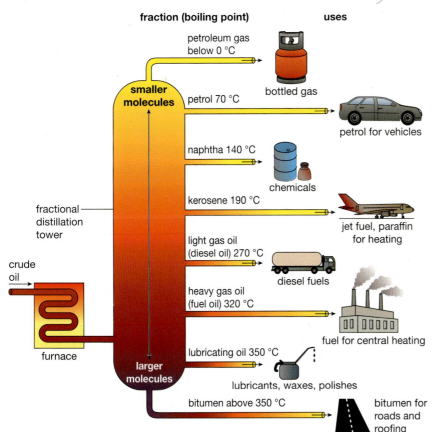

Crude oil is separated into fractions in the tower.

A fractionating tower at an oil refinery.

The diagram on the previous page shows crude oil being heated in a furnace. The hydrocarbons in crude oil then turn into gases and go into the fractionating tower, which is hottest at the bottom and coolest at the top. The different hydrocarbons are separated because they have different **boiling points**.

The tower is cooler towards the top. The gases condense and are taken out of the tower as they cool below their boiling point. Substances with bigger molecules condense in the hotter parts of the tower (at the bottom). Gases with smaller molecules travel all the way to the top where it is cooler.

This means that each 'fraction' contains molecules of similar sizes and boiling points, with similar forces between molecules. The higher boiling points are lower down the tower: the bigger the molecules are, the stronger the forces between the molecules.

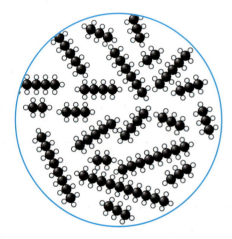

Crude oil is a mixture of hundreds of different hydrocarbons.

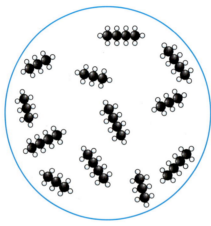

Small molecules go to the top of the column. These small molecules have weak forces between them so have very low boiling points.

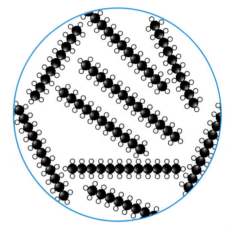

Further down the tower, the fractions have larger molecules with longer hydrocarbon chains. The forces between them are stronger, so they have higher boiling points.

Questions

1 a Look at the list of fractions that come out of the fractionating tower. Make two lists: 'Fractions used as fuels' and 'Fractions used to make other products'.

   b Alex says 'We need to conserve our supplies of crude oil or we will run out of fuels for our cars'. What would you say in response to Alex's point of view?

2 Petrol and lubricating oil are both crude oil fractions. Make a list of differences between the molecules in these two fractions.

Costs and benefits of using crude oil

Each fraction is still a mixture of molecules. Most of the fractions are processed further to be purified and are then used as fuels. The very large molecules are not very useful. Some of them are broken into smaller molecules, which can be used as a **feedstock** to make chemicals for products such as fibres and packaging. Only about 4% of crude oil is used in this way.

Crude oil is a fossil fuel and there is a finite supply. Some scientists think that oil will run out in the next 50 years. This would mean fuel would run out and there would be no oil to make the chemicals needed for production of consumer goods. The way we use crude oil today is not **sustainable**. The short-term benefits of using fuels need to be balanced and considered in terms of the costs of running out of hydrocarbons from crude oil in the near future.

B: Alkanes and molecular models

Alkanes in crude oil

Alkanes make up an important family of hydrocarbons. Alkanes can be used as fuels because they give out large amounts of energy when they burn. This is why the fractions from crude oil are so useful as fuels for vehicles, as well as providing energy in industry and generating electricity.

The table below shows the **molecular formula** and **structural formula** of the four smallest alkanes. C_nH_{2n+2}

Find out about

- the formulae and properties of alkanes
- homologous series
- intermolecular forces and changes of state
- representing molecules using formulae and models
- covalent bonds
- using dot-and-cross diagrams

Name	Molecular formula	Structural formula	Boiling point (°C)
methane	CH_4	H—C—H (with H above and below)	−162
ethane	C_2H_6	H—C—C—H (with H above and below each C)	−89
propane	C_3H_8	H—C—C—C—H (with H above and below each C)	−42
butane	C_4H_{10}	H—C—C—C—C—H (with H above and below each C)	0.5

The start of the alkane name tells you how many carbon atoms are in the alkane.

'meth' = 1

'eth' = 2

'prop' = 3

'but' = 4

The 'ane' ending identifies it as an alkane.

When the carbon atoms increase by one, the hydrogen always increases by two. You can see why by looking at the structural formula. The next alkane always has an extra

H
|
—C—
|
H

The lines represent bonds. C atoms always have four bonds. H atoms always have one bond.

As the molecules get larger, the boiling points increase.

The formulae of any alkane can be worked out using the idea that:

number of hydrogen atoms in formula = (number of carbon atoms × 2) + 2

The **general formula** for the alkanes can be written like this:

$$C_nH_{(2n+2)}$$

Key words

➤ alkanes
➤ molecular formula
➤ structural formula
➤ general formula

CH_4

The molecular formula shows how many carbon and hydrogen atoms are in each molecule.

The structural formula gives a 2D view of how the atoms are arranged.

The molecular and structural formulae of ethane.

Portable gas canisters contain butane.

Worked example: Predicting the formulae of alkanes

Butane contains four carbon atoms. Write down the formula of butane.

Step 1: Butane is an alkane. Write down the general formula for alkanes.

$C_nH_{(2n+2)}$

Step 2: Calculate the number of hydrogen atoms in butane.

number of hydrogen atoms = 2n + 2

number of hydrogen atoms in butane = (2 × 4)+2 = 10

Step 3: Write down the formula for butane.

butane – C_4H_{10}

Homologous series

The alkanes are a 'family' of compounds called a **homologous series.**

In a homologous series:

- As the molecules get bigger, each molecule has an extra CH_2 in its formula.

- All of the molecules in the series have the same general formula. For the alkanes this is $C_nH_{(2n+2)}$.

- The properties of all of the members of the series are similar. For example, alkanes are all very **flammable**.

- The properties show a trend as the molecules get bigger. For example, larger alkanes have higher boiling points and are less flammable.

Alkanes are only one example of a homologous series. There are many other homologous series of carbon compounds. Alcohols and acids are two other examples.

Key words

➤ homologous series
➤ flammable
➤ intermolecular forces
➤ state
➤ melting point

What happens when an alkane changes state?

When any substance in the solid state is heated, it melts to form a liquid. More heating changes the liquid to a gas. Energy is used to overcome the **intermolecular forces** when these changes of **state** happen.

The bonds *between* the carbon and hydrogen atoms in an alkane molecule are relatively strong compared to the intermolecular bonds between molecules of the alkane. The bonds between the atoms do not break. The solid, liquid, and gas all contain molecules of the alkane.

As the molecules in the alkanes get larger, the **melting points** and boiling points increase.

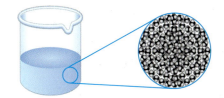

Molecules in liquid are close together. They move over each other.

- Larger molecules have a longer chain of carbon atoms.

- The forces between the molecules increase when the molecules are larger.

- When a solid alkane is heated, energy is used to overcome the intermolecular forces so that the molecules can move from their fixed positions and start to move over each other.

- When a liquid alkane is heated, energy is used to overcome the intermolecular forces so that the molecules can separate and move around very quickly.

- This is why larger molecules (with stronger intermolecular forces) have higher melting and boiling points.

- The bonds between carbon and hydrogen atoms do not break.

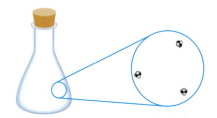

Molecules in a gas are far apart and move around very quickly.

Different ways of representing molecules

Chemists use different ways to represent molecules. Diagrams and models are useful to show the structures of molecules, but all of the representations are limited because they do not show what a 'real' molecule looks like.

	Methane	Ethane	
Molecular formula...	CH_4	C_2H_6	...shows the numbers of each type of atom in a molecule.
Empirical formula...	CH_4	CH_3	... shows the simplest ratio of each type of atom in a molecule.
Structural formula...	H | H—C—H | H	H H | | H—C—C—H | | H H	... shows the arrangement of bonds in two dimensions.
Ball-and-stick 3D model...			... shows the shape of the molecule. Bonds are shown as sticks.

Different representations of molecules give different information, but none of them show what a molecule really looks like.

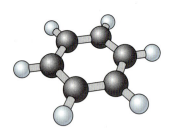

Benzene, C_6H_6, is a hydrocarbon in petrol.

Molecular and empirical formulae

Methane has the same molecular and **empirical formula**.

In methane, the ratio of carbon (C) to hydrogen (H) atoms in methane is 1:4. The molecular formula is CH_4.

This cannot be simplified any further, so the empirical formula and molecular formulae are the same.

In ethane, the ratio of carbon (C) to hydrogen (H) atoms in methane is 2:6. The molecular formula is C_2H_6.

The simplest ratio of C:H in ethane is 1:3.

The empirical formula of ethane is CH_3.

Worked example: Empirical formulae

Benzene is a hydrocarbon. Benzene has the molecular formula C_6H_6. What is the empirical formula of benzene?

Step 1: Work out the ratio of carbon to hydrogen for C_6H_6.

Ratio of C:H = 6:6

Step 2: Reduce to the simplest ratio.

Simplest ratio = 1:1

Step 3: Apply the simplest ratio to give the empirical formula.

Empirical formula: CH

Molecular and empirical formulae are useful in certain contexts but have limitations, as has every representation.

Empirical formulae are very useful for representing ionic compounds, where millions of ions are held together in a giant structure. The empirical formula shows the ratio of types of atoms in the compound, for example, the usual formula of sodium chloride is $NaCl$. This is actually an empirical formula. The formula shows that the sodium and chloride ions are combined in a ratio of 1:1. A crystal of salt will actually contain millions of ions joined together.

For alkanes, the empirical formula is useful for comparing the chemical composition of different types of molecule. The table below compares the advantages and limitations of molecular and empirical formulae.

Type of formula	Advantages	Limitations
molecular	• shows the numbers of each type of atom in the molecule • short and simple • useful when writing equations	• does not show the shape of the molecule • different molecules may have the same molecular formula
empirical	• shows the simplest ratio of types of atoms • useful for representing ionic compounds	• does not show the number of atoms in a molecule • does not show the shape of the molecule • different molecules may have the same empirical formula

Structural formulae and ball-and-stick models

A structural formula is a simple way of representing a molecule in two dimensions. The main use is to show the layout of atoms and bonds. Of course, real molecules are three dimensional. Three-dimensional models, such as **ball-and-stick models,** show the arrangements of the atoms in space so that you can see the overall shape and the angles between the bonds.

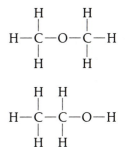

Both of these molecules have the same molecular formula, C_2H_6O, but they have different structures.

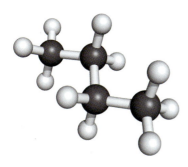

A ball-and-stick model of butane. Carbon atoms are shown as black balls, hydrogen as white balls.

Representation	Advantages	Limitations
structural formula	<ul><li>shows the numbers of each type of atom in the molecule</li><li>can be written easily on paper</li><li>shows the numbers of bonds around each atom and their arrangement in two dimensions</li></ul>	<ul><li>does not show the 3D shape of the molecule</li><li>does not show the correct bond angles</li></ul>
ball-and-stick model	<ul><li>shows the numbers of each type of atom in the molecule</li><li>shows the shape and bond angles</li><li>is an approximation for a 'real' molecule</li></ul>	<ul><li>both the sizes of the atoms and the lengths of the bonds are exaggerated</li><li>suggests that the electrons that make the bonds do not move</li></ul>

Covalent bonds

The formulae and models of molecules do not show how the atoms are bonded together. When non-metal atoms bond together to form molecules, they share electrons in their outer shells.

Chemists use computer models of molecule shapes to predict their reactions and behaviour. Simulations can predict how drug molecules will act in the body. This reduces both the cost and the time taken to develop new medicines.

The atoms are held strongly together by the attraction of their nuclei for the pair of electrons they share.

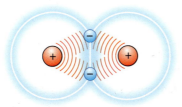

The atoms cannot move any closer together because the repulsion between the two positively charged nuclei will push them apart again.

A single covalent bond between two hydrogen atoms.

H_2

O_2

H_2O

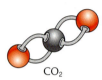

CO_2

Ball-and-stick models of H_2, O_2, H_2O, and CO_2. The molecules have a definite shape with fixed angles between the bonds.

Atom	Usual number of covalent bonds
H, hydrogen	1
Cl, chlorine	1
O, oxygen	2
N, nitrogen	3
C, carbon	4

The molecule of hydrogen is formed when the two hydrogen atoms share a pair of electrons. The nucleus of each atom is attracted to the electrons of the other. This pulls the atoms together. This is a **covalent bond**. Alkane molecules are also held together by covalent bonds. Carbon and hydrogen atoms share pairs of electrons.

A **single bond** is formed by two shared electrons.

The number of bonds that an element can form depends on the electrons in the outer shell of its atoms. The table in the margin gives the number of covalent bonds normally formed by atoms of some non-metal elements.

Sometimes atoms are bonded together by two bonds. This is a **double bond**. The two oxygen atoms in an oxygen molecule have a double bond between them. The bond between the carbon and oxygen atoms in carbon dioxide is also a double bond.

A double bond in a structural formula is shown by two lines like this:

$$O=C=O$$

Dot-and-cross diagrams

Dot-and-cross diagrams are used to show the shared electrons. For atoms, a full outer shell of electrons is a very stable arrangement. This explains why Group 0 elements (which have atoms with full outer shells) are very unreactive. When atoms join together to make molecules, they share electrons so that they have full outer shells.

● The first shell only holds two electrons, so a hydrogen atom (which has one electron in the first shell) only needs to share one more electron from another atom to make a full shell of two electrons.

● Other atoms need to share enough electrons to have eight electrons in their outer shell. When you draw diagrams with atoms with more than one shell of electrons (e.g., carbon) you usually only show the outer-shell electrons.

Hydrogen has one electron in its outer shell. It needs to share one electron to have a full shell. Forming one covalent bond means that hydrogen has two electrons and has a full shell.

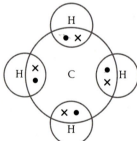

Carbon has four electrons in the outer shell. When it forms four bonds with hydrogen, it shares a total of four extra electrons (one from each hydrogen atom) giving a total of eight electrons. The outer shell is full.

Dot-and-cross diagram for methane. Carbon needs to share eight electrons to make a full shell.

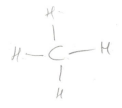

Dot-and-cross diagrams are a useful way of representing a covalent bond but they are limited because they show electrons in fixed positions as if they don't move. Electrons in atoms and molecules do not behave like this. They do not have fixed positions in space. Also, the dots and crosses imply that the electrons from each atom are different. Of course, they are all identical.

Properties of simple covalent compounds

Simple covalent substances are small molecules with covalent bonds.

- They have low melting and boiling points (they are often gases).
- They do not conduct electricity.

Dot-and-cross diagrams can be used to help explain the properties of simple covalent substances. The bonds between the atoms in the molecules are strong but there is very little attraction between molecules. This explains why they have low melting and boiling points, and often exist as gases. All of the electrons in the molecules are shared in the covalent bonds. There are no free electrons or other charged particles moving around so these substances cannot conduct electricity.

Hydrogen has a full shell when it shares two electrons.

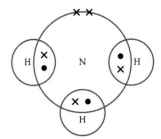

Not all of the electrons in the outer shell of the atom are always involved in bonding. In ammonia, NH_3, the nitrogen atom has five electrons in the outer shell. It forms bonds with three hydrogen atoms. Two of the electrons in the outer shell of nitrogen are not used in bonding.

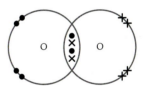

Oxygen atoms share four electrons, two from each atom. This makes a double bond.

Questions

1 Look at the table of boiling points for the alkanes. What state (solid, liquid, or gas) are these alkanes at room temperature (20 °C)? Explain your reasoning.

2 An alkane has 100 carbon atoms. What is its molecular formula?

3 **a** Describe the difference in arrangement and movement of methane molecules when it is a liquid and when it is a gas.
  b Explain why the boiling point of methane is lower than the boiling point of butane.

4 Look at the pictures of the ball-and-stick butane model. Use the model to help you write:
  a the molecular formula of butane
  b the empirical formula of butane
  c the structural formula of butane.

5 Look at the picture of the benzene molecule. Is benzene an alkane? Explain your reasoning.

6 Look at the models for hydrogen, oxygen, water, and carbon dioxide.
  a Draw structural formulae for each molecule.
  b Draw dot-and-cross diagrams for water and carbon dioxide.

7 Draw dot-and-cross diagrams for these molecules:
  - chlorine, Cl_2
  - hydrogen chloride, HCl
  - ethene, C_2H_4
  (Hint: there is a double bond between the carbon atoms.)

This 3D model of a butane molecule shows the atoms overlapping together. This is an improvement of the ball-and-stick model because it shows that the electron shells of atoms overlap. The disadvantage of this type of model is that you cannot see the bond angles.

C: Demand and supply of crude oil fractions

Find out about

- demand and supply of crude oil fractions
- cracking

Key words

➤ fossil fuel
➤ cracking
➤ catalyst
➤ alkene

Crude oil is in high demand to make petrol and diesel for vehicles. Fuels are used to transport people, raw materials, and manufactured goods.

Demand and supply

The demand for crude oil has increased over time. People need more oil to meet their energy needs for transport and industry, and also to meet their needs for manufactured products from oil. Look at the graphs below. The top one shows the demand for crude oil increases every year. Oil producers are increasing their supply of crude oil every year by taking more oil from the ground. However, the graph also shows that the gap between the supply and demand is increasing every year – there is not enough oil available for people's needs. Remember that crude oil is a **fossil fuel**. This means that it is in finite supply. Some forecasts predict that oil will run out in the next 50 years. Our current use of crude oil is not sustainable.

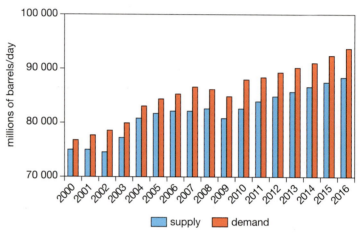

Graph showing world demand and supply of crude oil.

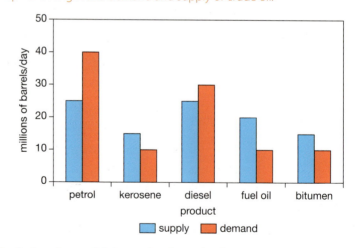

Graph showing world demand and supply of some crude oil fractions.

The second graph shows that the demand for each fraction in crude oil is not even. There is a high demand for some fractions (e.g., petrol for car fuel) and a lower demand for others (e.g., bitumen, which is used for road surfaces). The amounts of these fractions in crude oil do not 'match' the amounts people need. This means that people use more and more oil to meet demands for fractions such as petrol, while other fractions (such as bitumen) pile up or go to waste.

Petrol refineries solve this problem by processing the low-demand fractions to change them into high-demand fractions. They do this by chemically changing the molecules in the fractions.

Cracking

Fractions in low demand are processed in an oil refinery in a process called **cracking**. Cracking involves heating the fraction and passing it over a **catalyst**. There is more than one type of cracking process. Different processes use different temperatures and conditions, and produce different products.

- Cracking breaks large molecules into smaller molecules.
- Cracking re-arranges the bonds in molecules to give them different structures.

Cracking helps oil companies to make demand meet supply because fractions that are in lower demand can be chemically changed into molecules that are more useful.

Cracking produces molecules that are added to petrol, both to increase the amount of petrol available and to enhance its performance (to make it a more efficient fuel). A more efficient fuel produces more energy so less fuel is used per mile and the car produces fewer exhaust emissions. The energy used in cracking is offset by the benefits of making demand meet supply and by the usefulness of the products that are made.

Large molecules to small molecules

Cracking is used to produce molecules to increase fuel efficiency.

Structural formula equation to show hexane cracking.

When hexane is cracked it makes butane and ethene. Notice that:

- the number of carbon and hydrogen atoms stays the same when a molecule is cracked. Hexane contains six carbon atoms and 14 hydrogen atoms. The products, when their numbers of carbon and hydrogen are added up, also contain six carbon atoms and 14 hydrogen atoms
- one of the products (ethene) has a double bond between two carbon atoms. This is because each carbon atom in ethene needs to make four bonds around it
- ethene is an **alkene**. The 'ene' ending indicates the C=C double bond
- hexane and butane are alkanes – they are both hydrocarbons in which all the bonds are single bonds.

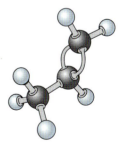

Propene is an alkene with a double bond between two of its three carbon atoms.

When petrol is cold, it is more difficult to ignite. Adding small molecules helps cars to 'cold start' in the winter.

Alkenes are used to make polymers, including the polyethene for these bags.

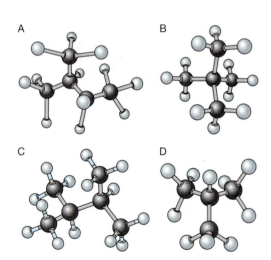

These molecules are all alkanes. However, the carbon atoms in these molecules are not in a straight chain.

Why are small molecules more useful?

The lighter fractions of crude oil are in highest demand. These fractions contain smaller molecules, which makes them useful:

- Small molecules have lower boiling points because they have weaker intermolecular forces. This makes them useful as fuels because they evaporate more easily so are more flammable and burn more easily. Winter petrol contains more small molecules because it makes it easier to start the car in cold weather.

- Small molecules with double bonds (alkenes) are very reactive. This makes them useful for making other products, such as polymers. Polyethene is a commonly used polymer made from ethene molecules.

Changing the shape of molecules

Another type of cracking changes the shape of the hydrocarbon molecules. Molecules with different shapes have different properties. The change in shape means the forces between the molecules are different, so properties such as melting points and boiling points change. Different shapes of hydrocarbon molecules are added to petrol to improve its performance. They are also used to make products such as solvents.

Questions

1 a Which crude oil fractions are in higher demand than supply?
   b Which fractions are in higher supply than demand?
   c Explain why some fractions have a much higher demand than others.

2 Suggest reasons why it is difficult to predict how long our supplies of oil will last.

3 Octane, C_8H_{18}, is an alkane. Write equations for what happens when octane is cracked to make two molecules. How many possible products of this cracking process can you recognise from this chapter? Name them.

4 a Which of the molecules in the margin (A, B, C, or D) has the same molecular formula as hexane, C_6H_{14}?
   b What is the empirical formula of this molecule?

5 a Write the molecular formulae for each of the four molecules.
   b Explain the advantages of using structural formulae and 3D models rather than molecular formulae for these molecules.

D: Alkenes

The alkene family

Alkenes are a family of hydrocarbons. Like the alkanes they are a homologous series. Alkenes are different from alkanes because alkenes always contain a double bond in their molecules.

Find out about

- the alkenes homologous series
- reactions of alkenes

The start of the name tells you how many carbon atoms are in the alkene.

'eth' = 2
'prop' = 3
'but' = 4
'pent' = 5

Every alkene contains a C=C double bond. Look closely at each carbon atom. You can see that each carbon atom has four bonds around it.

Name	Molecular formula	Structural formula
ethene	C_2H_4	
propene	C_3H_6	
butene	C_4H_8	
pentene	C_5H_{10}	

The 'ene' ending shows that it is an alkene and has a double bond between two carbon atoms.

The general formula for the alkenes is C_nH_{2n}. The formula shows that every alkene has twice as many hydrogen atoms as carbon atoms.

The smallest possible alkene is ethene. This is because two carbon atoms are needed for the molecule to have a C=C double bond.

Why are alkenes a homologous series?

Alkenes are a typical homologous series.

- As the molecules get bigger, each molecule has an extra CH_2 in its formula.
- All of the molecules in the series have the same general formula. For the alkenes this is C_nH_{2n}.
- The properties of all of the members of the series are similar. Alkenes all have similar reactions.
- The properties show a trend as the molecules get bigger. Larger alkenes have higher boiling points and are less flammable. This is because, like alkanes, larger molecules have stronger intermolecular forces between them.

Ethene is the smallest alkene. Ethene gas can be used to speed up the ripening of fruit before it is sold.

Reactions of alkenes

Alkenes all have similar reactions. Like all hydrocarbons, alkenes are very flammable and burn in **combustion** reactions to produce carbon dioxide and water.

| alkene | + | oxygen | → | carbon dioxide | + | water |

$$C_2H_4 + 3O_2 \rightarrow 2CO_2 + 2H_2O$$
ethene

Complete combustion produces carbon dioxide. **Incomplete combustion** may produce carbon monoxide (**CO**) or carbon particles.

The double bond makes alkenes much more reactive than alkanes. The usual reactions are **addition reactions**. During addition reactions, one bond in the double bond breaks and atoms 'add on' to the carbon atoms on either side. Only one type of product molecule is made.

When ethene reacts with bromine, atoms of bromine bond with the carbon and the double bond.

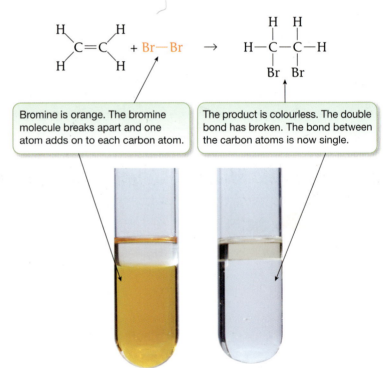

Bromine is orange. The bromine molecule breaks apart and one atom adds on to each carbon atom.

The product is colourless. The double bond has broken. The bond between the carbon atoms is now single.

When ethene reacts with water, ethanol is made.

Equation for the reaction of ethene with water.

This reaction needs heat and a catalyst. This reaction is used on a large scale to make ethanol in bulk for use in industry, where ethanol is used as a solvent and to make commercial products.

E: Alcohols and acids

What are alcohols?

Ethanol, C_2H_5OH, is the best known **alcohol**. This alcohol is made by fermentation. It is in all alcoholic drinks. Ethanol can also be made industrially by the addition reaction of water with ethene. Industrial ethanol is used as a **solvent**. Many cosmetic products, such as glues, windscreen washers, and perfumes, use ethanol as a solvent because it evaporates quickly.

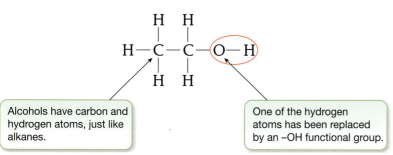

Alcohols have carbon and hydrogen atoms, just like alkanes.

One of the hydrogen atoms has been replaced by an –OH functional group.

Structural formula of ethanol.

Ethanol and other alcohols are flammable and burn to give carbon dioxide and water. This is partly because they contain a chain of carbon and hydrogen atoms, just like alkanes and alkenes.

Alcohols also contain an –OH group. This **functional group** is in all alcohols. There are many different types of functional group, for example, the –COOH functional group is in all **carboxylic acids**. Other than combustion, the hydrocarbon part of the molecule is not very reactive. It is the functional group that is responsible for the reactions of the alcohols.

Find out about

- alcohols
- carboxylic acids
- functional groups

Key words

➤ alcohol
➤ solvent
➤ functional group
➤ carboxylic acid

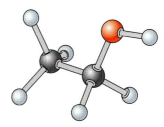

A model of ethanol shows the shape of the molecule and the bond angles. You can see that the C–O–H bond is bent at an angle of about 109°. Oxygen atoms in molecules are usually represented by red 'balls' in models.

The alcohols as a homologous series

Name	Formula	Fully displayed formula	Boiling point (°C)
methanol	CH_3OH	H | H—C—O—H | H	65
ethanol	C_2H_5OH	H H | | H—C—C—O—H | | H H	78
propanol	C_3H_7OH	H H H | | | H—C—C—C—O—H | | | H H H	97
butanol	C_4H_9OH	H H H H | | | | H—C—C—C—C—O—H | | | | H H H H	117

Perfumes and aftershaves use ethanol as a solvent. The ethanol evaporates to leave the perfume on the skin.

Testers check the quality of wine by measuring sugar content and pH. The pH check finds out if air has got into the wine to oxidise the ethanol.

Ant stings contain methanoic acid. Neutralising a sting with sodium bicarbonate from the kitchen cupboard helps to stop the pain.

Vinegar is a dilute solution of ethanoic acid.

- The 'start' of the names (meth-, eth-, prop-, and but-) are used to show how many carbon atoms are in each molecule.
- The **fully displayed formula** is a structural formula that shows all of the bonds, not only between the carbon and hydrogen atoms but also between the atoms in the functional group.
- The formulae are written with 'OH' on the end (rather than putting all of the hydrogen atoms together) to make it clear that the functional group is –OH. The general formula for the alcohols is $C_nH_{(2n+1)}OH$.

Properties of alcohols

You can compare each alcohol to the alkane with the same number of carbon atoms. This helps to show the difference that the –OH functional group makes to the properties.

1 Alcohols are all liquids at room temperature and their boiling points are much higher than alkanes (e.g., the boiling point of ethane is –89 °C and the boiling point of ethanol is 78 °C). This is because the –OH groups increase the strength of the intermolecular forces between molecules.

2 Like alkanes, alcohols burn to give carbon dioxide and water. Notice that alcohols already have an oxygen atom in the formula. They need less oxygen from the air than alkanes to burn completely. This makes them useful additives to fuels.

$$ethanol + oxygen \rightarrow carbon\ dioxide + water$$

$$C_2H_5OH + 3O_2 \rightarrow 2CO_2 + 3H_2O$$

3 Alcohols are good solvents. The –OH group means that, unlike most carbon compounds, they mix with water (water also has –OH in its structure).

4 Alcohols can be oxidised to make carboxylic acids. This happens when an extra oxygen atom from an oxidiser is added to the alcohol molecule.

An oxygen atom from an oxidiser is usually represented as [O].

Ethanoic acid has a –COOH functional group. It is a carboxylic acid.

Equation for oxidation of ethanol.

Alcohols can be oxidised by chemical oxidisers in practical experiments. Bacteria in the air can also oxidise ethanol. This reaction is useful for making food products such as vinegar, which contain carboxylic acids. However, it is a problem for wine and beer makers because the drinks are spoilt if the ethanol is oxidised. Carboxylic acids have a sour taste and smell.

Carboxylic acids

Carboxylic acids are another homologous series of compounds. They have a $-COOH$ functional group in their structure. Like all acids, they have a low pH and react with metals and carbonates to make salts. Carboxylic acids are less toxic, less reactive, and less corrosive than the acids usually used in the school laboratory. They are found in many food and cosmetic products.

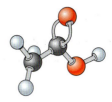

A model of ethanoic acid shows the shape of the molecule and the bond angles.

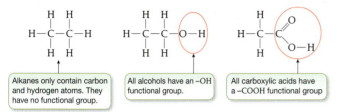

Alkanes only contain carbon and hydrogen atoms. They have no functional group.

All alcohols have an $-OH$ functional group.

All carboxylic acids have a $-COOH$ functional group

Fully displayed formulae of ethane, ethanol, and ethanoic acid to compare the functional groups.

Name	Formula	Fully displayed formula
methanoic acid	$HCOOH$	
ethanoic acid	CH_3COOH	
propanoic acid	C_2H_5COOH	
butanoic acid	C_3H_7COOH	

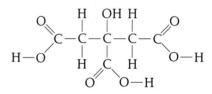

Lemons have a sharp taste because they contain citric acid. The structure of citric acid looks complicated but you can see that it contains carboxylic acid ($-COOH$) functional groups.

Questions

1 Look at the table of information about the alcohols. Why are alcohols a typical homologous series?

2 a A pentanol molecule has five carbon atoms. Draw a fully displayed formula for pentanol.

 b Write an equation to show the reaction that happens when pentanol is oxidised.

3 a Predict the trend in the boiling points of the carboxylic acids as the number of carbons increases.

 b The boiling point of methanoic acid is 101 °C. How does this compare with the boiling point of methanol? Predict how the boiling points of the first four members of carboxylic acids compare with alkanes and alcohols with the same number of carbon atoms.

 c A carboxylic acid has the formula C_4H_9COOH. Suggest its name. Use the names of the alkenes in the previous topic to help you.

 d Copy and complete this general formula for carboxylic acids: $C_nH_{(...)}COOH$.

Science explanations

C3 Chemicals of the natural environment

Our way of life depends on a wide range of products made from natural resources. The Earth's crust provides us with metal ores and crude oil, and our use of these impacts on the natural environment. Understanding the chemistry is fundamental to understanding the scale and significance of this human activity.

You should know:

- how the chemical bonds in metals explain the bulk properties of metals
- how the reactions of metals can be used to deduce an order of reactivity
- why the methods used to extract metals from their ores are related to their reactivity
- why electrolysis turns ions into atoms and splits ionic compounds into their elements
- how electrolysis is used to extract some metals from their ores
- **H** how to use half equations to show what happens to ions when electrolysis is used to extract metals
- how reduction and oxidation can be explained in terms of gain or loss of electrons, identifying which species are oxidised and which are reduced
- that crude oil consists mainly of hydrocarbons, of formula C_nH_{2n+2}, which are members of the alkane homologous series
- that crude oil is an important feedstock of the petrochemical industry and is a finite resource
- how and why fractional distillation separates the hydrocarbons in crude oil into fractions according to their chain length
- that covalent bonds occur when atoms share electrons and can be represented using dot and cross diagrams
- how the properties of materials depend on the covalent bonds they contain and their bond strengths in relation to intermolecular forces
- that useful materials are made by cracking long-chain molecules
- **S** the names, formulae, and structures of the first four members of the alkanes, alkenes, alcohols, and carboxylic acids
- that the characteristic reactions of organic compounds arise from their functional groups.

Ideas about Science

New technologies and processes based on scientific advances sometimes introduce new risks. Some people are worried about the effects arising from the extraction and use of metals. You should be able to:

- identify examples of risks that arise from mining and metal extraction
- suggest ways of reducing a given risk.

Some applications of science, such as the extraction and use of metals, can have unintended and undesirable impacts on quality of life or the environment. Benefits need to be weighed against costs. You should be able to:

- identify the groups affected by a given application, and suggest the main benefits and costs of a course of action for each group
- suggest reasons why different decisions on the same issue might be appropriate in different social and economic contexts
- identify and suggest examples of unintended impacts of human activity on the environment, such as the production of large volumes of waste by mining, mineral processing, and metal extraction based on low-grade ores
- explain the idea of sustainability and apply it to the methods used to obtain, use, recycle, and dispose of metals.

C3 Review questions

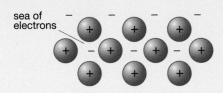

sea of electrons

1 Aluminium is a metal with a low density and a high electrical conductivity. It is used to make overhead power cables.

  a The diagram shows a model of the bonding in a metal.

   i What does the symbol ⊕ represent?

   ii Use this model to explain why metals have high melting points.

  H b Aluminium is extracted by electrolysis of molten aluminium oxide. Complete and balance the ionic equations to show what happens during the electrolysis of molten aluminium oxide.

$$...... + 3e^- \longrightarrow Al$$

$$......O^{2-} \longrightarrow O_2 + ...$$

  c Aluminium is a metal but aluminium oxide is an ionic compound. Solid aluminium metal and molten aluminium oxide conduct electricity in different ways.
  Describe and explain the differences.

2 Some students heated mixtures of metals and metal oxides.
  They looked for signs of a reaction.
  Their results are shown in the table in the margin.

	copper oxide	iron oxide	zinc oxide
copper		no reaction	no reaction
iron	reaction		no reaction
zinc	reaction	reaction	

  a Write down the three metals in their order of reactivity.

  b Zinc displaces iron from iron oxide, Fe_2O_3.

   The other product is zinc oxide, ZnO.

   Write a balanced equation for this reaction.

  H c Both copper and zinc ions have a charge of 2+.

   i Write half equations for the reaction of zinc with copper oxide.

   ii Which metal in this reaction is reduced and which is oxidised?

3 Zinc metal can be extracted from its oxide by heating with carbon. The equation for the reaction is:

$$ZnO + C \rightarrow Zn + CO$$

  a In the equation above:

   i name the element that is oxidised

   ii name the reducing agent.

  b Aluminium metal is extracted from its oxide by electrolysis. Explain why aluminium cannot be extracted from its oxide by heating with carbon.

4 The Vikings dug out small amounts of impure iron from peat bogs.

  The 'bog iron' contains iron oxide.

  They heated the bog iron in a charcoal fire to extract the iron.

  a What do we call a reaction where a metal oxide turns into a metal?

  b Several chemical reactions take place in the fire.

   In one reaction, iron oxide (Fe_2O_3) reacts with carbon to make iron and carbon dioxide.

   Write a balanced chemical equation for this reaction.

H **5** The diagram shows how copper is extracted from a waste heap using bacteria. The bacteria like acidic conditions so the waste heap is sprayed with sulfuric acid.

   a Write down two advantages and two disadvantages of using bacterial extraction of metals for this purpose instead of extraction with carbon.

   b An electric current is passed through the solution of copper sulfate.

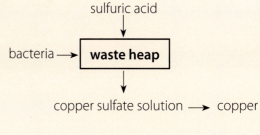

      i What is the name given to this process?

      ii What product would be found at:

         ● the anode?

         ● the cathode?

   c In 2015 it is more common for gold to be extracted using bacteria than copper.

      Suggest reasons why gold is extracted using bacteria and copper is not.

6 Crude oil is separated into fractions in a fractionating tower.

   The diagram shows crude oil entering the tower and two of the fractions coming out.

   These fractions are petroleum gas and liquid fuel oil.

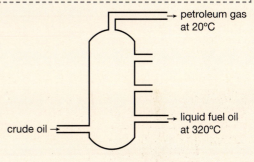

   a Petroleum gas and liquid fuel oil have very different boiling points.

      Describe the sizes of the molecules in petroleum gas and fuel oil, and explain how they are separated in the fractionating tower.

   b Some of the fractions of crude oil are made into more useful products by cracking.

      i Give two reasons why the products of cracking may be more useful than the fractions of crude oil.

      ii Heptane is changed into propane and ethene by cracking.

         Copy and complete the symbol equation to show how many molecules of propane and ethene are made when one molecule of heptane is changed by this process.

$$C_7H_{16} \longrightarrow \ldots\ldots C_3H_8 + \ldots\ldots C_2H_4$$

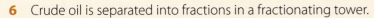

         heptane   propane   ethene

      iii The equation uses the molecular formula of ethene.

         Write down the empirical formula of ethene.

7 The graph shows the boiling points of five alkanes with different numbers of carbon atoms.

   a Estimate the boiling point of hexane, C_6H_{14}.

   b **i** What is the trend shown by the graph?

      ii Explain the trend shown in the graph.

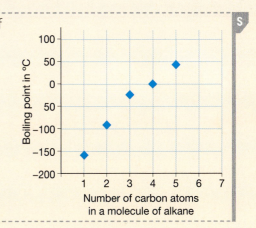

C4 Material choices

Why study material choices?

We all use materials and products that have been developed, tested, and modified by chemists. Materials are chosen for specific jobs because of their specific properties. Chemists have discovered ways to change the properties of some materials that occur in nature, and also to make completely new materials. By understanding the life cycle of a product, scientists can understand its impact on the environment and identify ways to reduce this impact.

What you already know

- An object is distinct from the material from which it is made.

- Chemicals can be elements or compounds.

- Chemicals can also be mixtures – two or more chemicals mixed together but not chemically combined.

- Different materials have different properties, and these properties make them suitable for particular uses.

- Properties of materials include hardness, solubility, transparency, conductivity (electrical and thermal), and response to magnets.

- The properties of simple molecular, ionic, and metallic substances are related to their bonding and structure.

- The temperature at which molecular materials melt or boil is related to the strength of the intermolecular forces.

- Chemical symbols and formulae can be used to represent elements and compounds.

- Word equations and balanced equations can be used to represent reactions.

The Science

Theories of structure and bonding in materials can be used to explain why different materials behave in different ways. This helps scientists to design new materials to meet a wide range of needs.

Polymers are an important class of material with a huge variety of uses. They occur naturally and can also be made synthetically by polymerisation reactions.

Ideas about Science

Scientists and engineers collect accurate data about the properties of materials so that valid comparisons can be made. This helps designers to make choices about which materials to use for a particular job. The environmental impact of materials and products is also an important consideration. To help assess this, scientists collect evidence about all the stages of the life cycle of products.

A: Choosing the right materials

Hockey sticks are made of wood or composites. These materials are stiff, strong, and lightweight.

Why are windows made of glass, while hockey sticks are not? Glass is ideal for windows because it allows light to pass through it. However, glass is **brittle**. If you used it to hit a fast-moving ball, it would shatter.

What would be a good material for a hockey stick? Steel is **stiff** and **strong**. There are over 6000 varieties of steel but they are all dense. So a steel hockey stick would be too heavy.

Hockey sticks are traditionally made from wood. This is because wood has the right combination of **physical properties**. It isn't as strong as steel but it is strong enough for a hockey stick. It is also much less dense than steel, so it is lighter.

Modern hockey sticks are made of **composite** materials. Composite materials consist of two or more materials mixed together. There is a very wide range to choose from. Composites can be stiff, strong, and very light. This makes them the ideal choice for a hockey stick.

When a designer is deciding how to make a product, they can choose which material to use according to which properties they require the finished item to have. As well as performance there are other factors to consider:

- Cost – is it affordable?
- Durability – how long will it last?
- Environmental impact – what damage might the material do during manufacture, use, or as waste at the end of its useful life?
- Versatility – how easily can the material be shaped or joined to other materials?
- Aesthetic appeal – does it look good?

Classes of material

There is a huge range of different materials to choose from when designing products. Most of these materials can be classed into four types:

- **ceramics**
- **polymers**
- **metals**
- composites

All these materials are chemicals. Some of the metals are pure chemicals, not mixed with anything else, but most of the materials we use are mixtures of chemicals. Each has its own range of properties with advantages and disadvantages, depending on how it is to be used.

Glass and clay ceramics

We use ceramics for mugs, plates, tiles, glass windows, bricks, and toilets. Ceramics are materials such as glass, brick, and pottery. Ceramics are useful because they are:

- strong in compression, so they don't give way when squeezed
- **hard**, even harder than metals.

Ceramics are strong in compression but weak in tension. This means that they break easily when stretched. They are also brittle, so they snap rather than change shape when stressed.

Polymers

We use polymers for making lots of different objects, including bags, clothes, window frames, and computers. Polymers include cotton, polyethene, polyurethane, and rubber. Polymers are useful because they are:

- **flexible**, so they change shape easily when forces act upon them
- **durable**, so they don't rot or rust easily and last a long time
- low **density**, so they are lightweight.

One disadvantage of polymers is that they are not very hard. They dent and scratch easily.

Metals

We use metals for an enormous number of products, including aircraft, cars, pipes, wires, jewellery, and sports equipment. Commonly used metals are iron, steel (which is an alloy), and aluminium. Metals are useful because they:

- are strong in tension and compression (large forces are needed to break them)
- are stiff, so they tend to keep their shape when forces act on them
- are **tough**, so they deform rather than break suddenly under stress
- are hard, so they don't scratch or dent easily
- are **malleable**, so they can be hammered into different shapes
- **conduct** electricity, so they can be used to make cables
- have high melting points, so they can be used for saucepans.

Because most metals have a high density, objects made of metal are usually heavy for their size. Aluminium has a low density compared with other metals. It is used to make products such as aeroplanes, where weight needs to be kept to a minimum.

Composites

Composites have a wide range of uses, from tennis rackets to helicopter rotor blades. Composite materials are made by combining two or more materials. This can result in a material with a more useful combination of properties. For example, a composite can be stronger and stiffer than any of the individual materials used to make it, while being as light as possible.

Glass is a ceramic. It is brittle, so it is poor at withstanding sudden impact and can shatter easily.

Synoptic link

See the appendix for properties of some examples of each category of material.

Key words

- ➤ brittle
- ➤ stiff
- ➤ strong
- ➤ physical properties
- ➤ composite
- ➤ ceramics
- ➤ polymer
- ➤ metal
- ➤ hard
- ➤ flexible
- ➤ durable
- ➤ density
- ➤ tough
- ➤ malleable
- ➤ electrical conductivity

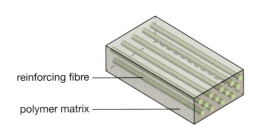

reinforcing fibre

polymer matrix

A model of a fibre-reinforced plastic. The fibres are embedded in a polymer matrix. The resulting composite is strong and lightweight.

Reinforced concrete is an example of a composite material. It has steel rods passing all the way through it. These give concrete a higher tensile strength.

Modern materials

Most composite materials consist of a main component, or **matrix**, with fibres or particles embedded in it. The matrix could be a polymer, metal, or ceramic.

Tennis racket designers want to design light, strong rackets with exactly the right amount of flexibility. Traditionally, tennis rackets were made from wood. Wood is a natural composite material. It consists of strong fibres held together in a tough matrix of lignin. The result is a material that is light, strong, and tough.

However, chemists have now developed synthetic composite materials with just the right combination of properties for a particular use. For tennis rackets, **fibre-reinforced plastics** are the perfect solution. The main fibres used are glass, boron, aramid (Kevlar), and carbon (graphite). The fibres are embedded in a polymer that can be moulded and set. This polymer matrix:

- holds the fibres in position
- transfers forces to the fibres
- protects the fibres from damage
- stops the fibres buckling under compression.

Combining materials at different scales

Reinforced concrete is a composite material widely used in building. Concrete is strong in compression but weak in tension. Inserting steel rods into the concrete increases its strength in tension. This is an example of a composite that combines materials on a large, or bulk, scale.

Other composites combine materials on a much smaller scale. Car tyres are made of a blend of natural and synthetic rubber embedded with tiny carbon particles, known as carbon black. These carbon particles are evenly spread throughout the rubber. They improve wear resistance so the tyres last longer.

Questions

1 Give three examples of each of these classes of material: ceramics, polymers, and metals.

2 Suggest the best material for making each of these items. Choose from ceramics, polymers, and metals and give a reason for your choice:
   a a football   **b** a hammer   **c** a cushion
   d a kitchen chopping board

3 Imagine tennis balls made from cotton, steel, or glass. Discuss the problems you might have using each type of ball.

4 Write a description of what a composite material is, for a student in Year 7. Use at least one example in your answer.

5 The bodies of most road cars are made from steel. The bodies of Formula One racing cars are made from composite materials, such as fibre-reinforced plastic. Suggest why composite materials are preferred for racing cars.

Key words

➤ matrix
➤ fibre-reinforced plastic

B: Testing materials

Scientists use words such as flexible, brittle, strong, and hard to describe materials. But there are times when more than a description is needed. Measuring the physical properties of materials quantitatively helps engineers to make better choices about which material to use for a particular job.

Collecting data

Manufacturers of sports equipment always want to improve their products. They test new materials to see if they are better than the current ones. To find out whether a new material would be good for making a bicycle frame, they would test:

- stiffness – by measuring how much it bends under a given force. This is called the **Young's modulus**. A stiff material has a high Young's modulus. A flexible material has a low Young's modulus

- tensile strength – by measuring the force that breaks it. The higher the force that breaks it, the stronger it is

- density – by measuring its mass and volume, and then dividing its mass by its volume. The higher its density, the heavier it will be for its size.

Find out about

- measuring the properties of materials
- using data to compare materials

Key word

➤ Young's modulus

The frames of racing bicycle need to have just the right combination of properties. Racing bicycle designers use data about different materials to decide which is best.

A material used to make bicycle frames should ideally be stiff and strong, with a low density. It is also important to know how much a material costs. The cost of materials often affects decisions about which material to use.

Here is some data about four different materials that might be used to make bicycle frames:

	Young's modulus (GPa)	Tensile strength (MPa)	Density (kg/m³)	Cost (£/kg)
Material 1	72	350	2780	£1.90
Material 2	205	670	7850	£0.60
Material 3	95	700	4500	£38.60
Material 4	320	930	1700	£25.00

Material 4 has the best combination of properties, but it is quite expensive. Material 4 might be suitable for a racing bicycle, while material 2 might be adequate for an affordable road bicycle.

Making a valid comparison

To decide whether one material is better than another for a particular use, engineers need to be able to make valid comparisons. This is so that they can be confident in their conclusions. A conclusion is **valid** if the procedures and tests used were suitable for the question being investigated.

To make a valid comparison, it is important to control all the factors that are not being tested but that may affect the results. For example, the strength of a material can be affected by temperature. If you measure the strength of one material at 20°C and another material at 30°C, you cannot make a valid comparison of the data. Sometimes engineers use an environmental chamber to control the temperature and humidity at which tests are carried out.

Measurements should also be accurate. The true value is what the measurement should really show, and **accuracy** is a measure of how close measurements are to the true value.

How accurate you need the data to be will depend on what you want to use the data for. For example, you might compare the strength of sewing threads by fixing them to a clamp and hanging 10 g masses from them, adding more and more until they break. This will tell you the strength to the nearest 10 g. You can improve the accuracy by repeating the test three or more times and calculating the mean. The mean is the best estimate of the true value. This is probably accurate enough to compare two very different sewing threads.

However, to compare two very similar sewing threads you would need much more accurate data. An engineer might use a machine that can gradually increase the force applied to a sewing thread by much smaller amounts. The data collected will be more accurate, but the equipment is much more expensive.

Questions

1   A sample of gold is measured.
    Its mass is 9.65 g and its volume is 0.5 cm³. Calculate its density.

2   An engineer is considering a new material for making a bicycle frame that is the same price as material 1 but is three times as stiff, twice as strong, and half as dense. Is this a better or worse choice than material 4? Use the data from the table on the previous page to explain your answer.

3   Give one factor that is controlled when sewing threads are tested.

4   An engineer took one measurement of the Young's modulus of a sample of polymer. Explain why this would not give you much confidence in the accuracy of the result.

C: Improving the properties of metals

Metals have properties that make them suitable for a huge variety of uses, from electrical cables to aeroplanes. All metals conduct electricity and most metals have high melting points and are strong. These properties can be explained by their structure and bonding.

There can, however, be disadvantages to using metals in their pure form. They do not always provide the best combination of properties for a particular use.

Skiers want ski poles that are strong and lightweight. Which metal would be best to use?

- Iron is very strong but it is heavy, rusts easily, and can be brittle.
- Aluminium has a low density, but it is quite soft and weak.

Neither iron nor aluminium is quite right for making a ski pole.

Find out about
- metals and alloys

Key word
➤ alloy

Synoptic link
You can learn more about the structure and bonding of metals in C3.1 *How can the properties of metals be explained?*

Ski poles need to be strong, so that they don't bend or break when forces are applied to them, and lightweight, so that they are easy to carry.

Alloys

Many of the metals we use today are not pure metals but alloys. An **alloy** is a mixture of elements, including at least one metal. By adjusting the proportions in the mixture, engineers can make a huge variety of alloys with properties that are just right for the intended use.

Alloys are often more useful than the pure metals they are made from because they are stronger, harder, less malleable, and less ductile.

Many modern ski poles are made from an alloy of aluminium. This alloy is stronger than pure aluminium and less dense than iron or steel.

Commonly used alloys

There is a huge range of alloys, each with slightly different properties. Each one is suitable for different uses. Some are very strong and hard, some resist corrosion, some are lightweight, and some make strong magnets. The table on the next page describes some common alloys.

Stainless steel is an alloy that is strong, very resistant to corrosion and has an attractive shiny appearance. This makes it an ideal material for cutlery.

Alloy	Composition	Key properties	Uses
mild steel	iron, plus carbon, chromium, manganese, vanadium	● hard ● durable ● fairly strong ● malleable ● inexpensive	building structures, car parts
stainless steel	iron, plus chromium, carbon, nickel, manganese, molybdenum	● very resistant to corrosion ● shiny appearance	jewellery, surgical instruments, cutlery
brass	copper, plus zinc	● malleable ● golden appearance	door locks and bolts, musical instruments
duralumin	aluminium, plus copper, magnesium, manganese	● fairly strong ● lightweight ● malleable	car and aeroplane parts
alnico	iron, plus aluminium, nickel, cobalt, copper, titanium	● magnetic	magnets in loudspeakers, pickups in electric guitars
amalgam	mercury, plus silver, tin, copper, zinc	● low melting point ● durable ● inexpensive	dental fillings

A model of an alloy

Scientists use models to describe what they have found out about the structure of alloys and to explain their properties.

The diagram on the left shows a model of an alloy. In this model the atoms are tiny spheres, packed closely together. The alloy has the structure of a metal but dotted throughout the main metal are atoms of other elements.

Unlike the atoms in a pure metal, the atoms in an alloy are not in a regular pattern. This is because the different atoms are different sizes. In a metal, the layers of atoms can slide over each other. However, the irregular arrangement in an alloy means that it is more difficult for the layers to slide over each other. This makes alloys harder than the pure metal.

This model can explain why alloys are harder than the metals that form them, but it has limitations. It cannot explain all the different properties of alloys. For example, it cannot explain why stainless steel is more resistant to corrosion than mild steel.

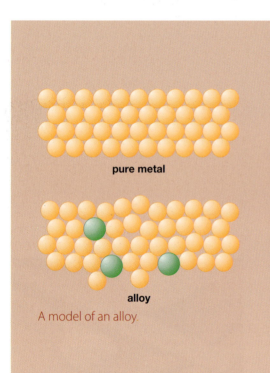

pure metal

alloy

A model of an alloy.

Questions

1 Choose, from the table above, a suitable alloy for making a container for chemicals. The container needs to be resistant to corrosion.

2 Use ideas about structure and bonding to explain why an alloy of aluminium is harder than pure aluminium.

A: Polymers all around us

Find out about
- natural and synthetic polymers
- polymer molecules
- discovering new polymers
- polymerisation reactions

Almost everywhere you look you will see objects made of **plastic**. A plastic material is one that can be moulded into shape. Plastics can be moulded into boxes, bottles, pipes, bristles, and fibres for making clothes.

Plastics are an example of what chemists call **synthetic polymers**. Synthetic refers to materials made by people. Polymers are made up of **long-chain molecules**. Synthetic polymers are usually made from crude oil or natural gas.

Key words
➤ plastic
➤ synthetic polymer
➤ long-chain molecule
➤ monomer
➤ polymerisation

Polymer molecules

Polyethene, nylon, and neoprene are all examples of synthetic polymers. Polymers can also be natural, such as cotton, leather, and wool. Whether they are synthetic or natural, polymers all have one thing in common: their molecules are long chains of repeating links.

Each link in the chain is formed from a smaller molecule. These small molecules are called **monomers**. They connect to one another to form the chain by a **polymerisation** reaction. 'Poly' means many and 'mono' means one, so a polymer is many monomers joined together.

Synoptic link

You can learn more about the processing of crude oil, including the formation of alkenes by cracking, in C3.4 *Why is crude oil important as a source of new materials?*

The discovery of polyethene

The 1930s was the decade of the first synthetic polymers. The world was a tense place and war was on its way. Governments were looking for scientific solutions to give them an advantage. This sped up many scientific developments, some of which used the new big idea: polymers. However, the first synthetic polymer was discovered by accident.

In 1933, Eric Fawcett and Reginald Gibson were working for Imperial Chemical Industries, investigating the reactions of gases at very high pressures. They put some ethene gas into the container and squashed it to 2000 times its normal pressure. However, some of the ethene escaped. When they added more ethene, they also let in some air.

Two days later, they found a white, waxy solid inside the apparatus. This was a surprise. They couldn't easily explain how the solid had formed. It took a leap of imagination for them to realise that, in some way, the small molecules of ethene gas had joined with each other to make bigger molecules. These bigger molecules made up the white solid.

They worked out that the new molecules were like chains. The chains were polymers made of ethene molecules that had joined together.

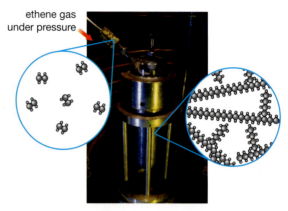

ethene gas under pressure

The original high-pressure container used by Fawcett and Gibson is on display at the Science Museum, London. The diagrams show what was happening to the small ethene molecules as they joined up in long chains to make polyethene. This process is called polymerisation.

At first they did not know exactly how the reaction had happened. After collecting more data they understood that oxygen in the air leaking into their apparatus had acted as a catalyst. The oxygen sped up what would otherwise have been a very, very slow reaction to join the ethene molecules together.

The polymer discovered by Fawcett and Gibson is polyethene. This is now commonly called polythene. A polyethene molecule is made from many ethene molecules joined together.

Natural polymers

Not all polymers are synthetic. Natural polymers are found in plants and animals and are essential to life. The human body is made up of many different natural polymers.

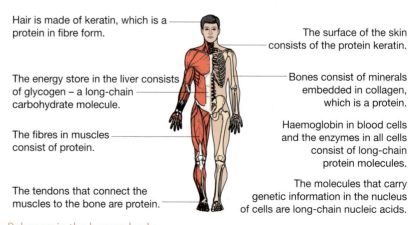

Hair is made of keratin, which is a protein in fibre form.

The energy store in the liver consists of glycogen – a long-chain carbohydrate molecule.

The fibres in muscles consist of protein.

The tendons that connect the muscles to the bone are protein.

The surface of the skin consists of the protein keratin.

Bones consist of minerals embedded in collagen, which is a protein.

Haemoglobin in blood cells and the enzymes in all cells consist of long-chain protein molecules.

The molecules that carry genetic information in the nucleus of cells are long-chain nucleic acids.

Polymers in the human body.

Proteins

Hair, skin, and muscle are built from **proteins**. The enzymes that control biochemical reactions are made of proteins. In your body, there are different proteins doing different kinds of job.

Proteins are polymers built by joining together monomers called amino acids. Each protein has a unique sequence of amino acids in its polymer chains.

Carbohydrates

Photosynthesis in the leaves of plants turns carbon dioxide and water into glucose. Glucose is a sugar, which belongs to the family of compounds called **carbohydrates**.

Glucose molecules have alcohol functional groups. Plants convert these to carbohydrate polymers such as:

- starch – an energy store
- cellulose – the polymer that makes up plant cell walls.

DNA

DNA is the molecule that carries the genetic code. It is made from four different monomers. These monomers are called nucleotides. Each nucleotide has three parts – a sugar group, a phosphate group, and one of four different bases. The sugar and phosphate groups form the backbone of the DNA polymer, alternating along the chain. The bases (adenine, cytosine, guanine, and thymine) are attached to this backbone. What matters in DNA is the order in which the different bases appear along the chain. This is the genetic code.

> ### Key words
> - protein
> - carbohydrate
> - DNA

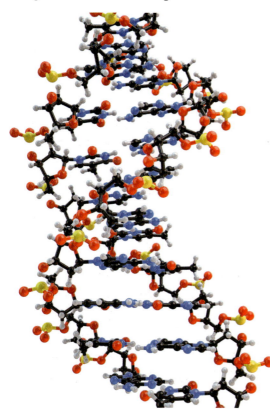

A model of a short length of DNA. DNA is a natural polymer made up of monomers called nucleotides. The long polymer chains wrap around each other in a double helix arrangement.

How to make a synthetic polymer

To make a synthetic polymer you first need to choose a suitable starting molecule, or monomer. This will be a small, reactive molecule. You then need to provide the right conditions for the monomers to react together and join up to form long chains.

Polymers can be formed by **addition polymerisation** or **condensation polymerisation**. The type of polymerisation reaction that happens depends on the atoms and bonds that make up the monomers.

Addition polymerisation

The simplest type of polymerisation is addition polymerisation. It is called this because the monomers 'add together' to form the long chains.

The reactive parts of monomers are called **functional groups**. There are many different functional groups. In addition polymerisation the functional group in the monomer has a double bond. Ethene is a hydrocarbon with two carbon atoms and four hydrogen atoms. It has a **double bond** between the carbon atoms. The double bond makes the ethene molecule reactive.

When the ethene molecules react you can think of the double bond between the carbon atoms 'opening up'. New bonds are then formed between the carbon atoms of each monomer molecule and a polyethene chain forms.

ethene monomers   polyethene

Thousands of ethene molecules react together to form polyethene. This is addition polymerisation. The ethene monomers add together to form the polymer. Nothing is left over.

This reaction can also be represented in a shorthand form like this, where *n* is a very large number:

$$nC_2H_4 \rightarrow$$

ethene   polyethene

The only product of an addition polymerisation reaction is a polymer. The monomers add together to form the polymer and nothing is left over.

Some examples of polymers formed by addition polymerisation are shown in the following table.

Monomer	Polymer	Uses of polymer
ethene H—C=C—H (H, H)	polyethene [—C—C—] H H / H H ₙ	• carrier bags • food packaging • cling film • washing-up liquid bottles • buckets • water pipes
chloroethene Cl—C=C—H (H, H)	poly(chloroethene) or PVC [—C—C—] Cl H / H H ₙ	• guttering • window frames • doors • food packaging • insulation for electrical cables • coating 'wipe-clean' fabrics
tetrafluoroethene F—C=C—F (F, F)	poly(tetrafluoroethene) or PTFE [—C—C—] F F / F F ₙ	• non-stick coating for frying pans • insulation for electrical cables • Gore-Tex for sports clothing • lining chemical reactors
propene CH₃—C=C—H (H, H)	poly(propene) [—C—C—] CH₃ H / H H ₙ	• carpets • car bumpers • drainpipes • food packaging • CD boxes • toys

Condensation polymerisation

Neoprene is another example of a synthetic polymer. It was discovered by Wallace Carothers, an American chemist. Like polyethene, neoprene was discovered by accident.

A worker in Carothers' laboratory left a mixture of chemicals in a jar for five weeks. When Carothers had a tidy-up, he discovered a rubbery solid in the bottom of the jar. Carothers realised that this new material could be useful and developed it into neoprene. This synthetic rubber first came on the market in 1931 and is still used today.

Carothers worked out that the polymer he had discovered was made from two different monomers. The different monomers had joined together as alternate links in the chain by a condensation polymerisation reaction. His team went on to invent nylon, another condensation polymer, as a synthetic replacement for silk.

Wetsuits are made of neoprene. Neoprene is a condensation polymer. It is made from two different monomers.

Condensation polymerisation is quite different from addition polymerisation, as the following table shows.

Addition polymerisation	Condensation polymerisation
There is one type of monomer molecule.	There are two types of monomer molecule.
Each monomer contains a C=C double bond.	Each monomer needs two functional groups. For example: ● alcohol (–OH) ● carboxylic acid (–COOH) ● amine (–NH$_2$)
The polymer is the only product.	As well as the polymer, a small molecule (e.g., H$_2$O or CO$_2$) is also formed each time a monomer joins the chain.

Nylon is a polyamide. A polyamide is formed from two different monomers, one with two carboxylic acid groups and one with two amine groups. When a carboxylic acid group in one monomer reacts with an amine group in another monomer, they join together and a water molecule is also formed.

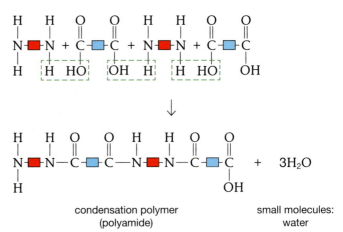

condensation polymer
(polyamide)

small molecules:
water

The molecules shown here are not displayed in full. The central part of each monomer is shown by a coloured block. This is so that the functional groups, which are involved in the reaction, can be seen more clearly. Amine monomers (red) react with carboxylic acid monomers (blue) to make polyamides. Water molecules also form.

condensation polymer (polyester)

small molecules: water

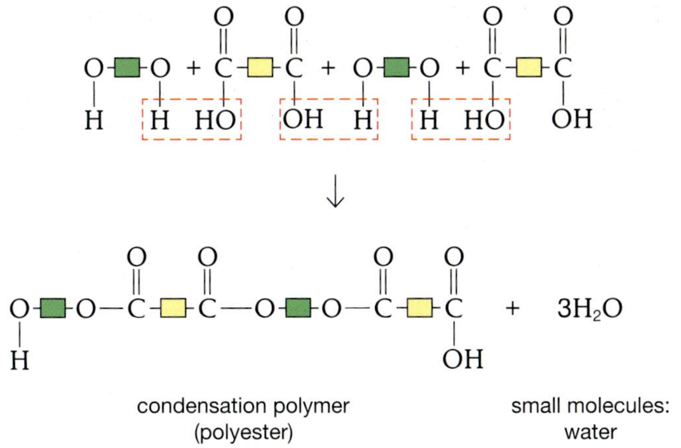

Alcohol monomers (green) react with carboxylic acid monomers (yellow) to make polyesters. Water molecules also form.

Polyester is another example of a condensation polymer. Polyesters are formed from two different monomers, one with two carboxylic acid groups and one with two alcohol groups. Again, water molecules are formed along with the polymer. Proteins are condensation polymers built from combinations of amino acids. Each protein consists of hundreds of amino acid monomers. DNA is also formed by condensation polymerisation, using just four different monomers.

Questions

1 What is a polymer and what is a monomer?

2 Give two examples of natural polymers and two examples of synthetic polymers.

3 Many scientists have made accidental discoveries. All of these words might be used to describe these scientists: lucky, skilful, foresightful, inventive, creative. Choose two of these words to describe the scientists that discovered polyethene. In each case, explain why you have chosen that word.

4 Name the monomers that form:
   a proteins
   b starch
   c DNA.

H 5 Write a short paragraph to explain the difference between addition polymerisation and condensation polymerisation.

6 Draw the polymer that would be formed from the monomer acrylonitrile:

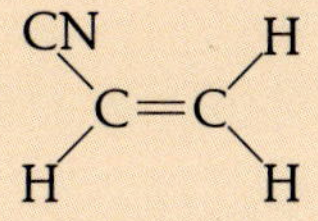

H 7 When glucose molecules join together to form carbohydrates, small molecules of water are formed. Explain whether this is addition or condensation polymerisation.

C4.3 How do bonding and structure affect the properties of materials?

A: Zooming in on materials

Find out about

- materials under the microscope
- molecules and atoms in materials
- models of molecules

Key words

➤ macroscopic
➤ microscopic
➤ fibres
➤ nanometre (nm)

A woollen jumper is very different from a silk shirt. The shirt is smoother and less stretchy than the jumper. They are both made from natural polymers but they are very different. Their properties depend on their make-up, from the large scale to the invisibly small:

- the visible weave of a fabric
- the microscopic shape and texture of the fibres
- the molecules that make up the polymer
- the bonds within and between the molecules.

The visible weave

The fabric of a silk shirt is tightly woven but even so it is possible to see the criss-cross pattern of threads. The fabric is hard to stretch because the strong threads are held together so tightly.

On the other hand, a knitted jumper is soft and stretchy. The loose stitches allow the threads to move around.

The weave and the stitches are visible to the naked eye. They are **macroscopic** features. However, the properties of a fabric also depend on invisible **microscopic** features.

Taking a closer look

A microscope can show details of the individual **fibres** in a fabric. Silk, for example, has smooth, straight fibres that slide across each other.

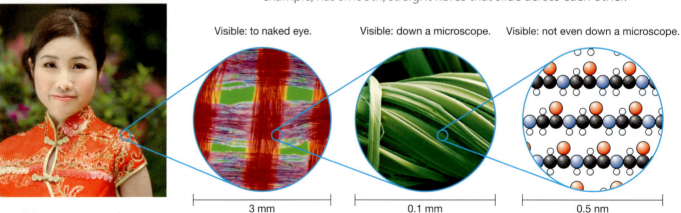

Visible: to naked eye. Visible: down a microscope. Visible: not even down a microscope.

3 mm   0.1 mm   0.5 nm

Levels of detail of the structure of silk. A millimetre is a thousandth of a metre. A micrometre is a thousandth of a millimetre. A nanometre is a thousandth of a micrometre.

Wool fibres have a rough surface that is covered in scales. The wool fibres tend to cling to each other within the thread. The scales also make the threads cling together.

The invisible world of molecules

It is difficult to look much further into the structure of materials without using microscopes of some kind. Scientists explain the differences between silk, wool, and other fibres by finding out about their molecules.

The properties of materials are related to the type of forces (bonds) in a material and how they are arranged. These might be forces between atoms or ions, or between molecules. The forces might be strong or weak. The atoms themselves do not have those properties.

Fibres are made up of polymer molecules. They are very long. Polymers have special properties because the molecules in them are so long. The length of a molecule affects how it interacts with the molecules around it.

Scientists measure the sizes of atoms and molecules in **nanometres** (**nm**). Some molecules, such as the small molecules in air, are smaller than 1 nm. The length of a polymer molecule can be 1000 nm or more. There are 1 000 000 000 nm in a metre. It takes a great leap of the imagination to think about things that are so small.

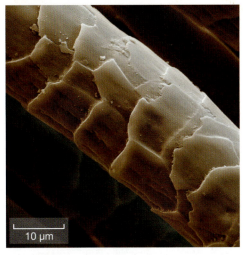
A wool fibre.

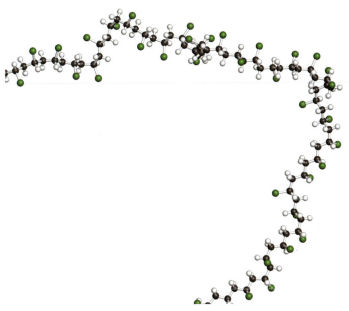
Computer-generated model of a PVC – polyvinyl chloride or poly (-chloroethylene) – molecule. No one knows what atoms and molecules look like. Scientists use models to help them understand the properties of materials. In this computer image the atoms of each element are colour-coded: carbon (black), hydrogen (white), and chlorine (green).

Questions

1 a List the following in order of size, starting with the largest: fibre, fabric, atom, thread, molecule.

   b Use the words in part **a** to write four sentences that describe the structures in order of size. The first sentence might be: 'Fabrics are made by weaving together threads.'

2 A polymer molecule is about 1000 nm long. An atom is about 0.1 nm across.

   a Estimate how many atoms there are along the polymer molecule.

   b How many polymer molecules, end to end, would fit into a millimetre?

B: Carbon creations

Find out about
- organic chemistry
- chains and rings
- long and short polymer chains
- using models to explain physical properties

A cotton plant: cotton is mainly cellulose. Cellulose is a natural organic polymer made up of chains of carbon atoms.

Carbon is the basis of all life on Earth. It is a special element because it can form a huge range of different compounds. There are more carbon compounds than there are compounds of all the other elements put together.

Carbohydrates, fats, proteins, and DNA are all polymers made up of a skeleton of carbon atoms supporting other atoms. These other atoms are often hydrogen, oxygen, nitrogen, and halogen atoms.

The chemistry of carbon is so important that it forms a separate branch of chemistry, called **organic chemistry**. The word 'organic' means 'living'. At first, organic chemistry was the study of naturally occurring compounds from plants and animals, but it now also includes synthetic compounds such as synthetic polymers, drugs, and dyes.

Chains and rings
Carbon forms so many different compounds because:

- carbon atoms can form chains and rings by joining to other carbon atoms
- carbon forms four strong covalent bonds. Two of these bonds can be used to make a chain or ring. The other two allow other atoms to join onto the chain. Very often these are hydrogen atoms.

Propane: a hydrocarbon with three carbon atoms in a chain.

Cyclohexane: a hydrocarbon with a ring of carbon atoms.

The chains can be very long, as in the polymer polyethene. A typical polyethene molecule may have 10 000 or more carbon atoms linked together. A polyethene molecule is still very tiny, but much bigger than a methane molecule. A methane molecule has just one carbon atom, bonded to four hydrogen atoms.

Families of compounds
To make sense of the huge variety of carbon compounds, chemists think in terms of families of carbon compounds. These families are also known as **homologous series**. Each family has a distinctive group of atoms.

Alkanes, alkenes, **alcohols**, and **carboxylic acids** are all families of carbon compounds. All the members of a family have similar chemical properties and the same general formula. They also have similar names. For example, the names of all the alkanes end in '-ane', all the alkenes end in '-ene', and all the alcohols end in '-ol'.

Alkanes

The alkanes are hydrocarbons. They are chains of carbon atoms with hydrogen atoms attached. They are well known because they are the compounds in fuels, such as natural gas and petrol. All alkanes burn in air, forming carbon dioxide and water. Alkanes do not react with acids or alkalis.

The first three members of the alkane series are methane, ethane, and propane.

methane, CH_4 ethane, C_2H_6 propane, C_3H_8

Alcohols

You can think of an alcohol as being an alkane with one of its hydrogens replaced with an –OH group. This is the reactive part of an alcohol and chemists call it the functional group. Alcohols burn in air and react with sodium and other alkali metals.

The first three members of the alcohol series are methanol, ethanol, and propanol.

methanol, CH_3OH ethanol, C_2H_5OH propanol, C_3H_7OH

Carboxylic acids

The functional group in carboxylic acids is –**COOH**. Like other acids, the carboxylic acids react with metals, alkalis, and metal carbonates.

The first three members of the carboxylic acids series are methanoic acid, ethanoic acid, and propanoic acid.

methanoic acid, HCOOH   ethanoic acid, CH_3COOH propanoic acid, C_2H_5COOH

Polymers – molecules big and small

A **hydrocarbon** is a compound of carbon and hydrogen only. Because carbon can form chains and rings, there are many different hydrocarbon compounds. But they do not all have the same properties.

Key words

➤ organic chemistry
➤ homologous series
➤ alkane
➤ alcohol
➤ carboxylic acid

Synoptic link

You can learn more about the alkanes as a homologous series in C3.4 *Why is crude oil important as a source of new materials?*

Synoptic link

You can learn more about functional groups, including carboxylic acids and alcohols, in C3.4 *Why is crude oil important as a source of new materials?*

Some are hard, while others are soft. Some are gases, while others are solids. This is because it is not the atoms themselves that have these properties. The properties of a hydrocarbon depend on the bonds within the molecules and the forces between the molecules.

Candle wax and polyethene are both polymers made up of long hydrocarbon chains. Wax is weaker, softer, and more brittle than polyethene. They are not the same because the molecules of candle wax are about 20 atoms long while the molecules of polyethene are about 5000 times longer. This difference in length is important. The physical properties of wax and polyethene depend on the length of the molecule, because this length affects the forces between molecules.

Two different bonds

Molecules of wax and polyethene are made of atoms. The bonds *between atoms* in the molecules are covalent bonds. These bonds are strong. This means it is very hard to pull a molecule apart. The molecules do not break when materials are pulled apart.

But the forces *between molecules* are very weak. It is much easier to separate molecules from one another because they can slide past each other. The forces between the molecules are called **intermolecular forces**.

Bending and breaking wax and polyethene

Compared with candle wax, polyethene is harder, stronger, and more flexible. Stretch or bend a candle and it cracks. This is because separating the small molecules is not difficult. The forces between the molecules are very weak and the molecules slip past each other quite easily.

Stretch or bend a lump of polyethene and it will not break easily. Its long molecules are all jumbled up and tangled. It is harder to make them slide over each other. The long molecules make polyethene harder, stronger, and more flexible than wax.

Warming and melting wax and polyethene

Some polymers, including wax and polyethene, can be softened by warming so that their shapes can be changed. This softening happens at different temperatures for different polymers. The softening temperature depends on the intermolecular forces between the chains.

Polyethene has a higher melting point than wax. The melting point of polyethene is about 120 °C, while for wax it is about 55 °C. This data supports the explanation that the intermolecular forces between wax molecules are weaker than the intermolecular forces between polyethene molecules.

Polyethene molecules are much longer than wax molecules. Intermolecular forces act all the way along the length of a molecule. There are more intermolecular forces between long polyethene molecules than between short wax molecules. More energy is needed to separate the polyethene molecules from each other so polyethene melts at a higher temperature than wax.

Wax seals were common in medieval times. They were used to show that a letter had not been opened. A seal is made with molten wax. When the letter is opened the wax cracks and breaks. This is because solid wax is brittle. The molecules of wax are about 20 atoms long.

The molecules of polyethene are similar to those of candle wax. But they are about 5000 times longer. Polyethene is much stronger and tougher than candle wax.

A polymer model

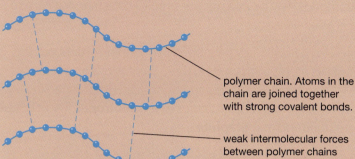

polymer chain. Atoms in the chain are joined together with strong covalent bonds.

weak intermolecular forces between polymer chains

This is a model that can be used to explain the melting point of polyethene compared with that of candle wax.

The model shows that hydrocarbon chains are attracted to neighbouring chains by intermolecular forces along their length.

So, for a longer chain there are more intermolecular forces between the chains.

This means more energy is needed to separate the chains. So, polyethene has a higher melting point than wax.

This model cannot explain everything about the properties of polymers. It has limitations. It cannot explain why different polymers with chains of the same length have different melting points.

Key words
➤ hydrocarbon
➤ intermolecular force

Questions

1 Pentane and butanol are carbon compounds.
   a Predict whether or not pentane will react with hydrochloric acid. Give reasons for your answer.
   b Predict whether or not butanol will react with sodium. Give reasons for your answer.

2 Bowls of pasta can be used as a model to explain the difference between wax and polyethene. One bowl contains cooked spaghetti. The other bowl contains cooked macaroni (or penne).
   a In the model, what represents a molecule?
   b Which kind of pasta represents wax and which represents polyethene?
   c Show how this model can help to explain why polyethene is stronger than wax.

3 Tests have shown that a hydrocarbon molecule that is 40 atoms long has a melting point of 84 °C. Does this increase or decrease your confidence in the explanation given in 'melting wax and polyethene' opposite? Explain your answer.

C: Structure and properties

Find out about

- allotropes of carbon
- giant covalent structures
- ionic compounds
- metals
- simple molecules
- the relationship between the structure of a material and its properties

Diamonds are transparent, and are cut so that they reflect light and sparkle.

Pencil lead is made of graphite and clay. As you slide the pencil across the paper, layers of graphite flake off and leave a mark.

Carbon structures

Carbon compounds may be polymers or simple molecules, but carbon as an element is quite different.

Carbon bonds to itself with covalent bonds. By forming four bonds it is able to arrange its atoms and bonds in three dimensions to create strong giant covalent structures. The atoms and bonds can be arranged in different ways to form different structures called **allotropes**. Two of these allotropes are **diamond** and **graphite**.

Non-identical twins

If you heat a diamond strongly, it will eventually burn and make carbon dioxide. Graphite will do the same. This is because both these minerals are made from carbon and nothing else. They both burn, but in many ways they could not be more different. How the carbon atoms are arranged and joined together in diamond and in graphite is very different. Diamond and graphite are good examples of how the behaviour of a solid depends on its structure and bonding. It is easy to picture the arrangement of atoms in diamond and graphite.

Diamond and graphite head-to-head

Property	Diamond	Graphite
hardness (1 = softest, 10 = hardest)	10	1–2
melting point (°C)	3550	3550
lubricating action	low	high
electrical conductivity	low	high

Diamond: a structure in three dimensions

In diamond, covalent bonds join each carbon atom to its four nearest neighbours. The four bonds are evenly spaced in three dimensions around each carbon atom, making pyramidal (tetrahedral) shapes. The arrangement is called a **giant covalent structure**, as it repeats over and over again until you have billions and billions of atoms covalently bonded together.

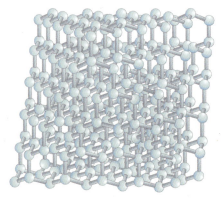

Model of the structure of diamond. Each carbon atom inside the structure is covalently bonded to four others in a huge 3D arrangement.

Graphite: a structure of flat sheets

Graphite also has a giant covalent structure. Covalent bonds join each carbon atom to its three nearest neighbours. The three bonds are evenly spaced in two dimensions, making flat sheets of hexagons that go on and on to include billions and billions of atoms. These flat sheets are stacked on top of each other in layers. Each atom has one outer-shell electron that has not been used to make a covalent bond. These electrons drift around freely in the gaps between the layers of atoms. They help stick the layers together, but only weakly.

A diamond drill bit. The tip of the drill bit is covered in a thin layer of diamonds. The extreme hardness of diamond allows it to cut through anything.

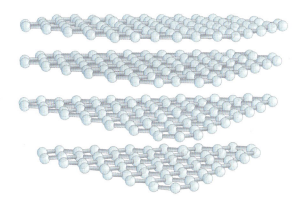

Model of the structure of graphite. Each carbon atom inside the structure is covalently bonded to three others, making flat sheets of hexagons with weak forces between them.

Different structures, different properties, different uses

Diamond and graphite both have high melting and boiling points. This is because many strong covalent bonds would need to be broken for them to melt. Strong covalent bonds in three dimensions make diamond the hardest known material. Drills used in mining have diamond tips that are hard enough to cut through rock.

In graphite, weak forces between the layers make it easy for them to slide over one another. This slipperiness makes graphite useful as a **lubricant**. The electrons drifting between the layers are free to move and take their charge with them. This allows graphite to conduct electricity – this is very unusual for a giant covalent structure. In contrast, diamond has no charged particles that are free to move, so it is an insulator.

Quartz: a diamond-like structure

Diamond and graphite are not the only materials that have giant covalent structures. Quartz is a form of silicon dioxide (SiO_2) commonly found in sand. It has a giant covalent structure similar to that of diamond. The strong bonds between the silicon and oxygen atoms make it a strong and rigid mineral. It is also hard, so it is suitable for use as an abrasive in sandpaper, for example.

5 cm

Pure quartz crystals are transparent and very hard.

Key words

- ➤ allotropes
- ➤ diamond
- ➤ graphite
- ➤ giant covalent structure
- ➤ lubricant

Polymers and giant covalent structures are just two types of chemical structure. There are others, such as:

- **ionic compounds**, for example, sodium chloride
- **simple molecules**, for example, oxygen
- metals, for example, copper.

Each of these has its own particular typical physical properties. To explain why they have these properties, materials scientists need to know something about their structure.

Ionic compounds

Crystals of sodium chloride are cubes. They are made up of sodium and chloride ions.

As a sodium chloride crystal forms, millions of Na^+ ions and millions of Cl^- ions pack closely together. The ions are held together very strongly by the attraction between their opposite charges. This is called ionic bonding, and the structure is called a giant ionic lattice. Unlike simple molecules, there is not an individual $NaCl$ molecule.

Because of the very strong intermolecular forces, it takes a lot of energy to break down the regular arrangement of ions. So $NaCl$ has to be heated to 801 °C before it melts, and to 14 138 °C before it boils.

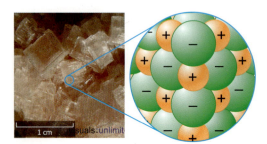

1 cm

Sodium chloride consists of sodium ions and chloride ions. These ions are not very reactive.

Sodium and chloride ions are charged particles. However, solid ionic compounds do not conduct electricity because the ions are not free to move. When an ionic compound is molten or dissolved in water, it can conduct electricity because its ions separate and move around.

Synoptic link

You can learn more about ionic bonding and structure in C2.3 *How do metals and non-metals combine to form compounds?*

Metals

Most metals have high melting points and are strong. Metals can be bent or pressed into shape. They bend without breaking. They are also malleable (aluminium sheet can be moulded under pressure to make cans) and ductile (copper can be drawn into wires).

The diagram shows the arrangement of atoms in copper, a typical metal. They are packed closely together in a regular arrangement. The structure carries on in all directions. It is a giant structure. Every atom inside the structure has 12 others touching it, the maximum number possible.

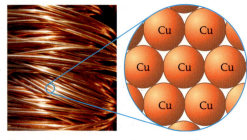

Copper has a giant metal structure. The atoms are packed together closely in a regular arrangement.

The atoms are held to each other by strong metallic bonding. Because the bonding is strong, copper is strong and difficult to melt. However, it is flexible, since the atoms are able to slide into new positions.

Metal atoms tend to share the electrons in their outer shell easily. In the solid metal, the atoms lose these shared electrons and become positively charged. The electrons, no longer held by the atomic nuclei, drift freely between the metal atoms. The attraction between the 'sea' of negative electrons and the positively charged metal atoms holds the structure together.

The electrons can move freely through the giant structure. This explains why metals conduct electricity well. When an electric current flows through a metal wire, the free electrons drift from one end of the wire towards the other.

Synoptic link

You can learn more about metallic bonding in C3.1 *How can the properties of metals be explained?*

Simple molecules

Most simple molecules are gases or liquids at room temperature.

All molecules have a slight tendency to stick together. For example, there is an attraction between one O_2 molecule and another O_2 molecule. But these intermolecular forces are very weak. This is why simple molecules have low melting and boiling points.

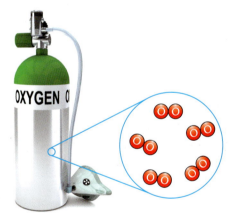

Oxygen gas consists of simple molecules of oxygen.

One way to picture this is that the molecules are moving so quickly that, when two O_2 molecules come close to each other, the attractive force between them is not strong enough to hold them together.

The forces inside molecules that hold the atoms together are very strong, many times stronger than the weak attractions between molecules. Small molecules such as O_2 do not split up into atoms except at very, very high temperatures.

Synoptic link

You can learn more about covalent bonds and intermolecular forces in C3.4 *Why is crude oil important as a source of new materials?*

Questions

1. Graphite is used to make some types of battery. Which property of graphite makes it suitable for this use?

2. Explain why graphite is soft but diamond is hard.

3. The melting point of polyethene is much lower than the melting point of diamond. Use ideas about structure and bonding to explain why this is.

4. Use ideas about structure and bonding to explain why copper conducts electricity but solid sodium chloride does not.

5. Carbon dioxide is a simple molecule with a boiling point of −57 °C. Titanium dioxide is an ionic compound with a boiling point of 2972 °C. Use ideas about structure and bonding to explain this difference.

A: Materials that are small and strange

Find out about

- the properties of nanoparticles
- carbon nanoparticles
- properties and uses of fullerenes, nanotubes, and graphene
- uses of nanotechnology
- benefits and risks of using nanotechnology

Bus companies, holidaymakers, and sports people are just some of the groups benefiting from recent scientific and technological advances. New products have been developed that rely on the properties of materials at a very, very small scale. These technologies are called **nanotechnology**.

At this small scale, materials often have very different properties from those of the same material at a bulk (large) scale. For example, they may have a different colour or melting point, or they may conduct electricity differently. Nanotechnology takes advantage of these unusual properties to make new products.

Not just small – very small

Nanotechnology is the use and control of structures that are called **nanoparticles**. These might be tiny particles of metals, salts, or ceramics, or they might be carbon structures. Some synthetic nanoparticles are made using specialist tools that build up new structures atom by atom. Others are made by chemical synthesis or by other techniques, such as etching.

Nanoparticles are measured in nanometres (nm) and are between 1 and 100 nm in at least one dimension (height, length, or depth).

A nanometre is a billionth of a metre or 0.000 000 001 m. You can use standard form to write this as 1×10^{-9} m. Standard form is a convenient way of representing numbers that are very small (or very large). For more about using standard form, see the *Maths skills* section at the end of this book.

It is difficult to understand just how tiny a nanometre is. One nanometre is about:

- the distance your fingernails grow in a second
- 1/80 000 the thickness of the average human hair
- the width of a DNA molecule.

Some bus companies use a nanoscale additive in diesel fuel. This reduces the amount of fuel used and the emissions from the vehicle, making the buses more efficient.

Modern sunscreens contain nanoparticles. This means that they are more transparent than older types of sunscreen when they are rubbed onto the skin.

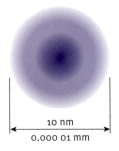

10 nm
0.000 01 mm

Illustration showing the size of a nanoparticle.

Imagine a nanoparticle that is a sphere with a diameter of 10 nm (0.000 01 mm). You could fit 100 000 of them across 1 mm on your ruler. Let's think about it another way…

If everything in the world were scaled up so that a 10 nm nanoparticle was the size of a football:

- a carbon atom would be about half a centimetre in diameter
- a glucose (sugar) molecule would be about the size of a 20p coin
- a red blood cell would be the size of a football pitch
- a cat would be about the same size as the Earth.

Properties of nanoparticles

If you wear a jumper and shirt both made of synthetic polymers, they sometimes seem to stick together. The force causing this 'stickiness' is similar to the electric force that allows a balloon rubbed against a jumper to stick to a wall.

Usually we do not notice the forces holding surfaces together, because they are very weak. But on the nanoscale they become very strong. These forces have the potential to be useful; they are what allow geckos to walk on ceilings. A gecko's feet have millions of tiny, nanoscale hairs, which give a huge surface area and provide the stickiness required to hold the gecko to the ceiling. This has inspired scientists in the US to design a robot that can walk up walls. They use a polymer that is similar to the bottom of the gecko's feet.

Geckos' feet have millions of nanometre-sized hairs that provide the 'stickiness' it needs to walk on ceilings.

Surface area and nanoparticles

For nanoparticles, the ratio of **surface area** to volume is very large. In a solid 30 nm particle, about 5% of the atoms are on the surface. In a solid 3 nm particle, about 50% of the atoms are on the surface.

The atoms on the surface tend to be more reactive than those in the centre. This means materials containing nanoparticles are often highly reactive or have unusual properties. It also means that they can be excellent **catalysts** for chemical reactions. Catalysts speed up chemical reactions and catalysed reactions occur on the surface of the catalyst.

Key words

- ➤ nanotechnology
- ➤ nanoparticle
- ➤ surface area
- ➤ catalyst

Worked example: Comparing surface-area-to-volume ratios

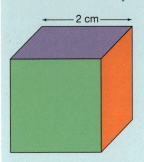
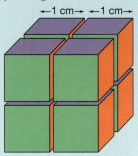

A cube (side 2 cm) and a stack of 8 cubes (side 1 cm)

Compare the surface-area-to-volume ratio of a 2 cm cube with that of a 1 cm cube.

Step 1: Write down the equation for the surface area of a cube.

surface area = 6 (faces) × area of one face = 6 × (edge)²

Step 2: Write down the equation for the volume of a cube

volume = length × breadth × height

Step 3: Write the surface-area-to-volume ratio for a 2 cm cube.

$$ratio = \frac{surface\ area}{volume} = \frac{6 \times 2\ cm \times 2\ cm}{2\ cm \times 2\ cm \times 2\ cm} = \frac{24\ cm^2}{8\ cm^3} = 3\ cm^{-1}$$

Step 4: Write the surface area to volume ratio for a 1 cm cube.

$$ratio = \frac{surface\ area}{volume} = \frac{6 \times 1\ cm \times 1\ cm}{1\ cm \times 1\ cm \times 1\ cm} = \frac{6\ cm^2}{1\ cm^3} = 6\ cm^{-1}$$

Step 5: Compare the values.

The smaller cube has twice the surface-area-to-volume ratio of the larger one.

A geodesic dome designed by the architect Richard Buckminster Fuller. This design inspired the scientists who discovered buckminsterfullerene.

Fullerenes, nanotubes, and graphene

Some of the most promising new nanotechnologies make use of carbon nanoparticles. For a long time, diamond and graphite were thought to be the only allotropes of carbon. We now know that there are others, with very different properties.

Fullerenes

Fullerenes were discovered in 1985 by a team of researchers interested in carbon found in stars. They were studying carbon reactions and were firing a laser at a graphite surface. They found that they had formed fragments made up of 60 carbon atoms (C_{60}). They tried to imagine how a cluster of 60 carbon atoms might be bonded together and were inspired by the designs of the architect Richard Buckminster Fuller. They called their newly discovered allotrope of carbon **buckminsterfullerene**.

Buckminsterfullerene is about 1 nm in diameter. Compare this with a water molecule, which is 0.16 nm across. Buckminsterfullerene has a structure made up of hexagons and pentagons and looks a lot like a football.

It soon became clear that a whole family of similar carbon nanoparticles was possible. This family is called the fullerenes. Because of their giant covalent structure, the fullerenes have high melting points and boiling points.

Fullerenes are unusual chemicals because they are hollow. This means other atoms can be trapped inside them. Researchers are studying the potential uses of fullerenes, including in treatments for HIV and for storing hydrogen for fuel-cell powered cars.

Graphene

If you have ever drawn with a pencil you have probably made **graphene**. Graphene consists of a sheet of carbon atoms. It has a structure similar to that of graphite, but it is just a single sheet, one atom thick.

Graphene is a sheet of carbon atoms, one atom thick. The sheet is almost one million times thinner than a human hair. Because it is so thin, on its own it is not visible to the human eye.

Scientists had known for some time that graphene existed but no one had been able to work out how to extract it from graphite. That was until 2003, when two researchers at the University of Manchester found some Scotch tape that had been used to clean a graphite stone. They noticed flakes of graphite were stuck to the tape.

They experimented to find out how thin the flakes on the tape could be made. They used more tape to peel the layers again and again until they were left with graphene one atom thick. They studied the properties of graphene and published their findings in 2004. Six years later they were awarded a Nobel Prize.

What makes graphene special is its unique combination of properties. Graphene is:

- the thinnest material known
- transparent
- the lowest-density material ever
- harder than diamond
- about 300 times stronger than steel
- a better conductor of electricity than copper
- very flexible.

Synoptic link

You can learn more about giant covalent structures in C4.3C *Structure and properties*.

Key words

➤ fullerenes
➤ buckminsterfullerene
➤ graphene

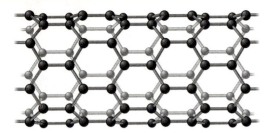

A model showing the structure of a nanotube. These have a very large surface area compared with their volume, so they make excellent catalysts to speed up reactions.

Graphene conducts electricity very well because the electrons, which in graphite hold the sheets together, are delocalised across the entire sheet.

Scientists are still researching possible uses for graphene and it is showing great potential. Some of the uses being investigated are in medicine, water purification, light bulbs, aircraft design, paints, food safety, mobile phone screens, electronic paper, and safer and more fuel-efficient transport.

Nanotubes

Carbon **nanotubes** are graphene sheets rolled into a tube and sealed at each end by a half fullerene.

Just like graphene, nanotubes are very strong and are excellent conductors of electricity. Their unusual electrical properties mean that they have found uses in electronic devices. Their strength makes them useful for reinforcing lightweight materials, for example, tennis rackets. They may also find uses as catalysts because of their large surface area. Even if the carbon itself doesn't act as a catalyst, atoms of a catalytic chemical could be spread atom by atom on the surface of the tube.

Nanotubes can be used to carry medicines into the body. The medicine stays wrapped up inside the nanotube until it reaches the place in the body where it is needed, for example, a tumour. This means that it is delivered safely without damaging other parts of the body.

Developing applications for these newly discovered materials requires great leaps of imagination by scientists and engineers.

Using nanotechnology

Products that make use of nanotechnology are becoming increasingly common. These include food packaging, sports gear, and clothing. You probably have some of these products at home.

Clothing

Silver is best known for use in jewellery because it is shiny and unreactive. However, for a long time it has been known that silver also has **antibacterial** properties. You can now buy socks that contain nanoparticles of silver. These kill the bacteria that lead to smelly feet and so help to keep feet smelling fresh.

Stain-resistant fabrics have been developed with tiny nanoscale hairs all over them. These help repel water and other liquids. If you spill a liquid on one of these stain-resistant fabrics it stays in droplets and doesn't spread out. All you have to do is wipe it off and it looks as good as new. This technology could reduce the amount of water and energy used in washing clothes.

Sunscreen

Many sunscreens contain particles of either zinc oxide or titanium oxide. These are white solids that absorb harmful ultraviolet light. In older sunscreens, the particles are relatively large and leave the skin looking white. More modern formulations use nanoparticles instead. These are transparent when they are rubbed into the skin. They give the same protection with a more natural appearance.

Self-cleaning windows

A company called Pilkington offers a product called Activ Glass, which is coated in nanoparticles. When light hits these particles, they break down any dirt on the glass. The surface is also hydrophilic ('hydro' means water; 'philic' means loving), which means that water falling on it spreads over the surface, helping to wash it.

Cleaner water

Chemists have developed special filters and processes that use nanoparticles to bond to pollutants in water. This means that they can be used to detect pollutants, such as arsenic and industrial solvents, and help remove them from the water.

Sports equipment

Shin-pads for footballers have been developed that use nano-structured plastics. They are lighter and thinner than other shin-pads, but still provide the same level of protection against kicks. Similar nanocomposite materials are now being used for hockey sticks, snowboards, fishing rods, and even bicycle frames.

'Double-core' tennis balls are now commonly used in professional tennis tournaments. They have an extra layer inside them made of clay particles of size 1 nm, mixed with rubber. This helps to slow down air escaping from the balls, keeping them inflated for longer.

Packaging

As well as being used in tennis balls, nano-sized clay particles are being added to plastics used to make food packaging and drinks bottles. This slows down the rate at which gases pass in or out of the packaging. Oxygen is kept away from foods so they spoil more slowly, and gases in fizzy drinks need to stay in drinks bottles so the drink doesn't go flat.

Clay nanoparticles are also added to plastics to increase their strength. This means less plastic needs to be used to make the food packaging, so reduces waste.

Is nanotechnology safe?

Nanoparticles have different properties from those of larger particles of the same material. This may mean that the nanoparticles have different effects on plants, animals, and the environment.

Different properties, different risks

Some doctors are concerned that nanoparticles are so small they may be able to enter the brain from the bloodstream. If this is true, it could mean some chemicals that are normally harmless become highly toxic at the nanoscale.

Exactly how all the various nanoscale substances differ from larger particles of the same material is not fully understood. There is more data about uses of nanoparticles than about possible health and environmental effects. This is because nanotechnology is very new and measuring these effects can take a long time and cost a lot of money.

Stain-resistant clothes were inspired by the nano-sized hairs on lotus leaves.

Some tennis balls use nanotechnology to stay inflated longer.

Key words

➤ benefits
➤ risk
➤ life cycle
➤ hazard
➤ exposure
➤ regulation

Balancing risks and benefits

When making decisions about different uses of nanotechnology, the **benefits** and **risks** must be balanced.

For example, the benefit of a towel containing nanoparticles of silver is that it stays fresher for longer and slows the growth of odour-causing bacteria. However, if the nanoparticles get washed out in the washing machine and end up in laundry waste water, they would ultimately end up in lakes and streams. Here they may or may not affect fish and other wildlife. They may or may not also have effects on human health. These risks can be difficult to measure.

Governments and companies make decisions about which products are put on the market and for what use, by analysing the benefits and risks. They need to consider the risks associated with the whole **life cycle** of a product, from manufacture to use and disposal.

All nanoparticulate materials are different and need to be analysed and tested separately. Just like ordinary materials, some are more harmful than others, and it may be safe to use some nanoparticulate materials in one situation but not another.

Hazards and exposure

A risk is made up of the **hazard** and **exposure**. A hazard is a potential source of harm, while the exposure is how much of the hazard you are going to come in contact with and in what way.

If a material is considered to be hazardous in some way but measures are put in place to reduce the exposure, then the risk will be quite low. For example, if a nanoparticle is found to be toxic but those handling it always use suitable protective equipment, the risk is low.

Regulation of nanotechnology

There is an ongoing debate about whether nanotechnology should be **regulated** by governments. At the moment there is no single law covering all the different uses of nanomaterials. But some groups and organisations think that there should be.

As mentioned in this section, nanotechnology has applications in many different areas, including food, cosmetics, chemical processes, electronics, and medicine. Each of these areas has its own laws and many of these are thought to be broad enough to cover nanoparticulate materials and their applications too. These laws often state what substance can be used in what way and how hazardous substances and products must be labelled.

However, some people think that these laws are not enough to protect health and the environment against the risks associated with some nanotechnologies, particularly engineered nanoparticles. They think that the existing laws need to be studied carefully and possibly changed. They would like to see detailed risk assessments covering the full life cycle of all new nanotechnology applications.

There are laws that determine how hazardous substances and products must be labelled, and how they can be used.

Towels treated with nanoparticles may stay fresher for longer, but what happens if the nanoparticles get washed out? Will they end up in lakes and streams, and how might they affect these ecosystems?

Questions

1 Write a description of what a nanometre is, for a student in Year 7.

2 a Calculate the surface area of a cube of side 0.01 cm. Write your answer in standard form.
  b Calculate the volume of a cube of side 0.01 cm. Write your answer in standard form.
  c Calculate the surface-area-to-volume ratio of a cube of side 0.01 cm.
  d How many times bigger is the surface-area-to-volume ratio of the cube of side 0.01 cm than that of the cube of side 2 cm?

3 Explain why nanoparticles have the potential to be good catalysts.

4 Explain why carbon nanotubes have the potential to be used in materials for making tennis rackets.

5 Explain why fullerenes and nanotubes have potential uses in drug-delivery applications.

6 Draw up a table to compare the structure and properties of graphene and graphite (see C4.3C).

7 Which properties of zinc oxide nanoparticles make them particularly suitable for sunscreens?

8 Give two ways in which nanotechnology products may be beneficial to the environment.

9 Silver has many different uses.
  a Which properties of silver make it good for making jewellery?
  b How are the particles of silver in socks different from the silver particles in jewellery?
  c Why will nanoparticles of silver be more effective at keeping socks smelling fresh than ordinary silver?

10 Explain why some people are concerned about the possible effects of nanotechnology on the environment.

11 Give one reason for not introducing new regulation for nanotechnologies.

C4.5 What happens to products at the end of their useful life?

A: Metals at the end of their useful life

Find out about

- rusting and corrosion
- oxidation reactions
- reduction reactions
- extending the life of metals **S**
- barrier methods for preventing corrosion
- using zinc or magnesium for sacrificial protection

Key words

➤ corrosion
➤ oxidation
➤ reduction

Iron and steel

More iron is produced every year than all other metals put together. This is because iron is cheap and can be used to make a wide range of useful alloys. Over 98% of iron produced is turned immediately into steel. The rest is used as cast iron or wrought iron.

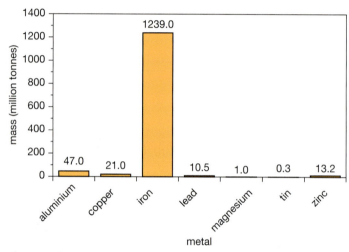

Worldwide annual production of metals in 2013. (Source: British Geological Survey.)

Steel is an alloy of iron with small amounts of carbon and other metals and non-metals. There are many different types of steel. Each type has slightly different properties that make it useful for different purposes. Most types of steel are strong, tough, durable, and easy to shape.

Steel is used to construct buildings, cars, and the machines and appliances we use every day (such as, fridges and washing machines). It is thought that there are more than 20 billion tonnes of steel currently in use. This is more than 2 tonnes for every person on Earth.

You might have noticed that when metals are exposed to the air and wet weather, they rust. Chemists call this **corrosion**. When iron or steel corrodes, bits flake off and it loses its strength. This can be dangerous if the metal is being used to keep a building up or a ship afloat.

What is corrosion?

Corrosion happens when a metal reacts with oxygen. When iron corrodes it reacts with oxygen from the air, in the presence of water, to form hydrated iron oxide.

Hydrated iron oxide is an orange-brown solid that is soft and crumbly. It flakes away from the surface of the iron, exposing the metal beneath to air and water. The metal beneath can then corrode and this continues until all the iron has corroded away.

Uses of iron.

Both oxygen and water are needed for iron and steel to corrode. The overall reaction is:

iron + oxygen + water $\longrightarrow$ hydrated iron oxide (rust)

$$4Fe(s) + 3O_2(g) + xH_2O(l) \longrightarrow 2Fe_2O_3 + xH_2O(s)$$

In the equation for the reaction, x is the amount of water in the hydrated iron oxide. This equation is not completely balanced because the amount of water can vary. It might be very wet or quite dry.

The corrosion reaction happens in two steps, each needing oxygen and water:

Step 1: formation of iron hydroxide

iron + oxygen + water $\longrightarrow$ iron hydroxide

$$2Fe(s) + O_2(g) + 2H_2O(l) \longrightarrow 2Fe(OH)_2(s)$$

Step 2: formation of hydrated iron oxide

iron hydroxide + oxygen + water $\longrightarrow$ hydrated iron oxide (rust)

$$4Fe(OH)_2(s) + O_2(g) + 2H_2O(l) \longrightarrow 2Fe_2O_3 + 6H_2O(s)$$

An oxidation–reduction reaction

Gain and loss of oxygen

When iron corrodes we say it is oxidised. **Oxidation** is the gain of oxygen – when iron corrodes, the iron atoms gain oxygen. We can see that the iron atoms gain oxygen by looking at the formula for hydrated iron oxide.

The opposite of oxidation is **reduction**. Reduction is the loss of oxygen.

H Gain and loss of electrons

Another way to explain oxidation is as the loss of electrons. When iron corrodes, the iron atom loses electrons.

In step 1, iron hydroxide is formed. This is an ionic compound. It is made up of Fe^{2+} ions and OH^- ions. The iron atoms each lose two electrons (e^-) to form the Fe^{2+} ions.

$$Fe \longrightarrow Fe^{2+} + 2e^-$$

The electrons that are lost during oxidation are accepted by the oxygen atoms to form the hydroxide ions (OH^-).

$$\tfrac{1}{2}O_2 + 2e^- + H_2O \longrightarrow 2OH^-$$

Reduction is the loss of oxygen, or gain of electrons. When iron corrodes the oxygen gains electrons, so we say it is reduced.

Remember: OIL RIG

Oxidation Is Loss of electrons, Reduction Is Gain of electrons.

When iron corrodes, it reacts with oxygen (from the air) and water, to form a crumbly red-brown solid.

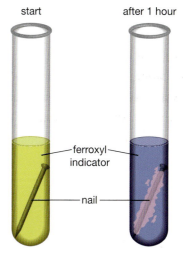

start   after 1 hour

ferroxyl indicator

nail

If a nail is left in a test tube of aqueous solution for an hour, it will begin to corrode. Yellow ferroxyl indicator can be used to detect the products of corrosion. The blue colour shows that Fe^{2+} ions are present and the iron has been oxidised to Fe^{2+}. The pink colour shows that hydroxide ions are present and that the oxygen has been reduced to OH^-.

Steel can last a long time if it is regularly treated to prevent corrosion. If it is not looked after and left to corrode, it stops being useful.

Chain-link fences are often used around parks and playgrounds. They are coated with plastic to prevent corrosion. The plastic coating is more expensive than paint, but it will last much longer. Painting would not be a good choice because the paint wears away over time and it would be a big job to repaint a long chain link fence.

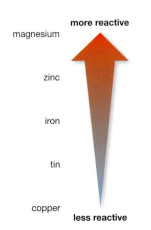

The reactivity series of metals. Iron is more reactive than tin, but less reactive than zinc or magnesium.

Preventing corrosion

Corrosion weakens metals. This can be dangerous and it can cost a lot of money to treat corrosion or replace corroded parts. This is why most items made of iron or steel are protected in some way to prevent or slow down corrosion. This can extend the useful life of metal products, so we can keep using them for longer.

By preventing corrosion, consumers save money by not having to replace a product so often. The environmental impact is also reduced because extracting metals to make new metal products uses a lot of energy and creates pollution.

Barrier methods

Most methods of preventing corrosion involve covering the surface of the metal with a layer of another material. These **barrier methods** keep air (oxygen) and water away from the metal surface. Without oxygen or water, corrosion cannot take place.

Examples of barrier methods are:

- painting
- plating
- coating with plastics
- coating with grease.

Painting is a good choice for protecting the steel bodywork of a car. Paint comes in many different colours, which can make cars more attractive or individual. A disadvantage of painting is that if the paint is scratched, the steel beneath can rust.

Plating is coating the metal with another metal. A 'tin' can is in fact made of steel with a thin layer of tin all over the surface. The tin is less reactive than steel. It does not corrode and will protect the steel, provided it is not scratched.

Bicycle chains and moving parts of an engine cannot be painted. This is because they are rubbing against other pieces of metal so the paint would wear away or flake off. They can be oiled and greased and this helps them move smoothly over each other as well as keeping out water and oxygen.

Chain-link fences may be coated with plastic to stop the steel wires from corroding. Again, this works well unless the plastic coating is scratched or broken.

Sacrificial protection

Sacrificial protection uses a more reactive metal to protect steel from corrosion. Galvanising is one type of sacrificial protection.

Metal dustbins are made of galvanised steel. This is steel plated with a thin layer of zinc. Zinc is higher in the reactivity series than iron. Zinc reacts with oxygen from the air to form zinc oxide. Unlike iron oxide, zinc oxide is a hard substance without cracks or pores. It stops oxygen and water from getting to the iron below. If it is scratched a little, it will simply react to make more zinc oxide, covering over the scratch.

With sacrificial protection, it is not always necessary to completely coat the steel surface. Ships and oil rigs are always in contact with water and sea air so are very susceptible to corrosion. They use blocks of magnesium or zinc bolted onto the outside of their hulls to protect them. The metal blocks are more reactive than the steel and so they corrode in preference to the steel.

When zinc is oxidised it forms Zn^{2+} ions:

$$\boxed{\text{H}}\quad Zn \longrightarrow Zn^{2+} + 2e^-$$

The electrons flow into the iron and are donated to any iron ions that might have formed. This turns the iron ions back into iron atoms, and stops the steel from corroding.

The zinc or magnesium blocks gradually corrode away but can easily be replaced. This is much less expensive than repairing a corroded hull.

Galvanised corrugated steel can last for decades before enough zinc is scratched away for the iron beneath to corrode.

This is the rudder and propeller of a fishing trawler. It has been painted, but for extra protection it also has blocks of a more reactive metal attached to it.

Questions

1 Think of an example of something that is made from steel and explain how the properties of steel make it suitable for this use.

2 What needs to be present for iron to corrode?

H 3 For each of the following equations say whether it shows oxidation or reduction:

   a $O_2 + 4e^- \longrightarrow 2O^{2-}$

   b $Cl_2 + 2e^- \longrightarrow 2Cl^-$

   c $Mg \longrightarrow Mg^{2+} + 2e^-$

4 When aluminium reacts with iodine, aluminium iodide (Al_2I_6) forms. Aluminium iodide is an ionic compound containing Al^{3+} ions and I^- ions. The equation for the reaction is:

$$2Al(s) + 3I_2(s) \longrightarrow Al_2I_6(s)$$

For this reaction state which element is oxidised and which is reduced and explain your answer.

5 Give two reasons why it is important to try and prevent corrosion.

6 Give one advantage and one disadvantage of painting as a way of preventing corrosion.

7 A piece of steel sheet coated with tin is scratched. A piece of steel coated with zinc is also scratched. Explain why the steel coated with tin will corrode, but the steel coated with zinc will not.

Key words

➤ barrier method
➤ plating
➤ sacrificial protection

B: From cradle to grave

Find out about

- the life of products from raw materials to disposal
- the impacts of the products we use

Key words

➤ life-cycle assessment (LCA)
➤ landfill
➤ biodegradable
➤ non-biodegradable

The products people buy and use affect the environment. Environmental scientists add up all the effects of a product from cradle to grave. A life-cycle assessment can show whether it is better to use a shopping bag made of a natural fibre or a bag made of plastic.

Lives or life cycles

Corrosion is what happens to metal products at the end of their useful life, but this is just one part of the life of a product. To understand fully the impact a product has on the environment and to be able to compare the impact of using different materials to make the same product, scientists need to think about the whole life of a product.

The life of a manufactured product has three distinct phases: cradle, use, and grave. At each phase in the life of a product, raw materials, water, and energy may be used:

● raw materials obtained and processed to make useful materials	**CRADLE**
● materials used to make the product	
● energy and water used in processing and manufacturing	
● energy used to transport the product to where it will be used	

● energy needed to use the product (e.g., electricity for a computer)	**USE**
● energy needed to maintain the product (e.g., cleaning, mending)	
● water and other chemicals needed to maintain it	

● energy needed to dispose of the product	**GRAVE**
● space needed to dispose of it	

Life-cycle assessment

Manufacturers can assess the use of raw materials, energy, and water at all stages in the life of their products. This **life-cycle assessment** (**LCA**) is part of legislation to protect the environment. The aim is to slow the rate at which humans use up natural resources that are not renewable.

An LCA involves collecting data about each stage in the life of a product. The assessment includes the use of materials and water, energy inputs and outputs, and environmental impact. An assessment of this kind can, for example, be used to compare the environmental impact of different types of food packaging.

In 2007 a company called Tetra Pak introduced a new kind of food carton that could be used to hold foods that would normally be stored in a 'tin' can (tin cans are actually made of steel). Cans are usually used to store food for long periods because the food can be sealed inside and then sterilised by heating. The new carton was designed so that it could be sterilised in a similar way. It was first used to store chopped tomatoes.

An LCA study was carried out by Tetra Pak to find out whether or not the cartons are better for the environment than tin cans. The table below compares the energy used in the production, transport, and disposal of the cartons and cans.

Tetra Pak carton	Energy (kWh/tonne tomatoes)	Solid waste (kg/tonne tomatoes)
materials production	1053.2	29.7
container production	60.7	1.2
transport from production to filler	29.8	0.1
sterilisation of contents	65.9	2.5
transport from filling to distribution centre	172.5	0.6
end of life	9.1	39.4
lifetime total	**1391.1**	**73.5**

Tin can	Energy (kWh/tonne tomatoes)	Solid waste (kg/tonne tomatoes)
materials production	1957.7	62.8
container production	332.7	15.7
transport from production to filler	9.2	0.0
sterilisation of contents	43.3	2.0
transport from filling to distribution centre	189.3	0.7
end of life	56.7	106.6
lifetime total	**2588.9**	**187.8**

The study showed that over their lifetime, the cartons use less energy and produce less solid waste than the cans. This is mainly because there is significantly more energy used to produce the materials for the cans than for the Tetra Paks.

The results of an LCA are often very dependent on what happens to a product at the end of its life. Manufacturers cannot always control how their products are disposed of. Some may be reused, recycled, or incinerated. Unfortunately, much polymer waste still ends up being tipped into holes in the ground. We call this **landfill**. Landfill is not sustainable because the materials cannot be used again. **Biodegradable** materials are decomposed by microorganisms, but most polymers and metals are **non-biodegradable**.

Cartons are an alternative form of food packaging to cans, but which has a lower impact on the environment?

Questions

1 Which stage in the life of a food carton or can uses the most energy?

2 Why is energy needed:
   a to produce the containers?
   b to transport the containers?
   c to sterilise the food?

3 Draw a bar chart to compare the energy use of cartons and tin cans.

4 Calculate the percentage reduction in energy use of a carton, compared with a tin can, over its lifetime using the values given for tomatoes.

5 Calculate the percentage reduction in solid waste of a carton, compared with a tin can, over its lifetime using the values given for tomatoes.

6 Give two reasons why it is not a good idea to put products in landfill once we have used them.

C: Incinerating, recycling, and reusing waste

Find out about
- ways of disposing of plastic waste
- open-loop recycling
- closed-loop recycling
- reusing plastic waste
- factors that affect decisions about recycling

Key words
➤ incinerator
➤ recycling
➤ PET

When you have finished a bottle or can of drink you probably put it in a recycling bin. This means that the materials go back into making another product.

There are many different options for dealing with waste that don't involve landfill. One way to dispose of polymer waste is to incinerate it. **Incinerators** burn polymer waste and use the energy released to generate electricity. This electricity can be used to provide power to people's homes. However, although energy is recovered the materials are lost.

Recycling

Recycling means that once the life of a product is over, the materials go back into another product. Soft drinks, including mineral water, are often sold in plastic bottles. These are usually made from **PET** (a polyester). PET bottles can be recycled in two different ways.

Open-loop recycling

Discarded PET soft-drinks bottles can be collected and fed through grinders that reduce them to flake form. The flake then passes through a separation and cleaning process that removes all foreign particles, such as, paper, metal, and other plastic materials.

The cleaned PET flake is sold to manufacturers, who convert it into a variety of useful products, such as, carpet fibre, moulding compounds, and non-food containers. Carpet companies can often use 100% recycled polymer to make polyester carpets. PET is also spun to make the fibre filling for pillows, quilts, and jackets.

Because the polymer has been flaked and may not be 100% pure, the resulting products are not as high quality as the original product. Once the new product has reached the end of its useful life it cannot be recycled again. It will probably end up in landfill. The PET can only be recycled once, then the loop is broken. This is called open-loop recycling.

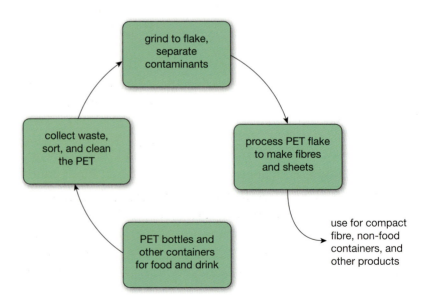

Open-loop recycling.

Closed-loop recycling

Recycling is at its best when the waste material can be used to manufacture the same product again with no loss in quality. With plastic waste this can be done by breaking down the waste into the monomers originally used to make the polymer.

Several companies have developed processes for depolymerising the PET in soft-drinks bottles, to turn the polymers back into monomers. The resulting monomers provide the raw material needed to make new polymer, for new bottles or other PET products.

This is called closed-loop recycling because the recycled material is as high quality as the original material. Manufacturers can keep recycling products in this way again and again, going round and round the loop.

Open-loop recycling is good because it cuts down the amount of raw material needed, and the amount of waste going to landfill. But it is not as good value as closed-loop recycling.

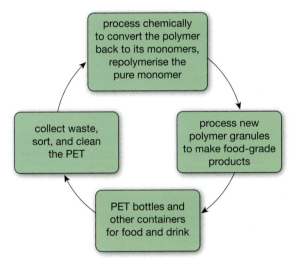

Closed-loop recycling.

Reusing waste

Once you have finished with a product the best thing you can do is reuse it. Many supermarkets now sell plastic carrier bags that are designed to be reused many times. These are sometimes called a 'bag for life'. Reusing a carrier bag is much better than throwing it away or even recycling it.

Is it always good to recycle?

We are encouraged to recycle aluminium cans because extracting aluminium from aluminium ores to make new drinks cans is very difficult. It takes a lot of energy, uses large volumes of water, and costs a lot of money. It takes less energy to turn aluminium waste into new products, and recycled aluminium is as good as new after reprocessing.

Bales of aluminium cans waiting to be recycled at a recycling plant.

While recycling can often be cost effective and reduce the environmental impact of a product's life cycle, this is not always the case. For example, mixed plastic waste can be very difficult to separate. The cost and energy needed to separate the waste material may be too high to make it worthwhile, or **viable**.

Decisions about whether or not to recycle, or to encourage others to recycle, depend on a variety of factors:

1 Are the raw materials (e.g., metal ores and crude oil) running out?

2 Is the waste material readily available?

3 How easy is it to collect and sort the waste material and how much does this cost?

4 How easy is it to remove impurities from the waste material?

5 How much energy is needed to transport and process the waste? Is this more or less than the amount of energy needed to transport and process the raw materials?

6 How much demand is there for products made from the recycled material?

7 What is the environmental impact of the recycling process?

Many of these factors will vary depending on local circumstances. This means that recycling may be viable in some circumstances but not others.

For example, the ease of collecting waste material depends on where it is being collected from. It is much more cost effective to collect recyclable waste from people's homes in cities, where they live closer together, than in rural areas. In rural areas, the rubbish trucks need to travel further to collect the same amount of waste.

Workers separating waste on a conveyor belt at a recycling facility.

Is recycling PET drinks bottles viable?

A life-cycle assessment can help with decisions about the viability of recycling. It can provide the data you need to help answer these questions.

- Crude oil is the main raw material used to make PET drinks bottles. Crude oil is a non-renewable resource. We will not be able to keep extracting it forever. As oil runs out, its price will go up. Higher oil prices will mean that making PET from crude oil becomes less cost effective than making PET from recycled waste.

- Most homes, businesses, schools, and hospitals now have regular recycling collections. There are also recycling bins in most town centres. Around 35% of PET plastic bottles in household waste are now collected for recycling. This means that there is a good supply of used PET drinks bottles.

- It is relatively easy to identify and separate out PET drinks bottles from mixed recycling waste, and technological advances, such as closed-loop recycling, mean that the recycled polymer material can be very high quality. The quality and range of potential uses mean that there is high demand for recycled PET.

- Recycling PET drinks bottles reduces both landfill and energy use. It is estimated that recycling 1 tonne of PET waste saves the equivalent of 1.5 tonnes of carbon dioxide emissions.

- The process of producing a PET plastic drinks bottle from crude oil, including processing, packaging, and transport, uses about 7 litres of water. Bottle manufacturers are constantly looking at ways to further reduce the amount of water used in the manufacturing process.

In most cases, recycling PET bottles is viable. Recycling protects natural resources, such as crude oil and metal ores, but will not be enough on its own to meet future demand for these resources. Scientists need to keep improving product life cycles at all stages.

Questions

1. a Explain, in a short paragraph, the difference between open-loop and closed-loop recycling.
    b Suggest one possible advantage and one possible disadvantage of each of these approaches to recycling.

2. Suggest examples of how to slow down the flow of materials from raw materials to waste. Include examples of:
    a reuse
    b recycling
    c recovering energy.

3. If a new process was developed that meant polymer waste could be separated quickly, easily, and cheaply, what effect would this have on the viability of recycling plastics? Explain your answer.

4. If the price of oil was to fall dramatically, what effect would this have on the viability of recycling plastics? Explain your answer.

Science explanations

C4 Material choices

Our society uses a large range of materials and products that have been developed, tested, and modified by the work of materials scientists. Materials used to make a particular product need to meet a specification that describes the properties the material needs to make it suitable for a particular use.

You should know:

- that one way of comparing materials is to measure their properties, such as, melting point, softening temperature (for polymers), electrical conductivity, strength (in tension or compression), stiffness, flexibility, brittleness, hardness, density, and ease of reshaping
- why it helps to have an accurate knowledge of the properties of materials when choosing a material for a particular purpose
- how the composition of alloys relates to their uses **S**
- that polymerisation is a chemical reaction that joins up small monomer molecules into long chains
- that DNA is a polymer
- **H** the basic principles of condensation polymerisation
- how the basic bulk properties of materials are related to the different types of bonds they contain, the intermolecular forces, and the arrangement of the bonds
- that the huge variety of organic compounds occurs due to the ability of carbon to form four covalent bonds, resulting in families of similar compounds, chains, and rings
- how the properties of polymers depend on the way in which the long molecules are arranged and held together
- the nature and arrangement of the covalent bonds in giant covalent structures
- how the properties of diamond and graphite can be explained by their structure and bonding
- the similarities and differences in the nature and arrangement of chemical bonds in ionic compounds, simple molecules, giant covalent structures, polymers, and metals
- that nanotechnology is the use and control of structures that are very small (1–100 nm)
- why nanoparticles of a material show different properties to larger particles of the same material
- how the properties of fullerenes and graphenes relate to their structures
- how the properties of nanoparticles are related to their uses
- that there are possible risks associated with nanoparticles
- the conditions that cause corrosion and how to explain methods of mitigation **S**
- how reduction and oxidation can be described in terms of loss or gain of oxygen, and also in terms of gain or loss of electrons **H**
- the principles behind the life-cycle assessment of a material or product.

Ideas about Science

Scientific explanations are based on data but they go beyond the data and are distinct from them. An explanation has to be thought up creatively to account for the data. In the context of the theories of structure and bonding you should be able to:

- recognise data, such as measures of the properties of elements and compounds, that is accounted for by explanations based on theories of structure and bonding.

Some applications of science, such as the development of new materials, can have unintended and undesirable impacts on quality of life or the environment. Benefits need to be weighed against costs. You should be able to:

- describe and explain examples of applications of science that have made significant positive differences to people's lives
- explain the idea of sustainability and relate it to life-cycle assessments of materials or products
- suggest reasons why different decisions on the same issue might be appropriate in view of differences in social and economic context
- identify and suggest examples of unintended impacts of human activity on the environment, such as the production of 'disposable' consumer goods.

Everything we do carries a certain risk of accident or harm. New technologies and processes can introduce new risks. In the context of developing new materials, you should be able to:

- identify examples of risks that have arisen from a new scientific or technological advance
- interpret and discuss information on the size of a given risk, taking account of both the chance of it occurring and the consequences if it did
- for a given situation:
  - identify risks and benefits to the different individuals and groups involved
  - suggest reasons for people's willingness to accept the risk
  - **H** distinguish between perceived and calculated risk
- discuss the public regulation of risk, and explain why it may in some situations be controversial.

C4 Review questions

1 The table shows some information about four different types of steel, A, B, C, and D.

Type of steel	Iron content (%)	Carbon content (%)	Tensile strength (psi)	Stretch before breaking (%)
A	99	0.3	80 000	20
B	99	0.1	43 000	48
C	98	0.4	115 000	8
D	98	0.2	78 000	25

a How does the data show that steel is an alloy? [S]

b How does the data show that steel contains elements other than iron and carbon?

c What conclusions can you make about the effect of carbon on the tensile strength of steel?

d A student tests the tensile strength of a 25 cm-long cable of steel D.
  Predict the length of the cable at its breaking point.

e Alex says that he thinks the stretch before breaking is linked to the carbon content of the steel. Do you agree? Give your reasons.

2 The diagrams show the structures of some monomers that can be used to make polymers. [S]

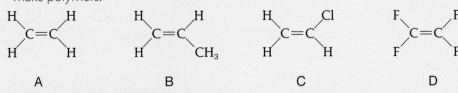

A   B   C   D

a Which **two** diagrams show monomers that are hydrocarbons?

b This diagram shows what happens when a polymer forms from monomer A.

$$n \times \quad \begin{array}{c} H \\ \\ H \end{array} C=C \begin{array}{c} H \\ \\ H \end{array} \quad \rightarrow \quad \left[\begin{array}{cc} H & H \\ | & | \\ C - C \\ | & | \\ H & H \end{array} \right]_n$$

Draw similar diagrams to show what happens when polymers form from each of the monomers B, C, and D.

c How do your diagrams show that all of these polymers are addition polymers?

d These two monomers react to make a condensation polymer. [H]

$$HO-\overset{O}{\overset{||}{C}}-\square-\overset{O}{\overset{||}{C}}-OH \quad + \quad OH-\square-OH$$

i Draw the **structure** of the repeating unit of the polymer.

ii Name the small molecule that is made in the reaction.

174

3   The table shows some properties of carbon dioxide and graphite.

	carbon dioxide	graphite
melting point (°C)	–57	3550
boiling point (°C)	–78	3800
electrical conductivity	does not conduct	good

a   Explain what is unusual about the melting point and boiling point of carbon dioxide compared to most substances.

b   Carbon dioxide and graphite have different structures.

Explain why carbon dioxide and graphite have different properties by comparing their structures.

carbon dioxide

4   a   Titanium dioxide nanoparticles are used in some sunblock creams. Because they are so small they do not reflect visible light.

i   Within what range is the diameter of a nanoparticle?

Choose the correct answer.

less than 1 nm   1–100 nm   100–500 nm   500–1000 nm

ii   Suggest one advantage of using nanoparticles in sunblock creams.

iii   Use ideas of risk and benefit to explain why people have different views about using sunblock creams that contain nanoparticles.

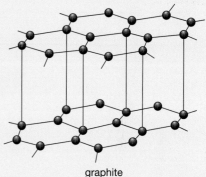

graphite

b   A book uses these mathematical expressions to describe nanoparticles:
size of nanoparticles > size of most molecules > size of atoms.
Explain what this expression means.

5   Some information about the life-cycle assessments (LCAs) for three types of disposable bags are shown in the table.

	Totals for 1000 bags for the whole LCA		
	paper (30% recycled fibre)	biodegradable plastic	polyethene
Energy use (MJ)	2620	2070	763
Fossil fuel use (kg)	23.2	41.5	14.9
Municipal solid waste (kg)	33.9	19.2	7.0
Greenhouse gas emissions (kg CO_2)	80	180	40
Fresh water use (litres)	4520	4580	260

a   Some people think that supermarkets should be banned from supplying polyethene bags.

They think that using paper bags or biodegradable bags would reduce harm to the environment.

Does the data in the table support this view?

b   The data in the table does not cover all aspects of the LCA of the bags.

Give two other factors that would need to be considered to compare the LCAs for the bags.

C5 Chemical analysis

Why study chemical analysis?

Analytical measurement of substances requires accuracy and attention to detail. It is essential to ensure that the many substances we use in our everyday lives do us good and not harm. Over £7 billion is spent each year on chemical analysis in the UK.

What you already know

- Filtration, distillation, and chromatography are techniques used to separate mixtures.
- Chemical reactions can be represented using formulae and equations.
- Ionic equations describe chemical changes by showing only the reacting ions in solution.
- Acids and alkalis react together in neutralisation reactions.
- The pH scale and indicators can be used to measure acidity and identify neutralisation.
- Acids react with alkalis to produce a salt plus water.
- The relative atomic mass of atoms can be found in the Periodic Table.

The Science

Chemists analyse substances to find out how much of each different element or compound they contain. This is important in making large-scale chemicals such as fertilisers, metals, and bulk chemicals and also in the quality control and analysis of drugs, personal products, and foods.

Chemists measure amounts and numbers of particles by measuring their masses or, for gases, their volumes. This enables chemists to calculate the amounts of reactants and products that are involved in reactions, and to analyse yields. Concentration measures the amounts of particles in solution. Titration is a type of quantitative analysis used to find out the concentration of a substance in solution.

Ideas about Science

Chemists need to be able to calculate the amounts of chemicals they plan to react together to make a product, both on a laboratory scale and in industry. This helps avoid using more resources than necessary, and reduces the waste to be disposed of. To ensure reliable analysis, standard procedures, such as titrations, are used. Titration procedures aim to deliver results as close as possible to the true value of the quantity measured, to ensure the accuracy, precision, repeatability, and reproducibility of the data.

A: Mixtures and pure substances

Find out about

- mixtures
- pure substances
- formulations

Examples of formulation products include paints, cosmetics, medicines, perfumes, and detergents.

Drinks manufacturers often use the word 'pure'.

Composition of the characteristic ingredients:		
Typical values mg/litre:		
Sodium	Na^+	13.2
Calcium	Ca^{2+}	29.1
Magnesium	Mg^{2+}	3.0
Chloride	Cl^-	31.1
Sulfate	SO_4^{2-}	42.7
Nitrate	NO_3^-	<0.5

Mineral water is not pure water.

Formulations

Many useful products contain **mixtures**. Mixtures consist of **pure substances** in the same state of matter, or in different states. A mixture is made by simply mechanically mixing pure substances in any proportion. **Formulations** are mixtures that are prepared according to a specific formula. Many consumer products such as medicines, cosmetics, detergents, and sunscreens are formulations. It is important that these products do not contain impurities that could be harmful to the user or have unwanted side effects.

A formulation usually contains one or more active ingredients, such as therapeutic drugs in medicines or pigments in paints. The other components in the mixture also have specific functions. For example, pharmaceutical companies add sweeteners and food flavours to some medicines to make them taste better. New formulations are usually made by trial and error, changing the proportions of the components until the desired properties are found. For example, a liquid may become more or less viscous depending on the exact proportion of the components.

Pure substances

When chemists talk about pure substances, they are referring to a substance that contains a single element (for example, oxygen) or a single compound (for example, water or carbon dioxide). This is in contrast to how the term 'pure' is often used in everyday life. For example, you will often see adverts for 'pure' fruit juice or 'pure' mineral water. This means that nothing else was added to the fruit juice or mineral water during manufacture. However, these substances are not 'pure' to a chemist, because both fruit juice and mineral water contain several different compounds, as you can see from the ingredients on the label.

Grades of purity

The reactions of acids with metals, oxides, hydroxides, and carbonates can be used to make useful salts. For uses such as food or medicines, these salts have to be made pure so that they are safe to be consumed.

Chemicals do not always have to be pure. Calcium carbonate, for example, is used in a blast furnace to extract iron from its ores. The iron industry can use limestone straight from a quarry. Limestone has some impurities but they do not stop it from doing its job in a blast furnace.

Suppliers of chemicals offer a range of grades of chemicals. In a school laboratory you might use one of these grades: technical, general laboratory, and analytical. The purest grade is the analytical grade. Technical-grade chemicals have relatively high levels of impurities.

Purifying a chemical is done in stages. Each stage takes time and money, and becomes more difficult. So the higher the purity, the more expensive the chemical is. Manufacturers therefore buy the grade most suitable for their purpose.

When deciding what grade of chemical to use for a particular purpose, it is important to know:

- the amount of impurities
- what the impurities are
- how they can affect the process
- whether they will end up in the product, and whether this matters.

Testing for purity

Pharmaceutical and cosmetic companies buy in many of their ingredients. Technical chemists working for the companies have to make sure that the suppliers are delivering the right grade of chemical.

Chemists test substances made in the laboratory and in manufacturing processes to check that they are pure. For instance, citric acid is often added to syrups such as cough medicines to control their pH. Technicians can check the purity of the acid using a procedure called **titration**, which measures the volume of alkali that it can neutralise.

Another way of assessing the purity of a substance is by testing its **melting point**. The melting point of a substance is the temperature range over which the first crystal of a solid just starts to melt and the last crystal completes its melting. Pure substances have a sharp melting point and can be identified by matching melting-point data to reference values. The presence of an impurity usually lowers the melting point and broadens the melting temperature range. This is because impurities cause defects in the crystalline lattice, making it is easier to overcome the intermolecular interactions between the particles.

Key words

- ➤ mixture
- ➤ pure substance
- ➤ formulations
- ➤ titration
- ➤ melting point

CALCIUM CARBONATE PRECIPITATED CP
QTY: 1kg BNO: C1042/R6 -708717

Assay	99%
Chloride (Cl)	0.005%
sulfate (SO$_4$)	0.05%
Iron (Fe)	0.002%
Lead (Pb)	0.002%

Laboratory-grade calcium carbonate. The term 'assay' tells you how pure the chemical is.

Synoptic link

You can learn more about titrations in C5.4C *Titrations*.

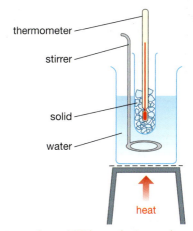

Pure ice melts at 0 °C, but salty ice melts between −5 °C and −2 °C. The greater the impurity, the bigger the difference from the melting point of the pure substance.

Synoptic link

You can learn more about the particle model and changes of state in C1.1 *How has the Earth's atmosphere changed over time, and why?*

Questions

1   Why is it important for manufacturers to test the purity of components before making up formulations?

2   Uses of sodium chloride (salt) include:
- flavouring food
- melting ice on roads
- saline drips in hospitals.

List these in order of the grade of sodium chloride required, with the purest first.

3   The melting-point range for an unknown white solid was found to be 112–116 °C. Compound A melts at 112 °C; compound B at 114 °C; and compound C at 116 °C; and compound D at 118 °C.

What does this information suggest about the white solid? Give reasons to support your answer.

B: Chromatography

Find out about

- principles of chromatography
- paper chromatography
- interpreting chromatograms

Key words

- ➤ mobile phase
- ➤ stationary phase
- ➤ equilibrium
- ➤ aqueous
- ➤ non-aqueous
- ➤ reference material
- ➤ solvent front
- ➤ locating agent

Chromatography can be used to:

1 separate and identify chemicals in a mixture

2 check the purity of a chemical

3 purify small samples of a chemical.

There are several types of chromatography. Paper chromatography, which can be done with some blotting paper and a solvent, is a cheap and simple method. Gas chromatography is an expensive method that involves high-precision instruments. All types of chromatography work on similar principles.

Principles of chromatography

Chromatography depends on the movement of a **mobile phase** through a fixed medium called the **stationary phase**. The analyst adds a small sample of the mixture to the stationary phase. As the mobile phase moves through the stationary phase, the chemicals in the sample move between the mobile and stationary phases.

For each chemical in the mixture there is a dynamic **equilibrium** as the molecules are distributed between the stationary phase and the mobile phase. If a chemical in the mixture is attracted more to the mobile phase, it moves faster. If a chemical is attracted more to the stationary phase, it moves more slowly. Since each chemical in the mixture is attracted differently, they move at different speeds and are separated.

- The chemical moves quickly if the position of equilibrium favours the mobile phase.

- The chemical moves slowly if the position of equilibrium favours the stationary phase.

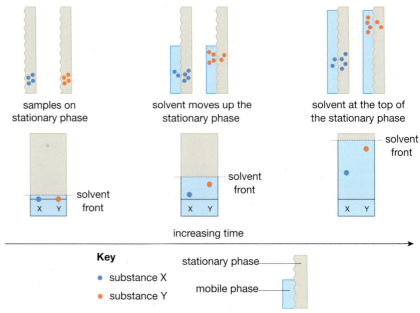

For each chemical there is a dynamic equilibrium as the molecules distribute themselves between the two phases. How quickly each chemical moves through the stationary phase depends on the position of this equilibrium.

Paper chromatography

Paper chromatography is used to separate and identify substances in mixtures. The technique does not require expensive instrumentation, but is limited in its use.

In paper chromatography the stationary phase is the paper, which does not move. The mobile phase is a solvent, which may be one liquid or a mixture of liquids. Substances are more soluble in some solvents than others. For example, some substances such as salts dissolve well in water, while others are more soluble in petrol-like, hydrocarbon solvents. With the right choice of solvent, it becomes possible to separate complex mixtures.

When a solute is dissolved in a solvent, a solution is formed. An **aqueous** solution is a solution in which water is the solvent. A **non-aqueous** solution is a solution in which something other than water is the solvent.

Practical procedure

The sample is dissolved in a solvent. This solvent is usually not the same as the mobile phase.

A small drop of the solution is put on the paper and allowed to dry, leaving a small 'spot' of the mixture.

If the solution is dilute, further drops are put in the same place. Each is left to dry before the next is added. This produces a small spot with enough material to analyse. The separation is likely to be poor if the spot spreads too much.

One way of identifying the chemicals in the sample is to add separate spots of solutions of substances suspected of being present in the unknown mixture. These are called **reference materials**.

The analyst adds the chosen solvent (the mobile phase) to a chromatography tank and covers it with a lid. After the tank has stood for a while, the atmosphere inside becomes saturated with solvent vapour.

The next step is to place the prepared paper in the tank, checking that the spots are above the level of the solvent.

The solvent immediately starts to rise up the paper. As the solvent rises, it carries the dissolved substances through the stationary phase. Covering the tank ensures that the solvent does not evaporate.

The chromatography paper is taken from the tank when the solvent gets near the top. The analyst then marks the position of the **solvent front**.

It is easy to mark the positions of coloured substances. All the analyst has to do is outline the spots in pencil and mark their centres before the colour fades. To locate colourless substances the chromatogram is 'developed' by spraying it with a **locating agent** that reacts with the substances to form coloured compounds.

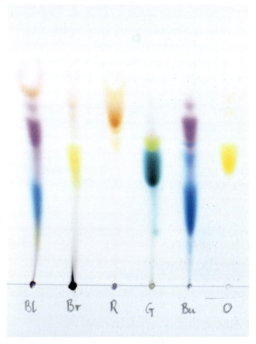

Chromatography plates must be spotted carefully. Small concentrated spots are needed. Their starting position should be marked.

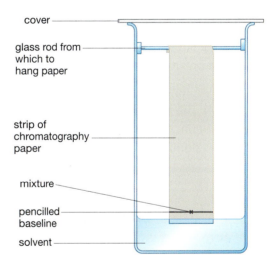

A chromatography tank. The sample spots on the paper must be above the level of the solvent.

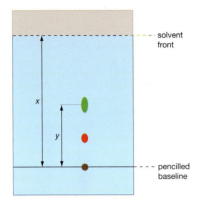

solvent front

x

y

pencilled baseline

Calculating the R_f.

Interpreting chromatograms

Chemicals may be identified by comparing spots with those from standard reference materials.

A chemical may also be identified by its **retardation factor** (R_f). This does not change, provided the same conditions are used. It is calculated, using the following formula, by measuring the distance travelled by the substance:

$$R_f = \frac{\text{(distance moved by chemical)}}{\text{(distance moved by solvent)}} = \frac{y}{x}$$

Key word

➤ retardation factor

Worked example: R_f values

Calculate the R_f value for the green spot in the diagram to an appropriate number of significant figures.

Step 1: Write down the equation you will use to calculate the R_f values.

$$R_f = \frac{\text{(distance moved by chemical)}}{\text{(distance moved by solvent)}}$$

Step 2: Measure the distance moved from the baseline by the solvent and the chemical.

distance moved by chemical = 1.9 cm
distance moved by solvent = 4.2 cm

Step 3: Substitute the quantities into the equation and calculate the R_f value.

R_f value does not have units as it is a ratio.

$$R_f = \frac{1.9 \text{ cm}}{4.2 \text{ cm}}$$
$$= 0.452$$

Step 4: The measurements were made to two significant figures, so give the R_f value to two significant figures.

$R_f = 0.45$ (2 significant figures)

Questions

1   Name two substances that dissolve better in water than in hydrocarbon (or other non-aqueous) solvents.

2   Why is it sometimes necessary to 'develop' a chromatogram? How can this be done?

3   What are reference materials used for?

4   Calculate the R_f value for the red spot in the diagram.

5   Paper chromatography is used to separate a mixture of a red chemical and a blue chemical. The blue chemical is more soluble in water while the red chemical is more soluble in the non-aqueous chromatography solvent.

   a   Sketch a diagram to show the chromatogram you would expect to form.

   b   Explain what the R_f value is and how it could be used to identify an unknown substance.

C: Separation and purification processes

The preparation of chemicals often produces impure products or a mixture of products. Separation processes in both the laboratory and in industry allow useful products to be separated from by-products and waste products. Chemists separate the components of mixtures using processes that use the different properties of the components, for example, state, boiling point, or solubility in different solvents.

Separation processes are rarely completely successful and mixtures often need to go through several stages or through repeated processes to reach an acceptable purity.

Find out about

● filtration
● crystallisation
● distillation

Filtration (Solid)

Filtering is a quick and easy way of separating a liquid, solution, or gas from a solid. It is one of the processes used in industrial water-treatment plants. Filters are substances such as paper, porous porcelain, a cloth, layer of sand, or charcoal through which a liquid or gas is passed to remove suspended particles or solids.

The solid or residue is caught by the filter while the liquid, or filtrate, passes through. In order to obtain a pure sample it is sometimes necessary to filter the mixture several times.

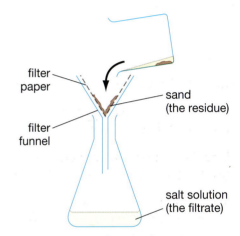

Filtering a mixture of sand, salt, and water in the laboratory.

Crystallisation

Crystals are formed by evaporating the water from a solution of a salt. The size of the crystals is related to the rate of evaporation of the water. If the water evaporates slowly, large crystals are formed. Rapid evaporation leads to the formation of much smaller crystals. The crystals of a pure product are often well formed and even in shape. Different colours seen in a crystal are usually an indication of the presence of impurities.

The Cave of Crystals is 290 m underground in Naica Mine, Mexico. The large crystals formed over millions of years in the geothermally heated, mineral-rich water that filled the caves. The calcium sulfate crystals were discovered after the water was pumped out of the mine. Above this cave is the Cave of Swords, where the crystals are smaller because the water cooled more rapidly.

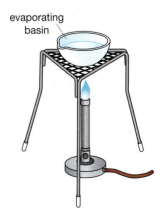

Salt crystals being prepared from a solution by crystallisation in the laboratory.

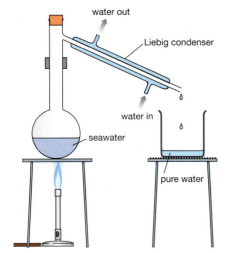

Simple distillation can be used to produce small amounts of pure water from seawater in the laboratory.

Synoptic link

You can learn more about fractional distillation of crude oil on an industrial scale in C3.4 *Why is crude oil important as a source of new materials?*

Key words

➤ distillation
➤ volatile
➤ fractional distillation

Distillation

Scientists use **distillation** to purify a solution by heating and cooling. First they heat the solution until it boils, when the more **volatile** liquid evaporates. The vapours are then condensed and collected, leaving a more concentrated solution behind in the reaction vessel.

Distillation is also a technique for separating a mixture of liquids by their boiling points. The boiling point of ethanol is 78.5 °C, while the boiling point of water is 100 °C. When a mixture of water and ethanol is heated to a temperature above 78.5 °C and below 100 °C the ethanol becomes a vapour, which is condensed and collected, leaving much of the water behind.

We also use this for cracking.

Fractional distillation is a process in which multiple simple distillations (e.g., boiling and condensing) can be carried out in a single piece of apparatus. It is dependent on each component of the mixture having a different boiling point. As the mixture is heated, the temperature rises and each fraction evaporates as its boiling point is reached. The larger the difference in boiling points of the components, the easier it is to collect a pure sample of the components. In practice it is very difficult to extract a pure component from the mixture.

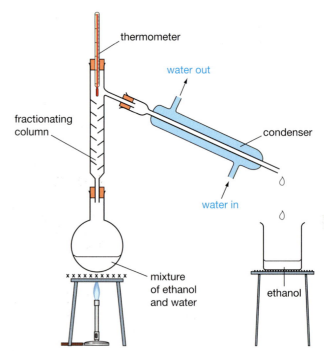

Fractional distillation can be used to separate a mixture of two or more liquids. This simple set-up is used in the laboratory.

Questions

1 Starting with a sample of rock salt, explain how you would obtain a sample of pure salt.

2 Look at the diagram showing the distillation of ethanol from an ethanol–water mixture. Explain why it is important to monitor the temperature during the process.

A: Chemical analysis

Chemical analysis is important. It is essential to ensure that the products we use in our everyday lives do us good and not harm. Toothpaste, detergents, medicines, cars, paints, fuels, and clothes are all subjected to chemical analysis at some stage of their production. Doctors often carry out analytical tests to determine the causes of illness, while forensic scientists do so to solve crimes.

Chemical analysis can be used to answer questions such as:

Find out about
- qualitative and quantitative methods of analysis
- collecting, storing, and analysing samples

- What is the chemical composition of this cosmetic cream?

- What minerals are contained in this ore?

- Is this water safe to drink?

- Does this soil contain the right nutrients for growing crops?

Analytical methods

Analysts use both qualitative and quantitative methods, depending on what they want to find out. A **qualitative** method can be used if the aim is to find out which chemicals are present in a material. This is its chemical composition. However, if the aim is to find out how much of each chemical is present, then a **quantitative** approach is required.

Each step of an analysis must be carried out as set down in the agreed standard procedures. Using these procedures, if an analyst carried out the same test twice they would expect to get the same results. If anyone else carried out the test they would also expect to get the same results.

Representative samples

Analysts work with **samples** of material that represent the bulk of the material under test. Rarely do they analyse the whole thing. How big the samples need to be depends on the analyses to be carried out.

Key words
➤ qualitative analysis
➤ quantitative analysis
➤ sample

Key words

➤ representative
➤ homogeneous
➤ heterogeneous

The composition of a **homogeneous** material is uniform throughout, like milk chocolate.

The composition of a **heterogeneous** material varies, like a chocolate caramel bar.

Bottled water is homogeneous. To test the water only a single sample is needed.

Water from a stream is likely to be heterogeneous. Samples may vary from one part of the stream to another.

The samples the analyst chooses must be **representative**. In other words, the samples should give an accurate picture of the material as a whole. This can be hard to achieve with an uneven mixture of solids, such as soil, but is much easier when the chemicals are evenly mixed in solution, as in urine.

Scientists have to decide:

● how many samples, and how much of each, must be collected to ensure they are representative of the material
● how many times an analysis should be repeated on a sample to ensure results are repeatable
● where, when, and how to collect the samples of the material
● how to store samples and transport them to the laboratory to prevent the samples from 'going off', becoming contaminated, or being tampered with.

Analysing water

Think about analysing water from two different sources. One is bottled water bought at a local supermarket. The other is from a local stream.

The bottled water is likely to be tested for dissolved metal salts. It is clear and there are no solids in suspension. The water is **homogeneous**, so only a single sample is needed. However, to check a batch of bottles, the analyst would take samples from a number of bottles. How much is needed depends on the test. There are no storage or transport problems. The bottle can be opened in the laboratory. This is a straightforward sampling task.

Water from the stream may be tested for a range of things, from the concentrations of dissolved chemicals to the number and variety of living organisms. Samples are likely to be **heterogeneous**. They may be cloudy and may vary from one part of the stream to another. The time of year when samples are collected will affect the composition of the water. Samples also need to be stored and taken back to the laboratory for analysis. This is a complex sampling problem that needs careful planning.

Questions

1 What is the difference between a quantitative and qualitative analytical method? Give an example of each.

2 Give two reasons why it is important for analysts to follow standard procedures.

3 Why must a sample be representative of the bulk material under test?

4 Suggest how an analyst should go about sampling when faced with the following problems. In each case, identify the difficulties of taking representative samples. Suggest ways of overcoming the difficulties.
   a measuring the concentration of chlorine disinfectant in a swimming pool
   b checking the purity of citric acid supplied to a food processor
   c monitoring the quality of aspirin tablets made by a pharmaceutical company
   d determining the level of nitrates in a farmer's field

B: Detecting ions in salts

Salts are found as solids, in minerals, and in solution, dissolved in the seas and oceans. Every litre of seawater contains about 35 g of dissolved salts. The most common salt is sodium chloride.

Salts are ionic compounds. They are made up of positively charged metal ions and negatively charged non-metal ions. Qualitative methods of analysis can be used to find out which ions form an unknown salt. These tests rely on each type of ion in the salt having special, characteristic properties.

Sodium chloride has a unique set of properties, such as melting point and solubility. But some of its properties are shared by all salts containing sodium ions, and others are shared by all salts containing chloride ions.

Flame colours

Flame tests can be used to identify metal ions in a salt. Flame tests are based on the principle that different metal ions give different colours to a flame when they are heated in it.

Early gas burners produced smoky and yellow flames. This would have made it difficult to see flame colours. But in 1855 Robert Bunsen, a German chemist, developed the type of burner that is still used today in laboratories all over the world. The great advantage of Bunsen's burner was that it could be adjusted to give an almost invisible flame. Bunsen first used his burner to blow glass, but he noticed that whenever he held a glass tube in a colourless flame, the flame turned yellow.

Robert Bunsen (1811–1899), who discovered the flame colours of elements with the help of his new burner.

Soon Bunsen was experimenting with different chemicals, which he held in the flame at the end of a platinum wire. He saw that different chemicals produced characteristic **flame colours**.

The table shows the flame colours produced by some metal ions.

Metal ion	Flame colour
sodium	bright yellow
potassium	lilac
calcium	orange-red
copper	blue-green

Metal ions in salts can be identified by comparing the flame colour with reference data, such as that given in the table.

Find out about

- tests for ions
- flame colours
- precipitation reactions
- H ionic equations

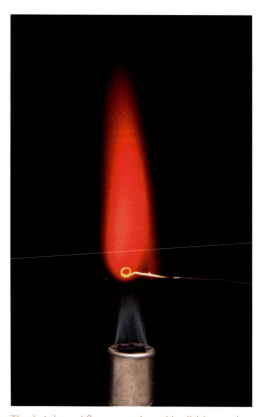

The bright red flame produced by lithium salts.

Key word

➤ flame colour

Precipitation reactions

Another method of identifying ions is by precipitation reactions. In a precipitation reaction two solutions are mixed and a solid forms. These reactions can be used to identify both metal and non-metal ions in solution. The tests rely on the different **solubilities** of different ionic compounds. Many ionic compounds are soluble in water, but some are not.

Shellfish use calcium carbonate to make their protective shells. At the surface of the sea, the concentrations of Ca^{2+} and CO_3^{2-} ions in the seawater are high enough for the calcium carbonate to precipitate. When shellfish die, their remains sink and form sedimentary limestone rocks on the seabed.

Sodium chloride is quite soluble in water over a wide range of temperatures. So a solution can contain quite high concentrations of Na^+ and Cl^- ions together. Calcium carbonate, however, has only a low solubility, so no solution can contain high concentrations of Ca^{2+} and CO_3^{2-} ions together.

To see what happens when you put high concentrations of Ca^{2+} and CO_3^{2-} ions together, you would start with two solutions: one with lots of a soluble calcium salt, and the other with lots of a soluble carbonate salt. Calcium nitrate and sodium carbonate would be good examples to use. You then mix the two solutions together.

Very quickly, huge numbers of Ca^{2+} and CO_3^{2-} ions cluster together to make solid crystals, which fall to the bottom of the solution as a solid white **precipitate**. The process is exactly the reverse of dissolving. The other two ions, sodium and nitrate, stay dissolved in the solution because sodium nitrate is a soluble salt.

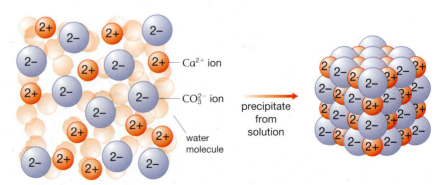

Ions of an insoluble salt cluster together and make a solid precipitate.

Predicting the precipitates

When two solutions of soluble salts are mixed, you can think of the metal ions changing places to form two new salts.

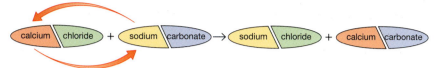

When solutions of salts are mixed, you can think of the metal ions changing places. Sodium chloride is soluble but calcium carbonate is insoluble.

If you know the solubilities of salts, you can predict whether either of these new salts is insoluble and will form a precipitate. This can help you to detect the ions present in a solution.

The table shows the solubilities of some salts and hydroxides.

Salt or hydroxide	Solubility
all nitrates	soluble
all salts of sodium and potassium	soluble
silver chloride	insoluble
silver bromide	insoluble
silver iodide	insoluble
barium sulfate	insoluble
calcium carbonate	insoluble
hydroxides of metals not in Group 1	insoluble

Testing for metal ions

One way to test for calcium ions in a solution would be to add a solution of sodium carbonate and see if a white precipitate of calcium carbonate forms. But there are other metal ions, such as zinc, that would also form a white carbonate precipitate. So the results would not be conclusive.

Several metal ions form insoluble hydroxide compounds, too. The ions of transition metals are special, though, because they are coloured. This gives their hydroxide precipitates characteristic colours that can be used to identify the metal ion in the original solution.

To carry out the test, add a small volume of a dilute solution of sodium hydroxide to a solution of the substance to be investigated and observe the colour of the precipitate. The table below shows the results of this test. In the test for Zn^{2+} ions, if there is excess sodium hydroxide the precipitate dissolves, leaving a colourless solution. The other hydroxide precipitates in the table are insoluble in excess sodium hydroxide.

The hydroxide precipitates $Fe(OH)_2$, $Fe(OH)_3$, and $Cu(OH)_2$.

Metal ion tested	Precipitate	Observations
copper, Cu^{2+}(aq)	$Cu(OH)_2$(s)	light blue (insoluble in excess NaOH(aq))
iron(II), Fe^{2+}(aq)	$Fe(OH)_2$(s)	green (insoluble in excess NaOH(aq))
iron(III), Fe^{3+}(aq)	$Fe(OH)_3$(s)	red-brown (insoluble in excess NaOH(aq))
calcium, Ca^{2+}(aq)	$Ca(OH)_2$(s)	white (insoluble in excess NaOH(aq))
zinc, Zn^{2+}(aq)	$Zn(OH)_2$(s)	white (soluble in excess NaOH(aq))

Key word

➤ ionic equation

The precipitates $AgCl(s)$, $AgBr(s)$, and $AgI(s)$.

The test for carbonate ions. Sodium carbonate reacts with ethanoic acid, producing bubbles of carbon dioxide.

H

Ionic equations

Chemists use **ionic equations** to summarise precipitation reactions like these. Ionic equations allow you to focus on the ions involved. When a solution of iron(III) nitrate is mixed with a solution of sodium hydroxide, only two of the four ions in the mixture make a precipitate. The ionic equation only shows the Fe^{3+} and OH^- ions:

$$Fe^{3+}(aq) + 3OH^-(aq) \rightarrow Fe(OH)_3(s)$$

It is important to include the state symbols in these equations.

To write an ionic equation for a precipitate reaction correctly, use the following steps:

1 Write the correct formula of the precipitate, with the state symbol (s), on the right-hand side of the arrow.

2 Write the ions that make the precipitate separately, with their charges and state symbols (aq), on the left-hand side.

3 If necessary, insert numbers of each of the ions on the left to show the ratio in which they join together. You can work this out from the formula of the precipitate.

Synoptic link

You can learn more about symbols, formulae, and ionic equations in C2.4 *How are equations used to represent chemical reactions?* and C3.2 *How are metals with different reactivities extracted?*

Testing for non-metal ions

Some tests for non-metal ions also use precipitation reactions.

Testing for halide ions

The ions chloride (Cl^-), bromide (Br^-), and iodide (I^-) from the halogen group each make insoluble precipitates with silver ions (Ag^+). Each precipitate has a characteristic colour.

H The ionic equation for the chloride ion (Cl^-) test is:

$$Ag^+(aq) + Cl^-(aq) \rightarrow AgCl(s)$$

Testing for sulfate ions

Sulfate ions (SO_4^{2-}) make a characteristic white precipitate with barium ions (Ba^{2+}).

H The ionic equation for the sulfate ion (SO_4^{2-}) test is:

$$Ba^{2+}(aq) + SO_4^{2-}(aq) \rightarrow BaSO_4(s)$$

Testing for carbonate ions

Carbonate ions make carbon dioxide gas when you add a dilute acid. This is not a precipitation reaction. Bubbles form, creating effervescence, and the gas can be tested to see if it makes limewater go cloudy. Carbon dioxide reacts with calcium hydroxide in limewater to make a milky white precipitate of calcium carbonate.

Tests for negatively charged ions

Ion tested	Test	Result
chloride, Cl^-(aq)	Acidify with dilute nitric acid, then add silver nitrate solution.	white precipitate, $AgCl$(s)
bromide, Br^-(aq)	Acidify with dilute nitric acid, then add silver nitrate solution.	cream precipitate, $AgBr$(s)
iodide, I^-(aq)	Acidify with dilute nitric acid, then add silver nitrate solution.	yellow precipitate, AgI(s)
sulfate, SO_4^{2-}(aq)	Acidify, then add barium chloride solution or barium nitrate solution.	white precipitate, $BaSO_4$(s)
carbonate, CO_3^{2-}(aq)	Add a dilute acid, such as dilute hydrochloric acid. If bubbles form, collect the gas and test with limewater.	effervescence on addition of acid; the gas, CO_2, turns limewater cloudy

Questions

1 Why was it important for Bunsen to have a burner with a colourless flame?

2 A flame test was carried out on an unknown salt. The flame colour was lilac. Which metal ion does the salt contain?

3 What would happen if you mixed solutions of potassium carbonate and calcium nitrate?

4 What would you expect to see if you added a solution of barium nitrate to an acidified solution of sodium sulfate?

5 Which tests would you carry out to distinguish between solutions of sodium chloride and sodium iodide? What results would you expect?

H 6 Write ionic equations for:
  a Ag^+ ions reacting with acidified I^- ions to make a precipitate of AgI
  b Cu^{2+} ions reacting with OH^- ions to make a precipitate of $Cu(OH)_2$
  c zinc nitrate reacting with sodium carbonate to form a zinc carbonate precipitate
  d barium nitrate reacting with potassium sulfate

7 Solution X gives a green precipitate when sodium hydroxide solution is added, and a white precipitate when acidified barium nitrate solution is added. What salt is dissolved in solution X?

8 Solution Y gives a white precipitate when sodium hydroxide solution is added, which dissolves in excess sodium hydroxide. Solution Y gives a cream precipitate when acidified and a solution of silver nitrate is added. What salt does Y contain?

C: Instrumental analysis

Find out about

- flame spectra
- emission spectroscopy
- interpreting charts from spectroscopy

Key words

➤ spectrometer
➤ emission spectroscopy
➤ electron shell

When chemists want to find out about the composition of a chemical they often use instruments called spectrometers. A **spectrometer** is a machine for recording and measuring a spectrum.

Emission spectrometers analyse the spectrum of light that is emitted from a hot sample. They can be used to check the purity of manufactured chemicals, to determine the composition of alloys such as steel, or to identify elements in distant stars. This type of analysis is called **emission spectroscopy**.

Flame spectra

Emission spectroscopy began with Robert Bunsen and his flame colours. Bunsen had been studying the flame colours produced by different metal ions. He thought that this might lead to a new method of chemical analysis, but he soon realised that it only works for pure compounds. It was hard to make any sense of the flames from mixtures. So he mentioned his problem to Gustav Kirchhoff, who was a professor of physics. Kirchhoff suggested looking not at the colour of the flames, but at their spectra.

Kirchhoff built a simple spectroscope, by putting a glass prism into a wooden box and inserting two telescopes at an angle. Light from a flame entered through one telescope. It was split into a spectrum by the prism and then viewed with the second telescope.

SPECTRA OF THE METALS OF THE ALKALIES & ALKALINE EARTHS.
From the Drawings of BUNSEN & KIRCHHOFF.

Emission spectra of the alkali metals as recorded by Robert Bunsen and Gustav Kirchhoff

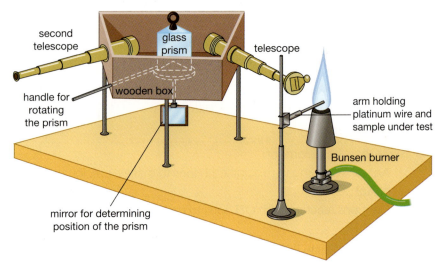

Kirchhoff's simple spectroscope.

Bunsen and Kirchhoff soon found that each element has its own unique spectrum when its light passes through a prism, and that each spectrum consists of a set of lines at particular wavelengths. With their spectroscope, they were able to record the line spectra of many elements. It was much easier to see differences between the spectra than differences between flame colours.

We now know how these unique spectra arise. When atoms of an element are heated in a flame, they gain energy. This forces electrons to move to higher-energy **electron shells** further from the nucleus. These electrons then drop back from outer electron shells to inner electron shells. Energy is transferred as they do so. Each energy transfer corresponds to a particular frequency of light in the spectrum. Only certain energy transfers are possible, so the spectrum consists of a series of lines.

Each element has a unique electron configuration, so its emission spectrum is also unique. This means that the spectrum of an unknown substance can be used to identify the elements it contains. The patterns and wavelength of the lines are compared with reference data from known elements.

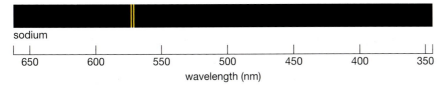

sodium

| | | | | | | | |
|650|600|550|500|450|400|350|

wavelength (nm)

The spectrum of sodium has just two lines very close together. Compare this with Bunsen and Kirchhoff's spectra.

Emission spectroscopy

Modern emission spectrometers heat a sample and then detect and record the exact wavelengths of light being emitted by the sample. They also interpret the results.

Using an emission spectrometer has many advantages over carrying out a simple flame test. It is:

An analyst using an emission spectrometer. This machine can be used to detect elements in samples of water, soil, air, food products, medicines, blood, and urine. It is very sensitive and can detect tiny amounts of an element in a sample.

- more sensitive – it can detect extremely small quantities of an element so only a very small sample is needed
- more accurate – unlike flame tests it does not rely on the judgement of the observer
- faster – it can carry out the analysis and interpret the results, usually very quickly, and can store the data.

The main disadvantage is:

- cost – it can be very expensive; emission spectrometers are not as freely available as the apparatus for a simple flame test.

Questions

1   What did Bunsen and Kirchhoff use to split the light from a flame into its spectrum?

2   A sample of an unknown salt is analysed in an emission spectrometer. This spectrum is obtained.
    Which metal ion does the salt contain?

3   Give one reason why a research chemist might use emission spectroscopy, rather than flame tests, to analyse chemical samples.

4   Why are emission spectrometers not more commonly used?

A: Does the mass change during a reaction?

Find out about

- mass changes and reactions
- the law of conservation of mass

Reactions and mass

If you mix ingredients to make a cake, the total mass of the finished cake is the same as the mass of all of the ingredients… or is it? If you measure the mass of a cake after it has been baked, it is usually *lighter* than the sum of the masses of the ingredients. Why? Cake ingredients contain water. When you bake a cake some of the water evaporates off. Therefore the cake has less mass.

Cake mixture looks wet. As the cake bakes, some of the water evaporates.

The same idea applies when thinking about chemical reactions. Some reactions appear to lose mass… but do they really?

As marble chips (calcium carbonate) react with dilute hydrochloric acid, the mass of the flask and contents goes down. The equation for the reaction helps to explain what has caused the apparent loss of mass.

| calcium carbonate | + | hydrochloric acid | → | calcium chloride | + | water | + | carbon dioxide |

$$CaCO_3(s) \ + \ 2HCl(aq) \ \longrightarrow \ CaCl_2(aq) \ + \ H_2O(l) \ + \ CO_2(g)$$

Carbon dioxide is a gas. It leaves the flask. If we collect the gas and add its mass to the final mass of the flask, we can see that the *total mass* does not actually change.

Other reactions appear to *gain* mass. For example, when magnesium burns in air, the final mass of the ash left behind is more than the mass of the magnesium at the start.

$$\text{magnesium} + \text{oxygen} \longrightarrow \text{magnesium oxide}$$
$$2Mg(s) \quad + \quad O_2(g) \ \longrightarrow 2MgO(s)$$

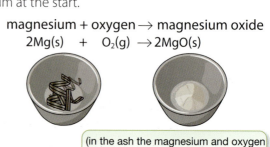

(in the ash the magnesium and oxygen atoms form a giant ionic lattice)

The white ash left behind after magnesium burns has a higher mass than the magnesium at the start.

Carbon dioxide gas escapes from the flask. The mass of the carbon and oxygen atoms is 'lost' from the flask so its mass goes down.

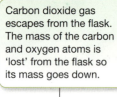

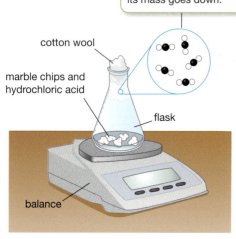

cotton wool

marble chips and hydrochloric acid

flask

balance

When marble chips react with dilute acid, the mass of the flask goes down.

This can be explained by using the particle model to think about what happens to the atoms in the reaction. The magnesium atoms react with oxygen atoms from oxygen in the air. The ash contains magnesium oxide. Overall the total mass of oxygen and magnesium atoms at the start is the same as the magnesium oxide at the end.

In magnesium oxide, the magnesium and oxygen form ions. Magnesium oxide is a giant ionic lattice with very large numbers of magnesium ions and oxide ions joined together.

Key words

➤ relative atomic mass
➤ relative formula mass
➤ law of conservation of mass

Putting numbers in

To find out whether there is an overall change in the total mass during a reaction, we compare the mass of the reactants with the mass of the products. Most reactions involve compounds reacting together. We can use the **relative atomic mass** to show the masses of the atoms in an element. When the equation includes compounds, we need to add up the masses of the atoms in the formula to find out the **relative formula mass** of each compound.

Worked example: Calculating the relative formula mass

What is the formula mass of CO_2?

Step 1: Write down the formula for the substance.

CO_2

Step 2: Use the Periodic Table to find out the relative atomic masses for each element.

relative atomic masses: $C = 12$, $O = 16$

Step 3: Add the atomic masses for every atom in the formula to find the relative formula mass.

1 carbon atom $\quad 1 \times 12 = 12$

2 oxygen atoms $\quad 2 \times 16 = 32$

relative formula mass of $CO_2 = (12 + 32)$
$\qquad\qquad = 44$

Answer:

relative formula mass of $CO_2 = 44$

Working out the relative formula masses helps to show that the total mass cannot change during a reaction. For example, when magnesium reacts with oxygen:

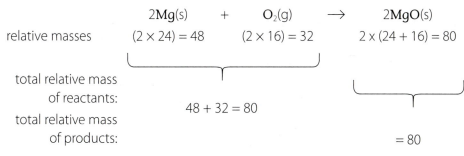

Overall there is no mass change in the total relative masses of the reactants and products. In a reaction the total mass stays constant. This law is called the **law of conservation of mass.**

Questions

1 Calculate the relative formula masses for these elements and compounds.
O_2 CO_2 H_2O CH_4 NaCl

2 Coal contains mainly carbon. When coal burns it loses mass.
Use ideas about relative mass to explain why the total mass does not change in this reaction. Use an equation in your answer.

B: The mole

Find out about

- how chemists measure amounts
- the mole
- calculations of amounts using the mole

Counting in chemistry

In high-street banks, coins are usually counted using a balance. The staff know the mass of £1 in 1p coins and so they weigh out that mass. They know that they have one hundred 1p coins to a value of £1. The same idea is used in automatic counting machines.

Atoms, molecules, and ions are too small to count individually. Chemists use the 'bank' method of counting particles by measuring the mass of large numbers of identical particles.

The mole

Chemists use the **mole** as an amount of particles. The number of particles in a mole is a very large number. It is 6.0×10^{23}. This number is also known as the **Avogadro constant**, because Amedeo Avogadro was the scientist who suggested the number originally. This number is very large – written out in full it is:

600 000 000 000 000 000 000 000

The easiest way to count large numbers of coins is by measuring mass.

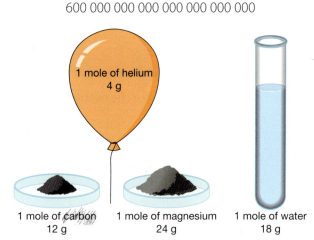

1 mole of helium 4 g

1 mole of carbon 12 g

1 mole of magnesium 24 g

1 mole of water 18 g

Each sample contains one mole of substance.

Conveniently, measuring a mole of a substance is easy. It is the relative formula mass of the substance in grams. This is why the number was chosen. It is easy to measure out of a mole of any element or compound if you know its relative formula mass.

Relative mass is measured relative to carbon. The definition of relative atomic mass uses carbon as the standard atom. The masses of other atoms are measured relative to a scale on which carbon has a mass of 12 g.

This means, for example, that a magnesium atom (relative mass 24) has exactly twice the mass of a carbon atom. So 24 g of magnesium contains the same number of atoms as 12 g of carbon.

A helium atom (relative mass 4) has only one-third the mass of a carbon atom, so 4 g of helium contains the same number of atoms as 12 g of carbon.

The box below gives the definitions of relative atomic mass and the Avogadro constant.

Relative atomic mass

The mass of an atom measured on a scale where an atom of carbon-12 has a mass of exactly 12.

Avogadro constant

The number of particles in exactly 12 g of carbon-12 (also known as the mole). The number is 6.0×10^{23}.

In C2.1B you saw that an element may have **isotopes**, which are atoms of different masses. Because of this, relative atomic mass is sometimes called 'average relative atomic mass'. The average relative atomic mass is the mean value for the atoms in a typical sample of the element, taking into account the relative amount of each isotope.

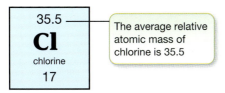

The average relative atomic mass of chlorine is 35.5

Chlorine has an average relative atomic mass of 35.5. There are two common isotopes of chlorine. Some chlorine atoms have a relative mass of 35.0; some have a relative mass of 37.0. The isotope with relative mass 35.0 is more common.

The mole and amounts of substances

In a bank, £2 of 1p coins has exactly double the mass of £1. This idea is used to measure amounts of elements and compounds. Two moles of carbon atoms have double the mass of one mole (2×12 g).

This idea can be written as an equation:

mass of substance (g) = number of moles × molar mass (g)

or in symbols: $m = n \times M$

You can use this equation for both elements and compounds. You may need to work out the relative formula mass before you can substitute it into the equation. The **molar mass** is the relative formula mass in grams.

Key words

➤ mole
➤ Avogadro constant
➤ isotope
➤ molar mass

Worked example: Calculating molar mass

What is the mass of two moles of water, H_2O?

Step 1: Write down the equation to calculate the mass of substance.

mass of substance (g) = number of moles × molar mass (g)

Step 2: Write down the formula for the substance.

H_2O

Step 3: Calculate the formula mass (see C5.3A) and then write down the molar mass.

formula mass of H_2O = $(2 \times 1) + 16 = 18$
molar mass = 18 g

Step 4: Calculate the mass.

mass of 2 moles of water = 2×18 g
= 36 g

Chemists can also work out the number of moles of a substance by rearranging the equation as follows:

$$\text{number of moles} = \frac{\text{mass of substance (g)}}{\text{molar mass (g)}}$$

or in symbols: $\quad n = \dfrac{m}{M}$

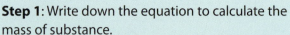

Worked example: Calculating the number of moles in a given mass of substance

A reaction uses 20 g of calcium carbonate. How many moles of calcium carbonate have been used?

Step 1: Write down the equation to calculate the mass of substance.

$$\text{number of moles} = \frac{\text{mass of substance (g)}}{\text{molar mass (g)}}$$

Step 2: Write down the formula for the substance.

$CaCO_3$

Step 3: Calculate the formula mass (see C5.3A) and then write down the molar mass.

formula mass of $CaCO_3$ = 40 + 12 + (3 × 16) = 100

molar mass = 100 g

Step 4: Calculate the number of moles in the given mass of the substance.

$$\text{number of moles} = \frac{20\ g}{100\ g}$$

= 0.2 moles

Ionic compounds and the mole

For elements and compounds, counting in moles usually means counting atoms or molecules. For example, 'one mole of helium' contains one mole of helium atoms and 'one mole of water' contains one mole of water molecules.

For ionic compounds it is slightly different. The giant structure of ionic compounds means that many millions of ions are joined together. One mole of an ionic compound is one mole of the formula unit. So 'one mole of sodium chloride ($NaCl$)' contains one mole of sodium ions (Na^+) and one mole of chloride ions (Cl^-).

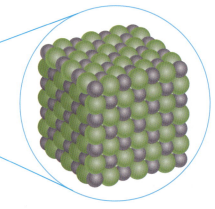

One mole of sodium chloride, $NaCl$, has a mass of 58.5 g. It contains one mole of Na^+ ions and one mole of Cl^- ions.

Counting particles using the mole in standard form

Remember that the relative formula mass of a substance always contains one mole of particles. One mole of a substance contains 6.0×10^{23} particles.

Helium	Water	Sodium chloride
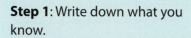		
1 mole of helium contains 6.0×10^{23} helium *atoms* (He)	1 mole of water contains 6.0×10^{23} water *molecules*.	1 mole of sodium chloride contains 6.0×10^{23} sodium chloride *formula units* (NaCl).
	The formula of water is H_2O so there are 3 moles of *atoms* in 1 mole of *molecules*.	The formula of sodium chloride is NaCl so there are 2 moles of *ions* in 1 mole of *formula units*.

Number of moles	Number of particles
1	6.0×10^{23}
2	$2 \times 6.0 \times 10^{23} = 12.0 \times 10^{23} = \mathbf{1.2 \times 10^{24}}$
3	$3 \times 6.0 \times 10^{23} = 18.0 \times 10^{23} = \mathbf{1.8 \times 10^{24}}$

In **standard form** there should be only one digit before the decimal point.

The numbers of particles should be shown in **standard form**.

- In standard form there should be only one digit before the decimal point.
- Every time you divide by 10, the value of the power of 10 decreases by one.
- Every time you multiply by 10, the value of the power of 10 increases by one

Worked example: Calculating the number of particles in a quantity of a substance

How many molecules of nitrogen, N_2, are in 2.5 moles? Give your answer in standard form.

Step 1: Write down what you know.

Avogadro constant = 6.0×10^{23}
number of moles = 2.5

Step 2: Calculate the number of particles.

number of particles = $2.5 \times 6.0 \times 10^{23}$
$= 15.0 \times 10^{23}$

Step 3: Convert to standard form. Divide (or multiply) by 10 until there is only one digit before the decimal point.

number of particles = 15.0×10^{23}
$= 1.5 \times 10^{24}$

Synoptic link

You can learn more about the particle model in C1.1 *How has the Earth's atmosphere changed over time, and why?*

Key word

➤ standard form

Questions

1 What is the mass of these substances?
   a 1 mole of argon atoms
   b 1 mole of chlorine molecules, Cl_2
   c 1 mole of sulfur dioxide molecules, SO_2
   d 4 moles of methane, CH_4
   e 3 moles of potassium fluoride, KF

2 How many moles of molecules are in:
   a 88 g of carbon dioxide?
   b 27 g of water?
   c How many molecules are in the samples in **a** and **b**? (Use the Avogadro constant.)

3 How many moles of oxygen *atoms* are in 64 g of oxygen molecules?

C: Calculations from equations

Find out about

- calculations based on equations
- calculating theoretical amounts of reactants and products
- limiting quantities
- how chemists calculate and measure yields S
- reasons for low yields
- using quantities to work out reactions

Key word

➤ limiting quantity

Amounts and equations

In industry, chemists need to know exactly how much of each reactant they need to use in a reaction to make a certain amount of product. This means that they can make exactly the amount of product they need, and that no reactants are wasted. This saves on cost and also saves having to separate unreacted chemicals from the product at the end. There is less waste to handle and to dispose of.

Chemists can work out the amounts and the masses of the reactants they need by looking at a balanced equation for a reaction. For example, making ethanol from fermentation is relatively expensive. Cheaper, synthetic ethanol can be made by reacting ethene (made by refining crude oil) with water.

The chemical industry makes large amounts of synthetic ethanol for use in a wide range of products. Many personal-care products, such as perfumes, aftershaves, mouthwashes, and hair-styling products, contain ethanol.

A balanced equation is used to work out the masses of ethene, water, and ethanol involved in the reaction to make ethanol.

	ethene	+	water	→	ethanol
	C_2H_4	+	H_2O	→	C_2H_5OH
	1 mole	reacts with	1 mole	to make	1 mole
	28 g	reacts with	18 g	to make	46 g

1 The equation shows that one molecule of ethene reacts with one molecule of water to make one molecule of ethanol.

2 Therefore, one mole of ethene molecules react with one mole of water molecules to make one mole of ethanol molecules.

3 Working out the relative formula mass of each substance gives the molar masses.

Notice that the law of conservation of mass gives a way of checking that the relative formula masses are correct. The total mass of the reactants and products is the same (46 g).

Making larger amounts

In the laboratory, chemists react small amounts of reactants together, but in industry reacting masses are likely to be in kilograms or even tonnes.

Chemists can scale up the masses in an equation in the same ratio to make larger amounts. For example, they can work out how to make larger amounts of ethanol from larger amounts of ethene.

If:	28 g of ethene react with 18 g of water to make 46 g of ethanol
scaling up would give:	28 kg of ethene react with 18 kg of water to make 46 kg of ethanol
and:	28 tonnes of ethene react with 18 tonnes of water to make 46 tonnes of ethanol

In practice, synthetic ethanol is made by passing ethene and water in the form of steam over a hot catalyst. The ethene is deliberately kept in excess. This means there is a bigger ratio of ethene to water than is needed for the reaction. There is not enough water to react with all of the ethene available, so the amount of ethanol made depends only on the amount of water. So the amount of water available is the **limiting quantity**. Adding more ethene does not make more ethanol, because some ethene is already left over.

The amount of water is limited because too much water stops the reaction working well. Water 'wets' the catalyst and makes it much less efficient. It is not a problem that some ethene is left over – the unreacted ethene is recycled back into the process so none is wasted.

Chemists in laboratories usually work with small masses of substances, measured in grams.

Chemists in industry control processes involving large amounts of substances, measured in kilograms or tonnes.

Calculating masses from equations

Chemists usually want to know one of two things about the masses of substances involved in a reaction:

- What mass of reactants reacts together to make a certain amount of a product?
- What mass of product can be made from a certain amount of reactant?

Both of these questions can be answered by doing calculations based on an equation for the reaction.

Food manufacturers scale up recipes to make large amounts of food. In the same way, chemists scale up laboratory reactions to make large amounts of chemicals on an industrial scale.

The box shows some rules for calculating masses of reactants and products from equations.

1 Write down the balanced symbol equation.

2 Write down a sentence about the number of moles of each substance. (You may find it helpful to put a ring around the substances for which you know the mass and those you need to calculate.)

3 Work out the relative formula mass of these substances.

4 Look at the units in the question and add units to your masses (g, kg, or tonnes).

5 Scale up to the known mass you were given in the question.

Calculating reacting masses

Chemists often want to know how much of each reactant to use to make a certain amount of product. This is similar to working out how much of each ingredient you need to make a cake. If you are expecting a lot of visitors, you might double up your recipe to make more cake. Chemists do the same type of calculation. They use the rules in the box above to scale up the quantities to the amounts they need.

Worked example: Calculating the quantity of reactant required

Calcium carbonate, $CaCO_3$, decomposes when it is heated to form calcium oxide and carbon dioxide.

What mass of calcium carbonate is needed to make 448 g of calcium oxide, CaO?

Note that you only need to do calculations for the substances mentioned in the question (not for carbon dioxide).

Step 1: Write the balanced equation. Identify the substances with known mass and the required mass.

calcium carbonate ⟶ calcium oxide + carbon dioxide

$CaCO_3$ ⟶ CaO + CO_2 Your 'known' mass in this question is CaO.

Step 2: Write a sentence in moles.

1 mole of $CaCO_3$ makes 1 mole of CaO

Step 3: Calculate the required relative formula mass.

$[40 + 12 + (3 \times 16)]$ = 100 $[40 + 16]$ = 56

Step 4: Write down the molar masses.

100 g $CaCO_3$ makes 56 g CaO

Step 5: Scale up to the formula mass to match the known mass.

To scale up, work out how much you need to make 1 g of known substance,

$\frac{100\ g}{56}\ CaCO_3$ makes 1 g CaO

then multiply to find the amount you want.

$448 \times \frac{100\ g}{56}\ CaCO_3$ makes 448 g CaO

Answer: **800 g of $CaCO_3$** makes 448 g of CaO

Calculating masses of products

Chemists also often want to know how much product they can make from a certain amount of reactant. The calculation 'rules' are exactly the same.

> **Worked example:** Calculating theoretical yield
>
> *In the blast furnace, iron is extracted from iron oxide, Fe_2O_3, by heating with carbon monoxide.*
>
> *What mass of iron can be made from 100 tonnes of iron oxide?*
>
> The equation shows there will be 2 moles of iron produced.
>
> **Step 1:** Write the balanced equation and identify the substances with known mass and the required mass.
>
> (Fe_2O_3) + $3CO$ → $(2Fe)$ + $3CO_2$
>
> **Step 2:** Write a sentence in moles.
>
> 1 mole of Fe_2O_3 makes   2 moles of Fe
>
> **Step 3:** Calculate the required relative formula masses.
>
> $[(56 \times 2) + (3 \times 16)]$ $[2 \times 56]$
>
> = 160   = 112
>
> **Step 4:** Write down the molar masses.
>
> 160 tonnes Fe_2O_3 makes   112 tonnes Fe
>
> **Step 5:** Scale up to the formula mass to match the known mass.
>
> To scale up, work out how much product 1 tonne of reactant will make,
>
> 1 tonne Fe_2O_3 makes $\frac{112 \text{ tonnes}}{160}$ Fe
>
> then multiply to find the amount you want.
>
> 100 tonnes Fe_2O_3 makes $100 \times \frac{112 \text{ tonnes}}{160}$ Fe
>
> **Answer:**
>
> 100 tonnes of Fe_2O_3 makes **70 tonnes of Fe**

In practice, the amount of product made from a certain mass of reactant is unlikely to exactly match the calculated value. This can be for many reasons. The reactants may not fully react, the reaction may be reversible, there may be impurities, and some reactants and products may be lost during the process. Chemists take all of this into account when they design large-scale processes.

Yield

Yield is a way of measuring how much product is made at the end of a reaction. The **theoretical yield** can be calculated before the reaction is carried out, from the equation and the amounts of reactants that are being added. It is the amount of a product that would be made if all the reactants were converted into product.

At the end of the reaction, chemists measure the **actual yield**. This is the mass of product collected at the end of the reaction, after separation and purification.

The **percentage yield** gives an indication of how the actual yield compares with the theoretical yield.

$$\text{percentage yield} = \frac{\text{actual yield}}{\text{theoretical yield}} \times 100\,\%$$

The higher the percentage yield, the more successful the process has been.

Key words

➤ theoretical yield
➤ actual yield
➤ percentage yield

Calculating percentage yield

To calculate the percentage yield, you need to work out the theoretical yield from the equation for the reaction. Usually, one of the reactants is added in excess, and one reactant is the limiting quantity. The yield depends on the limiting quantity. To calculate the theoretical yield, you need to use the method described earlier in this section.

Worked example: Calculating yield

A student reacted 8.0 g of copper oxide, CuO, with excess hydrogen gas to make copper.

He made 4.0 g of copper at the end of the reaction. What is the percentage yield?

Step 1: Work out the theoretical yield, based on the amount of copper oxide at the start (8.0 g), using the method from page 202.

$$CuO + H_2 \rightarrow Cu + H_2O$$

1 mole	makes	1 mole
(64 g + 16 g) = 80 g	makes	64 g
1 g	makes	$\frac{64\ g}{80\ g} = 0.8\ g$
8.0 g	makes	$0.8\ g \times 8.0 = \mathbf{6.4\ g}$

Step 2: Calculate the percentage yield.

$$\text{percentage yield} = \frac{\text{actual yield}}{\text{theoretical yield}} \times 100\%$$

$$= \frac{4.0\ g}{6.4\ g} \times 100\%$$

$$= \mathbf{62.5\%}$$

Why are actual yields less than theoretical yields?

Remember that the actual yield is a practical, measured value for the amount of product formed at the end of the reaction, separation, and purification. Actual yields may be less than theoretical yields for many reasons:

- Reversible reactions always have some reactants left at the end.
- The reaction may not have finished.
- Reactants or products may be 'lost' during transfer between containers or during the separation and purification process.

In industry, the conditions for some processes mean that the yields are low. Products can be separated out and the unreacted chemicals are recycled back into the process so raw materials are not wasted. It might take too long to wait for a complete reaction, which would make the process more expensive.

In the laboratory, information from yield calculations can be used to evaluate the success of a preparation method. Sometimes improvements to the method may give a higher yield. If a product has not been dried or completely separated from impurities, the measured actual yield may be higher than the accurate value. This is why it is important that products are purified and dried carefully at the end of a reaction.

Synoptic link

You can learn more about maximising industrial yields in C6.3 *What factors affect the yield of a reaction?*

In the laboratory, small amounts of reactants and products are lost at each stage of preparing a product.

Using amounts of products to work out reactions

Chemists usually measure the amounts of products of reactions to evaluate the success of a method. They can also use the amount of products that form to work out what is happening during a reaction. This is useful to check the equation for a reaction, or the formula of a reactant.

For example, when magnesium is heated it forms magnesium oxide. The magnesium gains mass when it reacts with oxygen gas from the air. The gain in mass can be used to work out both the chemical equation and the formula for magnesium oxide.

Analysing results

mass of magnesium at start	0.60 g
mass of magnesium oxide at end	1.00 g
calculated mass of oxygen gas gained in g	0.40 g

The number of moles of magnesium and oxygen that have reacted can be worked out using $n = \dfrac{m}{M}$

$$\text{number of moles of magnesium} = \frac{0.60 \text{ g}}{24 \text{ g}} = 0.025$$

$$\text{number of moles of oxygen gas} = \frac{0.40 \text{ g}}{32 \text{ g}} = 0.0125$$

The simplest ratio of moles can be worked out by dividing both numbers by the smallest value (0.0125).

$$\text{ratio of moles of magnesium:moles of oxygen} = 2:1$$

This tells us that two moles of magnesium react with one mole of oxygen gas, so we know that the left-hand side of the equation is:

$$2Mg + O_2 \longrightarrow$$

We can 'guess' that the formula of magnesium oxide must be **MgO** because oxygen gas contains two oxygen atoms, so there is one magnesium atom for each oxygen atom in the reaction. This can be confirmed by using the same data to do an empirical formula calculation.

$$2Mg + O_2 \longrightarrow MgO$$

So measuring the mass of reactants and products can also enable chemists to work out both equations and formulae.

Questions

1 What mass of ethanol can be made from the following amounts of ethene?

   a 14 g   **b** 56 g   **c** 112 tonnes

2 What mass of carbon dioxide is made when 50 tonnes of calcium carbonate is heated?

3 How much iron oxide needs to be heated to make 900 kg of iron? Give your answer to three significant figures.

4 In an experiment, the theoretical yield is 20.0 g and the actual yield is 8.0 g.

   a Work out the percentage yield.

   b Suggest ways that the experiment could be adapted to make the percentage yield higher.

5 Hydrogen, H_2, and iodine, I_2, react together to make hydrogen iodide, **HI**.

2 g of hydrogen were reacted with 300 g of iodine.

$$H_2 + I_2 \longrightarrow 2HI$$

   a Use the masses to work out the number of moles of hydrogen and iodine used.

   b Use your answer to explain how you can tell that hydrogen is the limiting quantity in the reaction.

   c What is the theoretical yield of hydrogen iodide from 2 g of hydrogen?

   d In an experiment, the actual yield was 200 g. What is the percentage yield?

D: Gases and the mole

Molar amounts of gases

The amounts of gases can be measured by mass, just as for solids and liquids. You have already seen how the reacting masses of ethene and carbon dioxide can be calculated from equations. However, it is quite difficult to measure amounts of gases by mass. Chemists working in laboratories and in industry also usually measure amounts of gases using volume.

The *mass* of one mole of gas depends on its formula and the mass of its atoms. However, the *volume* of a gas depends only on how many moles of gas there are. One mole of any gas has the same volume at the same temperature and pressure; that is, all gases have the same **molar volume**.

One mole of any gas has a volume of 24 dm³ at room temperature and pressure. 1 dm³ = 1000 cm³.

GCSE CHEMISTRY ONLY   HIGHER TIER ONLY

Find out about
- molar amounts of gases
- using molar volumes of gases
- calculations of amounts of gases based on equations

Key word
➤ molar volume

The petrol tank of an average motorbike holds just under 24 dm³: the same volume as a mole of any gas.

Because gases expand when they are heated and can be compressed if they are put under pressure, it is important that the volumes of gases are always measured under the same conditions. If the conditions change, the measured volumes of the gas will be different.

Calculations involving only gases

Calculations that involve only gases are very straightforward. You do not usually even need to work out the relative formula masses because the volumes can be compared directly from the equation.

For example, when methane burns:

$$\text{methane} + \text{oxygen} \longrightarrow \text{carbon dioxide} + \text{water}$$

$$CH_4(g) \; + \; 2O_2(g) \longrightarrow CO_2(g) + 2H_2O(l)$$

The equation shows us that:

1 mole of CH_4 reacts with 2 moles of O_2 to make 1 mole of CO_2 and 2 moles of H_2O

If we put molar volumes into the sentence…

24 dm³ of CH_4 reacts with 48 dm³ of O_2 to make 24 dm³ of CO_2

Under the ring on a gas cooker there are pipes with holes. This is to make sure enough air is available to react with the gas in complete combustion. Cookers that burn propane rather than methane have different sized holes because a different ratio of air : gas is needed.

We cannot use molar volume for water, because it is a liquid, not a gas, at room temperature and pressure.

The molar-volumes sentence tells us that whatever volume of methane is burned, the volume of oxygen needed will always be double the volume of methane. The volume of carbon dioxide made will always be the same as the volume of methane.

Worked example: Calculating using molar volumes

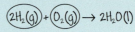

Hydrogen is used as a rocket fuel. Rockets are designed to mix hydrogen and oxygen in the correct proportions to react exactly, with no excess. What volume of oxygen is needed to burn 50 dm³ of hydrogen at room temperature and pressure?

Step 1: Write the balanced equation and identify the substances with known volume and the required volume.

$$2H_2(g) + O_2(g) \rightarrow 2H_2O(l)$$

Step 2: Write a sentence in moles.

2 moles of H_2 react with 1 mole of O_2

Step 3: For questions involving gases only, you can use a direct ratio of volumes to work out the unknown quantity.

Work out the ratio of moles.

hydrogen:oxygen = 2:1

Step 4: Apply the ratio to the known volume.

so 50 dm³ of hydrogen will react with **25 dm³ of oxygen**

Calculations involving both masses and volumes

Calculations that involve both masses and volumes need to be done by working out the numbers of moles involved in the reaction. These calculations are mainly of two types:

- You are given a *known mass* and asked to work out an *unknown volume*.
- You are given a *known volume* and asked to work out an *unknown mass*.

The box shows the steps in the calculation:

1 Start by spotting which substance is your known (the one for which you are given the mass or volume) and which is your unknown (the one you are working out).

2 Work out the number of moles of the known substance in the equation and use this to work out the number of moles of the unknown substance.

3 Convert moles to either masses or volumes (depending on the question).

The number of moles of a gas can be worked out using these equations:

volume (dm³) = number of moles × molar volume (24 dm³)

$$v = nV$$

which can be rearranged to give:

$$\text{number of moles} = \frac{\text{volume (dm}^3)}{\text{molar volume (24 dm}^3)}$$

$$n = \frac{v}{V}$$

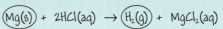

Worked example: Calculating the volume of gas produced

What volume of hydrogen is made when 2.4 g of magnesium react with hydrochloric acid?

Step 1: Write the balanced equation and identify the substances with known mass and the required volume.

$$Mg(s) + 2HCl(aq) \rightarrow H_2(g) + MgCl_2(aq)$$

Step 2: Write a sentence in moles.

1 mole of Mg   makes   1 mole of H_2

Step 3: Calculate the number of moles of known substances.

number of moles of Mg $= \dfrac{2.4\ g}{24\ g} = 0.1$ moles

relative formula mass of Mg = 24 g

Step 4: Use the balanced equation to calculate the number of moles of the product.

1 mole of Mg makes 1 mole of H_2

so

0.1 mole of Mg makes 0.1 mole of H_2

Step 5: Calculate the volume of gas from the number of moles.

volume of gas formed = number of moles × molar volume (24 dm³)

$= 0.1 \times 24$ dm³

$= 2.4$ dm³

Answer: **volume of hydrogen formed = 2.4 dm³**

Worked example: Calculating the mass of product

A student collects 120 cm³ of hydrogen from the reaction between magnesium and hydrochloric acid. What mass of magnesium chloride is made at the same time?

Step 1: Write the balanced equation and identify the substances with known volume and the required mass.

$$Mg(s) + 2HCl(aq) \rightarrow H_2(g) + MgCl_2(aq)$$

Step 2: Write a sentence in moles.

When 1 mole of H_2 is made, 1 mole of $MgCl_2$ is also made.

Step 3: First write down the volume of gas in standard units, then calculate the number of moles of known substances.

volume of $H_2 = 120$ cm³ $= \dfrac{120\ cm^3}{1000} = 0.12$ dm³

number of moles of $H_2 = \dfrac{0.12\ dm^3}{24\ dm^3} = 0.005$ moles

Step 4: Use the balanced equation to calculate the number of moles of the product.

When 1 mole of H_2 is made, 1 mole of $MgCl_2$ is also made.

So when 0.005 mole of H_2 is made, 0.005 mole of $MgCl_2$ is also made.

Step 5: Calculate the molar mass of product.

molar mass of $MgCl_2 = 24\ g + (2 \times 35.5\ g) = 95\ g$

Step 6: Calculate the mass of product from the number of moles.

mass of product = number of moles × molar mass

$= 0.005 \times 95\ g$

$= 0.475\ g$

Answer: **mass of magnesium chloride = 0.475 g**

Questions

1 Propane, C_3H_8, is used as a fuel in caravan cookers.
   a Use equations to show that a greater volume of oxygen is needed to burn 1 mole of propane than 1 mole of methane.
   b What volume of oxygen is needed to completely burn 100 cm³ of each of the two alkanes?

2 How many moles of gas are there in:
   a 48 dm³ of O_2?   b 60 cm³ of CO_2?
   c 24 cm³ of H_2?
   d Work out the masses of gas in a, b, and c.

3 What volume of hydrogen is formed when 2.3 g of sodium reacts with water?

A: Concentrations of solutions

Find out about

- how concentrations are measured in g/dm³
- how concentrations are measured in mol/dm³
- calculating concentrations and amounts

Key word

➤ concentration

Making solutions

Making accurate solutions of known **concentrations** of chemicals is an important task in any chemistry laboratory. Solutions are usually made by accurately measuring a known mass of dry solid using a precise balance. Laboratory balances typically measure mass to the nearest 0.001 g. The solid is then dissolved in a solvent (most often water) and diluted to a known volume.

A volumetric flask is usually used to make up solutions because it has a lower error value than a measuring cylinder. Many laboratories make up 'stock' solutions of chemicals at relatively high concentrations. The stock solutions can then be diluted down to make a range of solutions with more dilute concentrations.

The amount of solid that a laboratory technician needs to use to make up a solution to a particular concentration can be worked out using ideas about moles and masses.

A large part of the work of a laboratory technician is to make up accurate solutions of known concentration.

Measuring concentration in g/dm³

Concentration can be measured in g/dm³. This tells you how many grams of solid are in each dm³ of solution.

If you know the mass of a solid and the volume of a solution, the concentration can be worked out using this formula:

$$\text{concentration (g/dm}^3) = \frac{\text{mass (g)}}{\text{volume (dm}^3)}$$

or:

$$c = \frac{m}{V}$$

You can re-arrange this equation if you know the concentration you need, but don't know what mass to start with.

The units of concentration are g/dm³ because this formula for concentration is based on mass and volume.

Worked example: Calculating with concentration

A laboratory technician wants to make up 250 cm³ of a solution of sodium chloride of concentration 20.0 g/dm³.

What mass of sodium chloride should she use?

Step 1: Write down what you know, with the units. Convert to standard units.

volume of solution = 250 cm³

$$= \frac{250 \text{ cm}^3}{1000} = 0.250 \text{ dm}^3$$

required concentration = 20.0 g/dm³

Step 2: Write down the equation for concentration.

$$\text{concentration} = \frac{\text{mass}}{\text{volume}}$$

Step 3: Substitute the quantities into the equation.

$$20.0 \text{ g/dm}^3 = \frac{\text{mass}}{0.250 \text{ dm}^3}$$

Re-arrange the equation and calculate the mass. Include the units in your answer.

mass = 20.0 g/dm³ × 0.250 dm³

mass = 5.0 g

Measuring concentration in mol/dm³

Most concentrations are given in mol/dm³. This tells you how many moles of solute are in each dm³ of solution. Notice that the word 'moles' has been shortened to 'mol' in the unit.

If you know how many moles of solute have been dissolved, you can work out the concentration using this equation:

$$\text{concentration in mol/dm}^3 = \frac{\text{number of moles}}{\text{volume (dm}^3)}$$

or:

$$c = \frac{n}{v}$$

However, because it is usual to measure the number of moles by mass, you need to be able to convert between masses and moles. Then you can work out the concentration in mol/dm³ from a known mass, or vice versa.

Remember that:

$$\text{number of moles} = \frac{\text{mass (g)}}{\text{molar mass (g)}}$$

or:

$$n = \frac{m}{M}$$

You may need to re-arrange this equation if you are asked to work out the mass you need to make a particular concentration in mol/dm³.

Worked example: Calculating with concentration

What mass of solid sodium hydroxide, NaOH, is needed to make 400 cm³ of a solution of 0.50 mol/dm³?

Step 1: Write down what you know, with the units. Convert to standard units.

$$\text{volume of solution} = 400 \text{ cm}^3 = \frac{400 \text{ cm}^3}{1000} = 0.400 \text{ dm}^3$$

$$\text{required concentration} = 0.5 \text{ mol/dm}^3$$

Step 2: Write down the equation for concentration.

$$\text{concentration} = \frac{\text{number of moles}}{\text{volume}}$$

Step 3: Substitute the quantities into the equation.

$$0.50 \text{ mol/dm}^3 = \frac{\text{number of moles}}{0.400 \text{ dm}^3}$$

Re-arrange the equation and calculate the amount in moles.

$$\text{amount} = 0.50 \text{ mol/dm}^3 \times 0.400 \text{ dm}^3$$
$$= \mathbf{0.200 \text{ moles}}$$

Step 4: Work out the molar mass of NaOH.

$$\text{molar mass} = 23 \text{ g} + 16 \text{ g} + 1 \text{ g} = 40 \text{ g}$$

Step 5: Calculate the mass required from the amount in moles.

$$\text{mass} = 0.200 \times 40 \text{ g}$$
$$= \mathbf{8.0 \text{ g}}$$

Mass or moles?

Concentration is usually expressed in mol/dm³ because this indicates the number of moles of formula mass units in 1 dm³. A solution with a concentration of 1 mol/dm³ contains 1 mole of formula mass units.

So a solution of 1 mol/dm³ of **NaCl** contains 1 mole of **NaCl** in 1 dm³. There will be 1 mole of **Na⁺** ions and 1 mole of **Cl⁻** ions.

A solution of 1 mol/dm³ of **HCl** contains 1 mole of **HCl** in 1 dm³. There will be 1 mole of **H⁺** ions and 1 mole of **Cl⁻** ions.

Using mol/dm³ for concentration makes it easy to see if two solutions have the same concentration.

The masses you need to dissolve in solution to achieve the same concentration will be different if the molar masses of the compounds are different.

Questions

1. What is the concentration in g/dm³ and mol/dm³ of solutions containing:
   a. 49 g of H_2SO_4 in 250 cm³ of solution?
   b. 73 g of HCl in 500 cm³ of solution?
   c. 27 g of Na_2CO_3 in 200 cm³ of solution?

2. How much solid would be needed to make:
   a. 500 cm³ of a 25.0 g/dm³ solution of copper sulfate?
   b. 200 cm³ of a 0.1 mol/dm³ solution of calcium hydroxide, $Ca(OH)_2$?

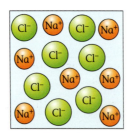

1 mol/dm³ **NaCl** (58.5 g/dm³)

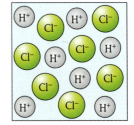

1 mol/dm³ **HCl** (36.5 g/dm³)

Solutions of concentration 1 mol/dm³ all contain 1 mole of relative formula units in 1 dm³ of solution.

A 1 mol/dm³ solution of **NaCl** contains 58.5 g/dm³ of **NaCl**.

A 1 mol/dm³ solution of **HCl** contains 36.5 g/dm³ of **HCl**.

It is not easy to see from the masses that these are the same molar concentration.

B: Acids, alkalis, and neutralisation

Which acids do we use in the laboratory?

Acids occur throughout nature. Very few substances are exactly neutral in solution and there are many examples of natural acids in plants and animals. Examples include citric acid in lemons and oranges and hydrochloric acid in our stomachs. Even the surface of our skin is slightly acidic.

The acids we use in the laboratory are often called 'mineral acids' because they are extracted from minerals rather than living things. The three most common laboratory acids are hydrochloric acid, sulfuric acid, and nitric acid.

Some people think that all acids are harmful. However, this is not the case; acids are routinely used in foods, medicines, and personal-care products. The hazard of an acid depends on the type of acid and its concentration. More concentrated acids may be corrosive, and even when very dilute, most laboratory acids cause skin irritation.

You already know that acids have pH values below 7 and react in **neutralisation** reactions. If you look at the table below you will see another feature that all acids have in common.

Acid	Formula	Ions in solution
hydrochloric acid	HCl	H^+ and Cl^-
sulfuric acid	H_2SO_4	$2H^+$ and SO_4^{2-}
nitric acid	HNO_3	H^+ and NO_3^-

All acids produce H^+ ions in solution.

Notice that the formula of sulfuric acid contains two hydrogen ions. If one mole of H_2SO_4 is dissolved in 1 dm^3 of water, the concentration of H^+ ions is 2 mol/dm^3. In the laboratory, sulfuric acid is usually used at a more dilute concentration than the other acids for this reason.

- Acids have a pH less than 7.
- Acids react with alkalis to make **salts**.
- Acids and alkalis react together in neutralisation reactions.
- Acids produce H^+ ions in solution.

Laboratory alkalis

Alkalis are soluble compounds that neutralise acidity. Alkalis are usually used as solutions. The most common alkali used in the laboratory is sodium hydroxide. Calcium hydroxide (limewater), potassium hydroxide, and ammonium hydroxide are also alkalis.

Common laboratory acids and alkalis.

Find out about

- common acids and alkalis
- ions in acids and alkalis
- neutralisation reactions
- ionic equations for neutralisation

Synoptic link

You can learn more about strong and weak acid chemistry in C6.1 *What useful products can be made from acids?*

Sulfur for making sulfuric acid is mined from active volcanoes.

Key words

➤ acid
➤ neutralisation
➤ salt
➤ alkali
➤ ionic equation
➤ state symbol

Alkali	Formula	Ions in solution
sodium hydroxide	NaOH	Na^+ and OH^-
calcium hydroxide	$Ca(OH)_2$	Ca^{2+} and $2OH^-$
potassium hydroxide	KOH	K^+ and OH^-

Alkalis produce OH^- ions in solution.

What happens during neutralisation?

Neutralisation reactions happen throughout nature and are important in the way that many personal-care products and medicines work. Toothpaste neutralises acids in the mouth. On an industrial scale, alkalis are used in neutralisation reactions to 'scrub' acidic gases to remove them from power-station emissions.

All reactions between acids and alkalis have some common features.

● Acids react with alkalis to make a salt and water.

$$acid + alkali \longrightarrow salt + water$$

● The type of salt depends on the acid and alkali used.

■ The positive (metal) ion in the salt comes from the alkali.

■ The negative ion in the salt comes from the acid.

The table opposite shows the negative ions formed from each acid.

Acid	Type of salt formed	Negative ion in salt
HCl	metal chloride	Cl^-
H_2SO_4	metal sulfate	SO_4^{2-}
HNO_3	metal nitrate	NO_3^-

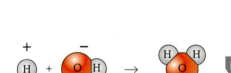

All neutralisation reactions involve hydrogen ions reacting with hydroxide ions to form water molecules.

Ionic equation for neutralisation

The **ionic equation** for neutralisation is the same for every acid–alkali reaction. This can be worked out from the symbol equation. It helps to include **state symbols** so that you can see which compounds contain ions in solution.

For example:

hydrochloric acid + sodium hydroxide $\longrightarrow$ sodium chloride + water

formulae: $HCl(aq) + NaOH(aq) \longrightarrow NaCl(aq) + H_2O(l)$

ions: $H^+(aq) + Cl^-(aq) + Na^+(aq) + OH^-(aq) \longrightarrow Na^+(aq) + Cl^-(aq) + H_2O(l)$

ionic equation for neutralisation: $H^+(aq) + OH^-(aq) \longrightarrow H_2O(l)$

The state symbol (l) shows that H_2O is a covalently bonded liquid; it does not split into its ions in solution.

The overall equation for every neutralisation reaction between an acid and an alkali is:

$$H^+(aq) + OH^-(aq) \longrightarrow H_2O(l)$$

The overall equation shows only the H^+ ions from the acid reacting with the OH^- ions from the alkali to make water.

Questions

1 Write word equations for reactions between these acids and alkalis.
   a hydrochloric acid and potassium hydroxide
   b nitric acid and sodium hydroxide
   c sulfuric acid and sodium hydroxide

2 Write symbol equations, with state symbols, for the reactions in question **1**.

H 3 Work out the ionic equation for the reaction between nitric acid and sodium hydroxide. Show your working.

C: Titrations

Why are titrations useful?

Many of the techniques that chemists use to analyse unknown samples are qualitative. **Qualitative analysis** identifies the substances in an unknown mixture. Testing gases, tests for ions in solution, and chromatography are all qualitative techniques.

Titrations are used to find out *how much* of a substance is in a sample, or to find its concentration. Analysis that measures amounts (or quantities) of a substance is called **quantitative analysis**. Titrations can be used to find out the concentration of a range of substances, for example, the vitamin C concentration in foods and drinks or the chlorine concentration in water or bleach. The most common type of laboratory titration involves reacting an acid with an alkali.

Titrations: the basic procedure

Suppose you wanted to know the concentration of an unknown sample of acid. A titration allows you to react the acid with a known concentration of alkali and measure exactly the amounts of the acid and alkali that react together. The data you collect can then be used to work out the concentration of the acid.

Using a titration to find the concentration of a dilute solution of acid

Find out about

- how to do titrations
- collecting and recording high-quality data
- H • judging the quality of titration data
- calculating concentrations S

Key words

➤ qualitative analysis
➤ titration
➤ quantitative analysis
➤ precision
➤ error
➤ pipette
➤ burette
➤ meniscus
➤ indicator
➤ end point

A pipette gives more **precise** results than a measuring cylinder for measuring volumes. There is less **error** in the measurement.

1 Use a **pipette** to measure out a fixed volume (e.g., 25 cm³) of acid solution and carefully transfer it into a flask.

2 Fill a **burette** with dilute alkali of known concentration. Make sure the bottom of the **meniscus** is on zero and that the tap is flushed out so that there are no air bubbles.

Air bubbles would make the volume reading incorrect.

3 Add a few drops of **indicator** to the flask and put the flask on a white tile under the burette.

Indicators change colour at different pH values. You need to use an indicator that changes colour when the acid is fully neutralised. The white tile helps you see the colour change.

Near the end point the indicator may change colour then change back again when swirled. Add the alkali drop by drop near the end point.

4 Add alkali from the flask a little at a time and swirl the flask. The **end point** is when the acid is just neutralised and the indicator changes colour.

5 At the end point, make a note of the volume of alkali that was added.

Chemists aim to get accurate titration volumes to within 0.1 cm³ of each other.

6 The first titration is a rough trial to give you an idea of the end point. Repeat the procedure, adding the alkali a drop at a time near the end point, until you have three volumes that closely agree.

• A pipette is used to measure the dilute acid into the flask. The burette is filled with dilute alkali. An indicator is added to the flask. Near the end point the alkali is added drop by drop.

Titrations: recording results

The table below shows some typical titration results. Notice that:

- you do not need to refill the burette every time. These readings sometimes start from where the last one ended.

- titration results should be recorded to (at least) one decimal place. If the reading is exactly 21.0 cm³, this should be recorded as '21.0' not '21'. This records the results *precisely*.

- the repeats should be close. If they are not, keep going until you have three close results. This makes sure the results are *valid*.

	Rough trial	Repeat 1	Repeat 2	Repeat 3
Burette reading at end (cm³)	22.0	43.1	22.0	42.1
Burette reading at start (cm³)	0.0	22.0	1.0	21.0
Volume added (cm³)	22.0	21.1	21.0	21.1

Titrations: aiming for accuracy

Quantitative analysis relies on the data being as high quality as possible. For a titration, this means that the concentration calculated needs to be as close to the true value as possible.

The results of a titration give high-quality data if the repeats are close together. If they are not, you may need to check if you are following the procedure properly and repeat your titration until you get values that are close to each other.

The systematic errors in titrations are kept to a minimum by having a detailed procedure and by using accurate measuring equipment.

- Pipettes and burettes have lower errors in their measurements than measuring cylinders. A typical burette is accurate to within ±0.05 cm³.
- Burettes have graduations to 0.1 cm³. This means that the readings are precise.
- Repeating readings until they give closely similar results makes sure the results are **repeatable**.
- Following a detailed procedure means that another person should be able to do the titration and get the same value for the concentration, or another method could be used to analyse the titration and also get the same concentration (it is **reproducible**).

Titrations: adapting the basic procedure

- The same procedure can be used to find out the concentration of an unknown alkali. This time you put the alkali in the flask and a dilute acid of known concentration in the burette.
- Solids can be used in titrations. The solid can be crushed and added directly to the flask. (This method can be used to analyse the amounts of acidic or basic substances in tablets, for example.)
- The indicator used must give a definite change at the end point. The usual indicator used is litmus, which is red in acid and blue in alkali, but there are many others.

Processing the data from titrations
Obtaining the best estimate of the volume added
The first stage in working out concentrations from titration data is to calculate the **mean value** for the volume added.

Key words

- repeatable
- reproducible
- mean value
- outlier

Litmus is a useful indicator to use in titrations because it has a definite colour change at the end point. Litmus is red in acid and blue in alkali.

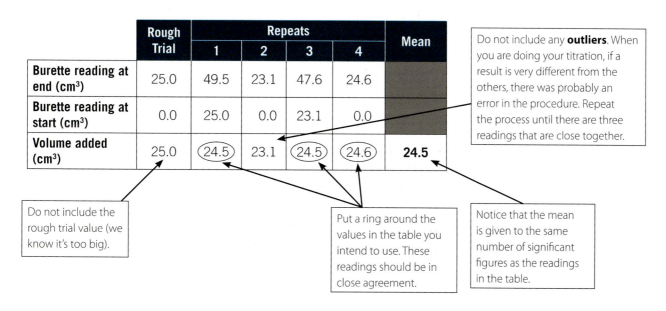

	Rough Trial	Repeats				Mean
		1	2	3	4	
Burette reading at end (cm³)	25.0	49.5	23.1	47.6	24.6	
Burette reading at start (cm³)	0.0	25.0	0.0	23.1	0.0	
Volume added (cm³)	25.0	24.5	23.1	24.5	24.6	**24.5**

Do not include any **outliers**. When you are doing your titration, if a result is very different from the others, there was probably an error in the procedure. Repeat the process until there are three readings that are close together.

Do not include the rough trial value (we know it's too big).

Put a ring around the values in the table you intend to use. These readings should be in close agreement.

Notice that the mean is given to the same number of significant figures as the readings in the table.

These students are recording and checking their titration results as they work. If there are outliers or if their readings show a large spread, they repeat their titrations until they record several readings close together.

In industry, automated equipment allows many titrations to be done at the same time. Repeatability checks and data recording are done electronically. This scientist is measuring the folic acid concentration in some drugs to check the dosage.

H S Calculating the concentration

- Use your titration results to work out the *number of moles added* from the burette.
- Use the equation for the reaction to work out the *number of moles in the flask*.
- Work out the *concentration* of the solution in the flask.

Worked example: Calculating concentration

20.0 cm³ of dilute hydrochloric acid was put in a flask. Sodium hydroxide solution (concentration 1.0 mol/dm³) was added from a burette.

HCl + NaOH → NaCl + H2O

The mean volume of sodium hydroxide used in the titration was 24.5 cm³.

What is the concentration of the acid? Give your answer to 3 significant figures.

Step 1: Write down what you know, with the units. Convert to standard units.

volume of NaOH added = 24.5 cm³ = $\dfrac{24.5 \text{ cm}^3}{1000}$ = 0.0245 dm³

concentration of NaOH = 1.0 mol/dm³

volume of HCl in flask = 20.0 cm³ = $\dfrac{20.0 \text{ cm}^3}{1000}$ = 0.0200 dm³

Step 2: Calculate the number of moles added.

number of moles = concentration × volume

n = 1.0 mol/dm³ × 0.0245 dm³ = 0.0245 moles NaOH

Step 3: Use the equation for the reaction to calculate the number of moles in the flask.

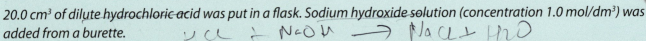

HCl + NaOH → NaCl + H2O

1 mole of HCl reacts with 1 mole of NaOH

0.0245 moles of NaOH neutralised the HCl in the flask

so

0.0245 moles of HCl reacts with 0.0245 moles of NaOH

Step 4: Calculate the concentration of the hydrochloric acid.

concentration = $\dfrac{\text{number of moles}}{\text{volume}}$ = $\dfrac{0.0245 \text{ moles}}{0.0200 \text{ dm}^3}$

= 1.225 mol/dm³

Answer:

concentration = 1.23 mol/dm³ (to 3 significant figures)

Worked example: Calculating concentration

25.0 cm³ of dilute potassium hydroxide solution was put in a flask. Dilute sulfuric acid (concentration 0.5 mol/dm³) was added from a burette.

The mean volume of sulfuric acid used in the titration was 19.8 cm³.

What is the concentration of the potassium hydroxide solution? Give your answer to 3 significant figures.

Step 1: Write down what you know, with the units. Convert to standard units.

volume of H_2SO_4 added = 19.8 cm³ = $\dfrac{19.8 \text{ cm}^3}{1000}$ = 0.0198 dm³

concentration of H_2SO_4 = 0.5 mol/dm³

volume of KOH in flask = 25.0 cm³ = $\dfrac{25.0 \text{ cm}^3}{1000}$ = 0.025 dm³

Step 2: Calculate the number of moles added.

number of moles = concentration × volume

n = 0.5 mol/dm³ × 0.0198 dm³ = 0.0099 moles of H_2SO_4

Step 3: Use the equation for the reaction to calculate the number of moles in the flask.

From the equation, the number of moles of KOH is twice the number of moles of H_2SO_4.

H_2SO_4 + 2KOH → K_2SO_4 + $2H_2O$

1 mole of H_2SO_4 reacts with 2 moles of KOH

0.0099 moles of H_2SO_4 neutralised the KOH in the flask

so

0.0099 moles of H_2SO_4 react with 2 × 0.0099 moles of KOH

0.0099 moles of H_2SO_4 react with 0.0198 moles of KOH

Step 4: Calculate the concentration of the potassium hydroxide solution.

concentration = $\dfrac{\text{number of moles}}{\text{volume}}$ = $\dfrac{0.0198 \text{ moles}}{0.0250 \text{ dm}^3}$

Answer:

concentration = 0.792 mol/dm³ (to 3 significant figures)

Notice that calculated values are often not whole numbers. Usually the concentration of the solution in the burette is chosen to be similar to the unknown concentration so that the titration values are a reasonable size. Titration results with very small volumes have a low percentage accuracy. On the other hand, titration results with very high volumes would involve refilling the burette, also introducing inaccuracy.

A 'common-sense' check is to compare your answer with the concentration of the known solution in the burette. As a very rough guide, the concentration of the unknown solution will usually be no more than double and no less than half that of the known solution.

Questions

1 Explain why universal indicator is not suitable for using in titrations.

2 A student does two titrations using the same acid and alkali, but gets very different readings. Read through the steps in the procedure. Suggest what might have gone wrong.

3 What is the difference between repeatability and reproducibility?

4 In a titration, a sample of 20.0 cm³ of an unknown concentration of sulfuric acid needed 22.0 cm³ of sodium hydroxide solution (concentration 0.2 mol/dm³) for neutralisation.
 a How many moles of sodium hydroxide were added?
 b Write an equation for the reaction.
 c How many moles of sulfuric acid were in the sample?
 d What is the concentration of the sulfuric acid?

Science explanations

C5 Chemical analysis

Chemical analysis is important in chemistry for the quality control of manufactured products and also to identify or quantify components in testing new products, mineral extraction, forensics, and environmental monitoring. Chemists need to identify both the substances that are present (qualitative analysis) and the quantity of each substance (quantitative analysis).

You should know

- the concept of a pure substance and how to identify a pure substance
- how different methods of chromatography separate mixtures as a mobile phase moves through a stationary phase
- why standard reference materials and locating agents are used in chromatography
- how to calculate and interpret R_f values
- how the processes of filtration, crystallisation, simple distillation, and fractional distillation are used in separating and purifying substances, and be able to give examples
- how chemists use precipitation reactions to detect which ions are present in an ionic compound **S**
- **H** how to use ionic equations to describe precipitation reactions
- how to carry out a flame test and interpret the results
- how to interpret data from emission spectroscopy
- the advantages of using instrumental methods of analysis
- **H** how to do calculations using the Avogadro constant and the mole
- how to use the balanced equation for a reaction to work out the quantities of chemicals to use in a synthesis
- how to calculate the theoretical yield **S**
- the relationship between molar amounts of gases and their volumes
- how to calculate the concentration of a solution in mol/dm^3 or in g/dm^3
- how chemists use ionic theory to explain why acids have similar properties, and how alkalis neutralise acids to form salts
- how to carry out a titration and interpret the data.

Ideas about Science

When carrying out analysis of substances, chemists have to take into consideration sampling techniques as well as practical procedures and safety. When describing procedures for both quantitative and qualitative analysis, you should be able to:

- suggest appropriate equipment, materials and techniques, justifying the choice with reference to the precision, accuracy and validity of the data that will be collected

- suggest an appropriate sample size and/or range of values to be measured, and justify the suggestion

- construct a clear and logically sequenced strategy for data collection

- identify hazards associated with the data collection and suggest ways of minimizing the risk.

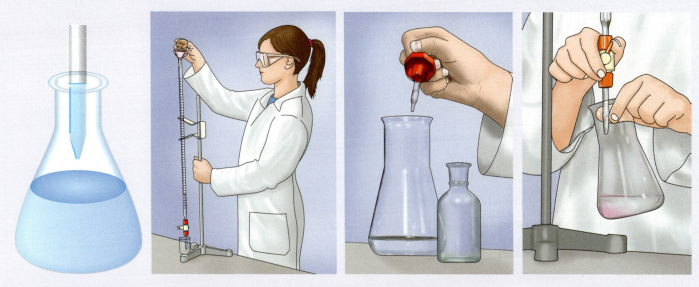

A pipette is used to measure the dilute acid into the flask. The burette is filled with dilute alkali. An indicator is added to the flask. Near the end point the alkali is added drop by drop.

C5 Review questions

Product
a solid product mixed with a solution
a liquid product mixed with another liquid
a liquid product which contains dissolved impurities
a soluble product dissolved in a solution

Method	Aspirin melting temperature in °C
1	137
2	138–140

1   At the end of a reaction, chemists often need to separate a product from a mixture.

   a   Which method of separation is needed to separate each product given in the table?
      Choose from this list.

      filtration   crystallisation   simple distillation   fractional distillation

   b i Joe makes some aspirin using two different methods.
         He measures the temperature at which the aspirin melts. He does this for aspirin made using each method.
         His results are shown in the table.

         Which method has made the purest aspirin?

         Explain how you can tell.

      ii Joe decides to look up some secondary data to help him to assess the purity of his aspirin.
         Explain how Joe should use secondary data and his results to check the purity of his aspirin.

2   Eve is a geologist. She collects samples of four minerals, A, B, C, and D, from a cliff face.

   The table shows the metal compounds that the minerals contain.

   Eve uses these solutions to test the metal samples:

   ● dilute sodium hydroxide solution

   ● dilute hydrochloric acid

   ● dilute silver nitrate

   ● dilute barium nitrate.

   a   How could Eve show that minerals A and B both contain copper ions?

   b   How could Eve show which of the minerals C and D contains calcium ions and which contains zinc ions?

   c   How could Eve show that mineral A contains carbonate ions and mineral B contains sulfate ions?

   d   Eve also uses emission spectroscopy to identify the metal ions in the minerals. Evaluate the use of emission spectroscopy compared to the use of solutions to identify ions in a mineral.

   e   Eve wants to identify all of the metal compounds in the cliff face. The cliff face is over a kilometre long and has many layers. How should she take samples to make sure that her samples are representative of the whole cliff?

Mineral	Metal compound in mineral	Formula
A	copper carbonate	$CuCO_3$
B	copper sulfate	$CuSO_4$
C	calcium chloride	$CaCl_2$
D	zinc chloride	$ZnCl_2$

3   Reema reacts 0.5 g of calcium carbonate with excess dilute acid. She measures the mass change during the reaction.

   The mass of the flask plus contents goes down during the reaction.

   a   Explain why the mass of the flask plus contents went down.

   b   Reema works out the number of moles of calcium carbonate she uses in the reaction.

   i Calculate the relative formula mass of calcium carbonate, $CaCO_3$.

   Use the Periodic Table to find the relative atomic masses that you need.

   H ii Calculate the number of moles of calcium carbonate in 0.5 g.

4 Jake investigates what happens when iron forms rust (iron oxide).
   He measures the mass of 10 new iron nails. He leaves the nails in damp
   conditions until they appear to be completely rusted.
   He dries them thoroughly and then measures their mass.
   These are his results.

   mass of 10 new, dry iron nails = 5.6 g
   mass of 10 rusted, dry iron nails = 7.2 g

   This is the equation for the reaction that happens when iron rusts.

   $$2Fe + 3O_2 \rightarrow Fe_2O_3$$

   a Use the equation to calculate the theoretical mass of the rusted iron nails.
      Use the Periodic Table to find the relative atomic masses that you need.
      Give your answer to two significant figures.

   b Has all the iron in the new nails reacted to form rust?
      Use the value you calculated in part **a** to justify your answer.

   c Explain why it is important that Jake dried the nails before measuring
      their mass.

5 Liz uses a series of titrations to find out the concentration of sodium hydroxide
   in a solution.
   She uses 25.0 cm³ of the sodium hydroxide solution each time.
   She does a set of titrations and works out the mean volume of hydrochloric
   acid that reacts with 25 cm³ of the sodium hydroxide solution. She then does
   another set of titrations using sulfuric acid.
   The table shows her results.

Acid used in titration	hydrochloric acid	sulfuric acid
Concentration in mol/dm³	0.1	0.1
Mean volume used in titration in cm³	24.9	12.5

   a Write a symbol equation for the reaction between sodium hydroxide and
      hydrochloric acid.

     b 25.0 cm³ of sodium hydroxide was used in each titration. Use your equation
      and the titration results to estimate the concentration of the sodium
      hydroxide. Justify your answer.

   c The hydrochloric acid and the sulfuric acid both have a concentration of
      0.1 mol/dm³.

      i   What is the formula for each acid?

      ii Use the formulae to explain why the mean volumes used are different.

     iii Liz does a further titration using nitric acid, HNO_3, of concentration
           0.2 mol/dm³. Predict the volume of nitric acid needed in the titration.
           Justify your answer.

C6 Making useful chemicals

Why study useful chemicals?

We use chemicals to grow and preserve our food, treat disease, and decorate our homes. Many of these chemicals do not occur naturally: they are synthetic. Developing useful products such as drugs to treat diseases or fertilisers to grow our food depends on chemists. Initially the chemicals are made in a laboratory so they can be tested to make sure that they are safe and have the desired properties before moving to large-scale production.

What you already know

- Chemical changes result in the formation of new materials.
- Chemicals can be represented using formulae and their reactions can be represented by equations.
- Alkalis neutralise acids to make salts.
- Acids react with metals to produce a salt plus hydrogen gas and with an alkali to produce a salt plus water.
- Indicators are used to determine the pH of a substance, which is a measure of its acidity or alkalinity.
- A catalyst increases the rate of a chemical reaction without being used up.
- An energy change takes place during a change of state.
- An exothermic process gives out energy to the surroundings.
- An endothermic process takes in energy from the surroundings.

The Science

The reactions of acids and bases are used to synthesise useful products in the laboratory and on an industrial scale. An understanding of the properties and characteristics of acids and bases is important in ensuring that the desired products are made. Chemists manage the rate of the reaction taking place by controlling experimental conditions including temperature, pressure, concentration, and surface area of the reactants and through the use of catalysts.

Some chemical reactions, such as the reaction between nitrogen and hydrogen to produce ammonia, are reversible, meaning they can react in both directions and eventually come to chemical equilibrium. Chemical equilibria are important considerations when determining the optimum conditions in important industrial processes such as the Haber process to ensure a good balance between yield, rate of reaction, and economic viability. The manufacture of fertilisers is used as a context for considering how chemists reach decisions about optimum processes for large-scale production of bulk chemicals.

Ideas about Science

Chemists make sure that they use the right grade of chemical for a reaction. Technical chemists test chemicals from suppliers to check for purity. They take measurements and make sure that the data they collect is as accurate as possible. They can then make the best estimate of the true value of the purity. Chemists use data and interpret graphs to make predictions about the yield of a chemical product as well as determining other properties of the chemical.

Chemists have theories to explain observed data and answer key questions about chemical reactions. The particle nature of matter is used to explain the rate of reaction and predict what will happen if variables such as temperature, pressure, or concentration are changed. Computer models may be used to simulate these changes to help chemists agree upon the optimum conditions for an industrial process.

Science-based technologies often improve the quality of life but may also harm the environment. For example, fertilisers are needed to meet the food demands of an increasing world population but at the same time overuse of fertilisers can lead to eutrophication of rivers and lakes. Therefore, benefits must be weighed against harms. The chemical industry is changing to become more sustainable by looking at alternative feedstocks and developing new catalysts, including biocatalysts.

A: Acid reactions

Find out about

- everyday products that are made from acids
- reactions of acids
- practical procedures for making salts
- formulae of salts

Key words

➤ acid
➤ bulk chemical
➤ fine chemical
➤ neutralisation
➤ salt

Many products that you use every day, including cleaning products, pharmaceutical products, and food additives, are based on the chemistry of **acid** reactions. The chemical industry converts raw materials, including acids, into these useful products. It makes **bulk chemicals**, such as fertilisers, on a scale of thousands or even millions of tonnes per year and **fine chemicals**, such as drugs, on a much smaller scale.

Many acids are part of life itself. Biochemists have discovered the citric acid cycle. This is a series of reactions in all cells. The cycle harnesses the energy from respiration for movement and growth in living things.

Citric acid is a solid at room temperature. It dissolves in water and is found in citrus fruits such as oranges and lemons. The human body processes about 2 kg of citric acid a day during respiration. Citric acid and its salts are added to food to prevent them reacting with oxygen in the air, and to give a tart taste to drinks and sweets.

 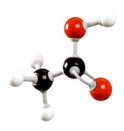

Acetic acid (ethanoic acid) is a liquid at room temperature. It is the acid in vinegar. Most white vinegar is just a dilute solution of acetic acid. Brown vinegars have other chemicals in the solution that give the vinegar its colour and flavour. Most microorganisms cannot survive in acid, so vinegar (E260) is used as a preservative in pickles.

Sulfuric acid, H_2SO_4, is manufactured from sulfur, oxygen, and water. The pure, concentrated acid is an oily liquid at room temperature. The chemical industry in the UK makes about 2 million tonnes of the acid each year. The acid is essential for the manufacture of other chemicals, including detergents, pigments, dyes, plastics, and fertilisers.

Many pharmaceutical products are made from acids, including aspirin, which is produced by reacting 2-hydroxybenzoic acid and ethanoic anhydride (which come from ethanoic acid).

Reactions of acids

Acids react in **neutralisation** reactions with metals, hydroxides, and carbonates. All neutralisation reactions produce **salts**, which have a wide range of uses and can be made on an industrial scale. For example, magnesium chloride is used as a component in fertiliser and in the production of paper and textiles.

Acids with metals

Acids react with metals to produce salts. The other product is hydrogen gas.

$$\text{acid} + \text{metal} \longrightarrow \text{salt} + \text{hydrogen}$$

For example:

$$2HCl(aq) + Mg(s) \longrightarrow MgCl_2(aq) + H_2(g)$$

Not all metals will react in this way. Recall the list of metals in order of reactivity (see C3). Metals that come below lead in the list of order of reactivity do not react with acids, and even with lead it is hard to detect any change in a short time.

Acids with metal oxides or hydroxides

An acid reacts with a metal oxide or hydroxide to form a salt and water. No gas forms.

$$\text{acid} + \text{metal oxide} \longrightarrow \text{salt} + \text{water}$$
$$\text{(or hydroxide)}$$

For example:

$$2HCl(aq) + MgO(s) \longrightarrow MgCl_2(aq) + H_2O(l)$$

The reaction between an acid and a metal oxide is often a vital step in making useful chemicals from ores.

Acids with carbonates

Acids react with carbonates to form a salt, water, and carbon dioxide gas, which can be observed as bubbles.

$$\text{acid} + \text{metal carbonate} \longrightarrow \text{salt} + \text{water} + \text{carbon dioxide}$$

Geologists can test for carbonates by dripping hydrochloric acid onto rocks. If they see any fizzing (caused by the CO_2), the rocks contain a carbonate. This is likely to be calcium carbonate or magnesium carbonate.

The word equation is:

hydrochloric acid + calcium carbonate ⟶ calcium chloride + water + carbon dioxide

The balanced equation is:

$$2HCl(aq) + CaCO_3(s) \longrightarrow CaCl_2(aq) + H_2O(l) + CO_2(g)$$

This is a foolproof test for the carbonate ion. So the term 'the acid test' is used to describe any way of providing definite proof.

Factors to consider when making magnesium sulfate

Magnesium sulfate is a salt produced by the chemical industry. It has many uses, including as a nutrient for plants.

The process of making magnesium sulfate (or any other soluble salt) on a laboratory scale illustrates the stages in a chemical synthesis.

Choosing the reaction

Any of the characteristic reactions of acids can be used to make salts:

- acid + metal ⟶ salt + hydrogen
- acid + metal oxide or hydroxide ⟶ salt + water
- acid + metal carbonate ⟶ salt + carbon dioxide + water

Magnesium metal is relatively expensive because it has to be extracted from one of its compounds. So it makes sense to use either magnesium oxide or carbonate as the starting point for making magnesium sulfate from sulfuric acid.

In the method suggested in C8P *Making a sample of a salt* an excess of solid is added to make sure that all the acid is used up. This method is only suitable if the solid added to the acid is either insoluble in water or does not react with water.

Carrying out a risk assessment

It is always important to minimise exposure to risk. You should take care to identify hazardous chemicals. You should also look for hazards arising from equipment or procedures. This is a **risk assessment**.

Formulae of salts

All salts are ionic compounds and the formulae can be worked out from knowledge of the ions present. The formula of sodium chloride is $NaCl$ because there is one sodium ion (Na^+) for every chloride ion (Cl^-). There are no molecules in table salt, only ions.

Remember that all compounds are overall electrically neutral. Some non-metal ions consist of more than one atom. The table opposite shows some examples of common ions. In the formula for magnesium nitrate, $Mg(NO_3)_2$, the brackets and subscript 2 around the NO_3 show that two complete nitrate ions appear in the formula.

Synoptic link

You can learn more about the full procedure for making a salt in C8P *Making a sample of a salt*.

Testing for carbonate using hydrochloric acid.

Synoptic link

You can learn more about writing formulae and balanced symbol and ionic equations in C3.2 *How are metals with different reactivities extracted?*

Positive ions			Negative ions		
Charge	Name of ion	Symbol	Charge	Name of ion	Symbol
1+	ammonium	NH_4^+	1–	hydroxide	OH^-
	sodium	Na^+		nitrate	NO_3^-
	potassium	K^+		chloride	Cl^-
2+	calcium	Ca^{2+}	2–	sulfate	SO_4^{2-}
	magnesium	Mg^{2+}		carbonate	CO_3^{2-}
	zinc	Zn^{2+}		oxide	O^{2-}

Key word

➤ risk assessment

Questions

1 A pattern can be etched onto a zinc plate using hydrochloric acid. The hydrochloric acid reacts with the zinc, forming soluble zinc chloride, $ZnCl_2$. Write a word equation and a balanced symbol equation for the reaction.

2 Magnesium hydroxide, $Mg(OH)_2$, is an antacid used to neutralise excess stomach acid, HCl. Write a word equation and a balanced symbol equation for the reaction.

3 There is a volcano in Tanzania, Africa, where the lava contains sodium carbonate, Na_2CO_3. The cooled lava fizzes when it reacts with hydrochloric acid. Write a word equation and a balanced symbol equation for the reaction.

4 Limescale forms in kettles where hard water is heated. Limescale consists of calcium carbonate. Three acids are often used to remove limescale: citric acid, acetic acid (in vinegar), and dilute hydrochloric acid. Which acid would you use to descale an electric kettle? Explain your answer.

5 Read about the production of magnesium sulfate.
   a Write a balanced equation for the reaction of magnesium carbonate with sulfuric acid.
   b Why does the mixture of magnesium carbonate and sulfuric acid froth up?
   c Why is it important that the magnesium carbonate is added in excess?
   d Identify the impurities removed during the purification stages of making the salt.

6 Write down the formulae of these salts:
   a potassium nitrate
   b magnesium carbonate
   c sodium sulfate
   d calcium nitrate

B: How strong is an acid?

Find out about

H
- strong and weak acids
- the pH scale
H
- concentrated and dilute acids

Key words

➤ pH
➤ concentration
➤ indicator
H
➤ strong acid
➤ weak acid
➤ dilute acid

Indicators and the pH scale

As you learnt in C5, all acids in solution produce hydrogen ions. The **pH** of a solution is a measure of the **concentration** of the hydrogen ions in that solution.

Indicators change colour to show whether a solution is acidic or alkaline. Blue litmus turns red in acidic solutions and red litmus turns blue in alkaline solutions. Special mixed indicators, such as universal indicator, show a range of colours and can be used to estimate pH values.

pH values can also be measured electronically using a pH meter with an electrode that dips into the solution. The meter can be read directly from the display, or it may be connected to a datalogger or computer.

The term pH appears on many cosmetic, shampoo, and food labels. The pH scale is a number scale that shows the acidity or alkalinity of a solution in water. Most laboratory solutions have a pH in the range of 1–14.

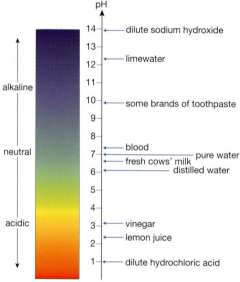

The position of some common acids and alkalis on the pH scale mapped against the universal indicator colour range.

A pH meter can be used to measure pH values.

Hydrangea flowers contain natural indicators – they are blue if grown on acidic soil and pink on alkaline soil. Note that this is the opposite of the litmus colours.

H Strong and weak acids

Some acids ionise completely to give hydrogen ions when they dissolve in water. For example:

$$HCl(g) + water \longrightarrow H^+(aq) + Cl^-(aq)$$

Chemists call these **strong acids**. Hydrochloric, sulfuric, and nitric acids are all strong acids.

Other acids, such as carboxylic acids, ionise to produce hydrogen ions to a lesser extent when dissolved in water than the strong acids. Only a small proportion of the molecules ionise, so not all the hydrogen is released as ions into the solution. This type of acid is called a **weak acid**. For example, ethanoic acid:

$$CH_3COOH(aq) \rightleftharpoons CH_3COO^-(aq) + H^+(aq)$$

Since the H^+ ion is responsible for the acidic properties, weak acids still show the characteristic reactions of acids with metals, alkalis, and metal carbonates, but their reactions are slower than those of strong acids.

The symbol $\rightleftharpoons$ is used in equations describing reversible chemical reactions. You will learn more about these reactions in C6.3.

Weak acids and strong acids of the same concentration have different pH values because the concentration of $H^+(aq)$ ions is different.

The relative reactivity of magnesium ribbon with hydrochloric acid and ethanoic acid.

Synoptic link

You can learn more about concentration of solutions in C5.4 *How are the amounts of chemicals in solution measured?*

Dilute and concentrated acids

Both strong and weak acids can be prepared at a range of concentrations. Concentration refers to the amount of substance, in moles, in a given volume. A **dilute** sample of an acid contains relatively few hydrogen ions when compared to a more concentrated sample of the same acid. This means that the dilute sample will have a higher pH value.

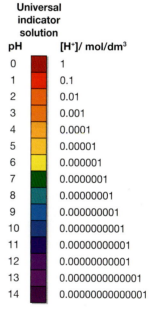

Universal indicator solution

pH	$[H^+]/ mol/dm^3$
0	1
1	0.1
2	0.01
3	0.001
4	0.0001
5	0.00001
6	0.000001
7	0.0000001
8	0.00000001
9	0.000000001
10	0.0000000001
11	0.00000000001
12	0.000000000001
13	0.0000000000001
14	0.00000000000001

Linking the pH scale and hydrogen ion concentration.

If this represents the hydrogen ions in a dilute acid...

...then this represents the hydrogen ions in a concentrated acid.

Comparing the relative amounts of the hydrogen ions in a dilute and concentrated acid.

A pH value of 1 means that the concentration of $H^+(aq)$ in the solution is 0.1 mol/dm³.

A pH value of 2 means that the concentration of $H^+(aq)$ in the solution is 0.01 mol/dm³.

A pH value of 3 means that the concentration of $H^+(aq)$ in the solution is 0.001 mol/dm³.

So as the $H^+(aq)$ concentration increases by a factor of 10, the pH value decreases by 1. The pH scale is a useful indication of the acidity, rather than writing the concentration of the hydrogen ions present.

Questions

1 Explain what is meant by the term pH.

H 2 What is the difference between a strong acid and a weak acid?

3 Draw a particle diagram to show how the pH of a strong acid and a weak acid at the same concentration differs.

4 Suggest why the pH scale is more convenient than describing a solution by its molar concentration of hydrogen ions.

C6.2 How do chemists control the rate of reaction?

A: Factors affecting the rate of reaction

Find out about

- collision theory
- activation energy
- catalysts
- enzymes

Some chemical reactions seem to happen in an instant. An explosion is an example of a very fast reaction. Other reactions take time – seconds, minutes, hours, or even years. Rusting is a slow reaction and so is the rotting of food.

It is a chemist's job to work out the most efficient way to synthesise a chemical. It is important that the chosen reactions happen at a convenient speed. A reaction that occurs too quickly can be hazardous. A reaction that takes several days to complete is not practical because it ties up equipment and people's time for too long, which costs money.

Powdered antacids react faster in water than a tablet. Milk standing in a warm kitchen goes sour more quickly than milk kept in a refrigerator. Changing the conditions alters the rate of these processes and many others.

Factors that affect the rate of chemical reactions are:

- the concentration of reactants in solution – the higher the concentration, the faster the reaction
- the **surface area** of solids – powdering a solid increases the surface area in contact with a liquid, solution or gas, and so speeds up the reaction
- the temperature – a 10 °C rise in temperature can roughly double the rate of many reactions
- the **pressure** – increasing the pressure of a gaseous reaction increases the rate of reaction
- **catalysts** – these are chemicals that speed up a chemical reaction without being used up in the process.

An explosion is an example of a very fast reaction.

Rusting is an example of a very slow reaction.

one big lump (slow reaction)

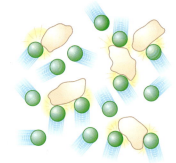

several small lumps (fast reaction)

Breaking up a solid into smaller pieces increases the total surface area. This increases the amount of contact between the solid and the solution, making the reaction happen faster.

Collision theory

Chemists have a theory to explain how the various factors affect reaction rates.

The basic idea is that particles, such as molecules, atoms, and ions, can only react if they bump into each other. Imagining how these particles collide with each other leads to a theory that can account for the effects of concentration, temperature, pressure, surface area, and catalysts on reaction rates.

Collision theory states that when molecules collide, some bonds between atoms can break while new bonds form. This creates new molecules.

Molecules are in constant motion in gases, liquids, and solutions. There are millions upon millions of collisions every second. Most reactions would be explosive if every collision led to a reaction. But only a very small proportion of all the collisions are successful and actually lead to a reaction. These are the collisions in which the molecules are moving with enough energy to break bonds between atoms. This minimum amount of required energy is the **activation energy**. Activation energy is like an energy hill that the reactants have to climb before a reaction will start. The higher the hill, the more difficult it is to get the reaction started. The collisions between molecules have a range of energies. Head-on collisions between fast-moving particles are the most energetic.

Synoptic link

You can learn more about activation energy in C1.2 *Why are there temperature changes in chemical reactions?*

Any change that increases the number of successful collisions per second has the effect of increasing the rate of reaction.

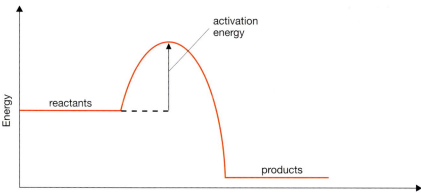

Energy-level diagram showing the activation energy of a reaction. The size of the activation energy is usually less than the energy needed to break all the bonds in the reactant molecules because new bonds start forming while old bonds are breaking, returning the energy to the mixture.

Key words

➤ surface area
➤ pressure
➤ catalyst
➤ activation energy

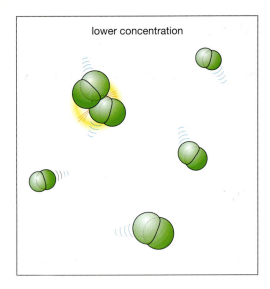

lower concentration

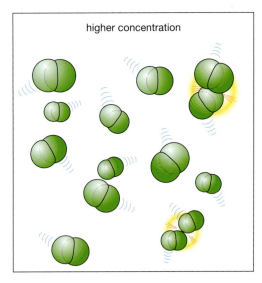

higher concentration

Molecules have a greater chance of colliding in a more concentrated solution. More frequent collisions means faster reactions. Reactions get faster if the reactants are more concentrated.

Concentration

Increasing the concentration of solutions of dissolved chemicals increases the frequency of collisions. The same small proportion of the collisions will be successful, but now there are more of them. So there will be more successful collisions.

Temperature

Increasing the temperature increases the kinetic energy of the particles. This increases the rate of collisions and, more significantly, increases the energy available to the particles to overcome the activation energy and react.

Pressure

Increasing the pressure increases the rate of collisions. For example, by pressing the plunger of a sealed syringe you are increasing the pressure. In this case you now have the same number of gas particles in a smaller volume, which effectively means that the concentration of the gas has increased.

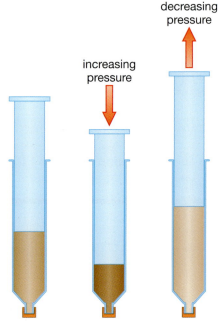

decreasing pressure

increasing pressure

There are more collisions in a given time when you increase the pressure of a gas. Decreasing the pressure of a gas leads to fewer collisions in a given time.

Surface area

Earlier in this section, you learnt that breaking up a solid into smaller pieces increases its surface area. Increased surface area means that there are more atoms, molecules, or ions of the solid available to react. This speeds up the reaction by increasing the frequency of successful collisions with particles in liquid, solution, or gas.

Catalyst

Using a catalyst provides an alternative route for a chemical reaction with a lower activation energy. The energy of the reactants and the energy of the products does not change, but the activation energy is smaller. This means a higher proportion of collisions have enough energy to cross the activation-energy barrier.

Catalysts in industry

The use of a catalyst can reduce the economic and environmental cost of an industrial process, leading to more sustainable 'green' chemical processes. Modern catalysts can be highly selective. This is important when reactants can undergo more than one chemical reaction to give a mixture of products. With a suitable catalyst it can be possible to speed up the reaction that gives the required product, but not speed up other possible reactions that create unwanted **by-products**.

Better catalysts

Catalysts are essential in many industrial processes. They make many processes economically viable. This means that chemical products can be made at a reasonable cost and sold at affordable prices.

Research into new catalysts is an important area of scientific work. This is shown by the industrial manufacture of ethanoic acid from methanol and carbon monoxide. This process was first developed by the company BASF in 1960 using a cobalt compound as the catalyst at 300 °C and at a pressure 700 times atmospheric pressure.

About six years later, the company Monsanto developed a process using the same reaction but a new catalyst system based on rhodium compounds. This ran under much milder conditions: 200 °C and 30–60 times atmospheric pressure.

In 1986, the petrochemical company BP bought the technology for making ethanoic acid from Monsanto. It has since devised a new catalyst based on compounds of iridium. This process is faster and more efficient because iridium is cheaper and less of the catalyst is needed. Iridium is even more selective so the amount of ethanoic acid produced (yield) is greater and there are fewer by-products. This makes it easier and cheaper to make pure ethanoic acid and there is less waste.

Synoptic link

You can learn more about endothermic and exothermic reactions and energy-level diagrams in C1.2 *Why are there temperature changes in chemical reactions?*

The manufacture of ethanoic acid from methanol and carbon monoxide uses a catalyst to speed up the reaction.

methanol + carbon monoxide → ethanoic acid

$$CH_3OH(g) + CO(g) \rightarrow CH_3COOH(g)$$

➤ enzyme
➤ active site
➤ denatured
➤ green chemistry
➤ biocatalyst
➤ fermentation
➤ feedstock

Enzymes

Enzymes are proteins that catalyse processes in living organisms. An enzyme is a long protein molecule that is folded to produce a specific shape called an **active site**. The reactant molecules fit into the active site where the reaction takes place.

Due to the nature of the active site, each enzyme controls one particular reaction and works within a narrow temperature and pH range. If the shape of the active site changes so that the reactant no longer fits, the enzyme is **denatured**, meaning that it no longer works. This can happen at high temperatures or extreme pH values. Just like catalysts, enzymes can be used again and again.

Enzymes in biological processes

Enzymes play an important role in many biological processes, such as digestion. For example, the enzyme amylase found in saliva starts to break down starchy food in our mouths, while the enzyme protease breaks down proteins into amino acids in the small intestine. The enzyme catalase found in the liver is essential to avoid a build-up of hydrogen peroxide, which is potentially harmful to cells, by catalysing the following reaction.

$$\text{hydrogen peroxide} \xrightarrow{\text{catalase}} \text{water} + \text{oxygen}$$

$$2H_2O_2\ (l) \longrightarrow 2H_2O(l) + O_2(g)$$

Synoptic link

You can learn more about enzymes in biological processes in B3.1 *What happens during photosynthesis?*

Enzymes in industrial processes

One aim of **green chemistry** is to make processes more energy efficient. New processes are being developed that run at much lower temperatures. One way of doing this is to use **biocatalysts**. These are the enzymes produced by microorganisms. Remember that enzymes operate within a limited temperature range, above which they are denatured and no longer work. So, if a process can be designed to use an enzyme at a lower temperature than a more traditional route, the energy demand will be lowered.

Petrochemical route

benzene from crude oil

Several steps take place at 260 °C and 40× atmospheric pressure.

Uses phosphoric acid, oxygen, sulfur dioxide, iron ions, and hydrogen peroxide.

catechol → vanillin

Biosynthetic route

glucose sugar

Glucose is dissolved in water at 37 °C.

Uses enzymes produced by *E. coli* bacteria.

Comparing two routes to the flavouring agent vanillin.

Fermentation

The production of ethanol is an important industrial process. Ethanol can be produced by three different routes: **fermentation**, biotechnology, and chemical synthesis. Most of the world's alcohol, 93%, is produced by the traditional method of fermentation of sugar with yeast. Enzymes present in the yeast play an important role in the reaction. Ethanol produced by this method is mainly used as a fuel, with smaller amounts used in alcoholic beverages and the chemical industry.

Common **feedstocks** for the fermentation reaction are sugar cane, sugar beet, corn, rice, and maize. Large areas of land are needed to grow the crops, and only some parts of the plants can be fermented. The parts that cannot be fermented are used to make animal feeds and corn oil. Recent developments mean that more plant material can be fermented and agricultural waste, paper-mill sludge, and even household rubbish can be used for fermentation.

Cellulose polymers from the feedstock are heated with acid to break them down into simple sugars such as glucose. Glucose is then converted into ethanol and carbon dioxide. This reaction is catalysed by enzymes found in yeast:

$$glucose \rightarrow ethanol + carbon\ dioxide$$

$$C_6H_{12}O_6 \rightarrow 2C_2H_5OH + 2CO_2$$

The optimum temperature for the fermentation reaction with yeast is in the range 25–38 °C. At lower temperatures the rate of reaction is too slow, and at higher temperatures the enzymes are denatured. Enzymes are also affected by pH. This is because changes in pH can make and break bonds within and between the enzymes, changing their shape and therefore their effectiveness.

The concentration of ethanol solution produced by the fermentation process is limited to between 14 and 15% ethanol. If the ethanol concentration rises higher than 15%, it becomes toxic to the yeast, which is killed, and the fermentation stops.

Sugar cane is a feedstock for the production of ethanol by fermentation.

Bioethanol is the most widely used green car fuel in the world.

Enzymes play an important role in the manufacture of cheese.

Enzymes in everyday products

Enzymes have many uses in everyday products including food production, detergents, biological washing powders, and medical testing. For example, enzymes play an important role in the production of cheese. The enzyme rennet coagulates the milk to make solid casein. As the cheese matures, enzymes break up some of the proteins into their amino acids. This gives the cheese its characteristic flavour. The ripening process takes several weeks; cheeses such as cheddar take up to nine months to mature. Scientists are still looking for new ways to try and make the enzymes speed up the reaction. After all, time is money to the manufacturers.

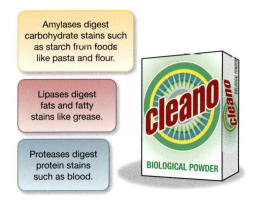

Amylases digest carbohydrate stains such as starch from foods like pasta and flour.

Lipases digest fats and fatty stains like grease.

Proteases digest protein stains such as blood.

Use a biological washing powder and save energy by washing at lower temperatures.

Enzymes are now widely used in the manufacturing of the fabrics that your clothing, furniture, and other household items are made from. Increasing demands to reduce pollution caused by the textile industry has caused the industry to turn to biotechnology to try and resolve some of its problems. In many instances the use of enzymes has been the answer.

Questions

1 Suggest reasons why it is important to develop industrial processes that run at lower temperatures and pressures.

2 Using collision theory, explain why increasing the concentration of the reactants increases the rate of reaction.

3 Adding a catalyst to a reaction mixture means that the activation energy for the reaction is lower. Explain why this speeds up the reaction even if the temperature does not change.

4 Copy the energy-level diagram from earlier in this section and add in a new line to show the effect of adding a catalyst.

5 Explain why an enzyme only works within a narrow temperature and pH range.

6 What are the advantages and disadvantages of using enzymes in industrial processes? Explain your answer.

7 Explain why the concentration of ethanol solution made by fermentation will never reach 16%.

B: Measuring the rate of reaction

Your pulse rate is the number of times your heart beats every minute. The production rate in a factory is a measure of how many articles are made in a particular time. Similar ideas apply to chemical reactions.

Chemists measure the **rate of a reaction** by finding the quantity of product produced or the quantity of reactant used up in a fixed time. Manual timing of reactions is only suitable for reactions that are over in a few minutes (or more). Modern methods using lasers and dataloggers allow chemists to measure the rates of reactions that are over in less than one million of a billionth of a second (one femtosecond).

Different procedures are used in different circumstances – there are details of some methods in C8.

Interpreting rate-of-reaction graphs

The apparatus in the diagram below was used to investigate the effect of changing the conditions of the reaction of zinc metal with sulfuric acid.

$$Zn(s) + H_2SO_4(aq) \rightarrow ZnSO_4(aq) + H_2(g)$$

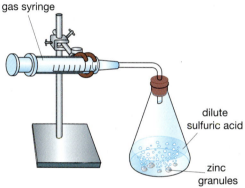

gas syringe

dilute
sulfuric acid

zinc
granules

Apparatus used in an investigation into the effect of changing the conditions on the reaction of zinc metal with sulfuric acid.

The red line on the graph plots the volume of hydrogen gas against time using zinc granules and 50 cm³ of dilute sulfuric acid at 20 °C. The reaction gradually slows down and stops because the acid concentration falls to zero.

The effect of concentration

Line A on the graph shows the result of using acid that was half as concentrated while leaving all the other conditions the same as in the original set up.

The investigator added 50 cm³ of this more dilute acid. Halving the acid concentration lowers the rate at the start. The final volume of gas is cut by half because there was only half as much acid in the 50 cm³ of solution to start with.

Find out about
● measuring rates of reaction
● interpreting rates graphs

Key word
➤ rate of reaction

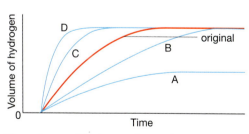

The volume of hydrogen formed over time during an investigation into the factors affecting the rate of reaction of zinc with sulfuric acid. The investigator used the same mass of zinc each time. There was more than enough metal to react with all the acid.

Key words

➤ gradient
➤ tangent

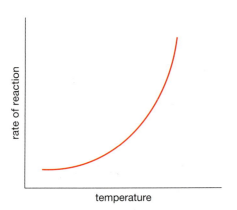

The graph shows how, for most reactions, the rate of reaction increases with temperature.

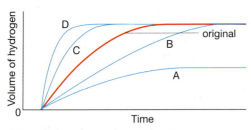

The volume of hydrogen formed over time during an investigation into the factors affecting the rate of reaction of zinc with sulfuric acid. The investigator used the same mass of zinc each time. There was more than enough metal to react with all the acid.

The effect of surface area

Line B on the graph shows the result of using the same excess of zinc metal in fewer, larger pieces. All other conditions were the same as in the original set up. Fewer and larger lumps of metal have a smaller total surface area so the reaction starts more slowly. The amount of acid is unchanged and the metal is still in excess so that the final volume of hydrogen is the same.

The effect of temperature

Line C on the graph shows the result of carrying out the reaction at 30 °C while leaving all the other conditions the same as in the original set up. This more or less doubles the rate at the start. The quantities of chemicals are the same so the final volume of gas collected is the same as it was originally.

The effect of adding a catalyst

Line D on the graph shows what happens when the investigation is repeated with everything the same as in the original set up but with a catalyst added. The reaction starts more quickly. Catalysts do not change the final amount of product, so the volume of gas at the end is the same as before.

Determining the rate of reaction

The rate of reaction is determined from the experimental data. We can calculate the average rate of a reaction from the time taken to make a fixed amount of product by dividing the amount of product by the time.

For example, in the reaction between zinc and sulfuric acid if a volume of 10 cm³ of hydrogen gas is produced at 50 s:

$$\text{average rate of reaction} = \frac{\text{volume of hydrogen produced}}{\text{time}}$$

$$= \frac{10\ \text{cm}^3}{50\ \text{s}}$$

$$= 0.2\ \text{cm}^3/\text{s}$$

To calculate the rate of reaction at a specific time in the reaction we need to plot and interpret a graph. The **gradient** of a graph showing the change in a variable over time is equal to the rate of change at that time. If the graph is a curve we will need to draw a **tangent** to the curve in order to find the gradient.

Synoptic link

You can learn more about finding the gradient of a graph in the *Maths skills* section at the end of this book.

Worked example: Calculating rate of reaction

During the reaction between magnesium and sulfuric acid a student collected the following data. What is the rate of hydrogen gas production at 100 s?

Time (s)	0	10	30	60	90	120	150	180
Volume of hydrogen gas (cm^3)	0	10	22	38	50	57	60	60

Step 1: Plot a graph of volume of gas against time. Choose suitable scales, with the volume of hydrogen produced on the vertical axis and the time on the horizontal axis.

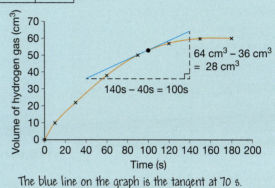

$64\ cm^3 - 36\ cm^3 = 28\ cm^3$

$140s - 40s = 100s$

Step 2: Calculate the rate by finding the gradient of the tangent at 100 s.

The blue line on the graph is the tangent at 70 s.

● Draw the tangent at 100 s. The tangent just touches a curve and has the same gradient as the curve does at that point.

The line rises from 36 cm^3 to 64 cm^3 between 40 s and 140 s.

● To find the gradient of a straight line:

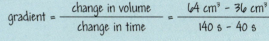

gradient = $\dfrac{\text{change in volume}}{\text{change in time}}$ = $\dfrac{64\ cm^3 - 36\ cm^3}{140\ s - 40\ s}$

■ Choose any two points on the line.

■ Draw a right-angled triangle with the line as the hypotenuse.

= $\dfrac{28\ cm^3}{100\ s}$ = 0.28 cm^3/s

■ Use the scale on each axis to find the changes in values.

rate of reaction = 0.28 cm^3/s (to 2 significant figures)

Questions

1 For each of these reactions, what method could be used to measure the rate of reaction? (Hint: Look at the states of the reactants and products.)

 a $CaCO_3(s) + 2HCl(aq) \rightarrow CaCl_2(aq) + CO_2(g) + H_2O(l)$

 b $Zn(s) + H_2SO_4(aq) \rightarrow ZnSO_4(aq) + H_2(g)$

 H **c** $Na_2S_2O_3(aq) + 2HCl(aq) \rightarrow 2NaCl(aq) + SO_2(g) + S(s) + H_2O(l)$

2 How is it possible to control conditions to speed up these changes?

 a the setting of an epoxy glue

 b the cooking of an egg

 c the conversion of oxides of nitrogen in car exhausts to nitrogen

3 When investigating the effect of temperature on a chemical reaction, why is it important to keep all other conditions the same?

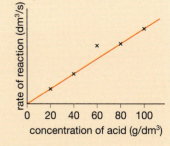

H **4** The effect of concentration on the rate of reaction between magnesium and hydrochloric acid was investigated. The results were plotted on the graph.

 a Is there a correlation? If so, describe it.

 b Which result is an outlier? Suggest a reason why this result is different to the expected value.

A: Reversible reactions and equilibrium

Find out about

- reactions that go both ways
- factors affecting the direction of change
- chemical equilibrium
- dynamic equilibrium
- factors affecting the position of equilibrium
- Le Chatelier's principle

Synoptic link

You can learn more about calculations of yields in C5.1 *How are chemicals separated and tested for purity?*

Industrial processes are managed to get the best yield as quickly and economically as possible. Chemists select the conditions, such as temperature, pressure, and concentration that give the best economic outcome in terms of safety, maintaining conditions and equipment, and energy use.

Some reactions go only in one direction. For example, the reactions that happen to a raw egg in boiling water cannot be reversed by cooling the water. When methane burns in air, carbon dioxide and water are produced. It is then almost impossible to turn the products back into methane and oxygen. These changes are known as **irreversible reactions**, and give a high **yield**.

The reactions in some processes are reversible. This can be problematic in industry because the reactants never completely react to make the products, meaning that yield can be potentially very low.

In C5.3D you learnt how to calculate the percentage yield for a reaction from the expression:

$$\text{percentage yield} = \frac{\text{actual yield}}{\text{theoretical yield}} \times 100$$

In **reversible reactions** reactants are often wasted and have to be separated out from the products; this increases the cost as extra purification stages have to be added into the industrial processes.

Reversible changes of state

Melting and evaporating are familiar reversible processes. Heating turns liquid water into steam, but the liquid re-forms as steam condenses on cooling.

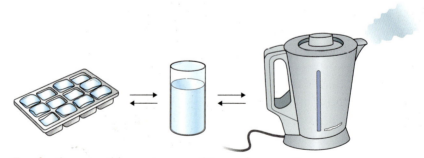

Two familiar reversible processes: melting and evaporation.

Heating turns ice into liquid water:

$$H_2O(s) \rightarrow H_2O(l)$$

Ice re-forms if the liquid cools to 0 °C or below:

$$H_2O(l) \rightarrow H_2O(s)$$

Combining these two equations gives:

$$H_2O(s) \rightleftharpoons H_2O(l)$$

Many chemical reactions are also reversible. A **reversible reaction** can go forwards or backwards, depending on the conditions. The direction of change may vary with the temperature, pressure, or concentration of the chemicals.

Temperature and the direction of change

Heating decomposes blue copper sulfate crystals to give water and anhydrous copper sulfate, which is white:

$$CuSO_4 \cdot 5H_2O(s) \rightarrow CuSO_4(s) + 5H_2O(l)$$

Add water to the white powder after cooling, and it changes back into the hydrated form. As it does so it turns blue again and gets very hot:

$$CuSO_4(s) + 5H_2O(l) \rightarrow CuSO_4 \cdot 5H_2O(s)$$

Temperature also affects the direction of change in the formation of ammonium chloride. At room temperature ammonia gas and hydrogen chloride gas react to form a white solid, ammonium chloride:

$$NH_3(g) + HCl(g) \rightarrow NH_4Cl(s)$$

Gentle heating decomposes ammonium chloride back into ammonia and hydrogen chloride:

$$NH_4Cl(s) \rightarrow NH_3(g) + HCl(g)$$

Clouds of ammonium chloride form where ammonia and hydrogen chloride gases meet above concentrated solutions of the two compounds.

Key words
➤ irreversible reaction
➤ yield
➤ reversible reaction

Concentration and the direction of change

This equation describes the reaction between iron and steam:

$$3Fe(s) + 4H_2O(g) \rightarrow Fe_3O_4(s) + 4H_2(g)$$

The change from left to right (from reactants to products) is the forward reaction. The change from right to left (from products to reactants) is the backward reaction.

The forward reaction is favoured if the concentration of steam is high and the concentration of hydrogen is low.

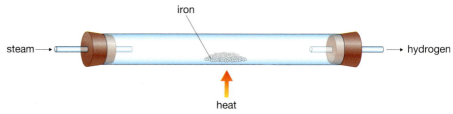

The forward reaction.

The backward reaction is favoured if the concentration of hydrogen is high and the concentration of steam is low:

$$Fe_3O_4(s) + 4H_2(g) \rightarrow 3Fe(s) + 4H_2O(g)$$

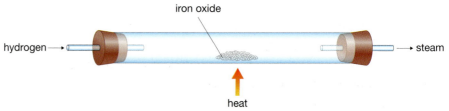

The backward reaction.

Equilibrium

Reversible changes often reach a state of balance, or **equilibrium**. A solution of litmus in water at pH 7 is purple because it contains a mixture of the red and blue forms of the indicator. Similarly, melting ice and liquid water are at equilibrium at 0.8 °C. At this temperature, the two states of water coexist with no tendency for all the ice to melt or all the water to freeze.

When reversible reactions are at equilibrium, neither the forward nor the backward reaction is complete. Reactants and products are present together and the reaction appears to have stopped. Reactions like this are at equilibrium. Chemists use a special symbol in equations for reactions at equilibrium: $\rightleftharpoons$

So at 0.8 °C: $H_2O(s) \rightleftharpoons H_2O(l)$

At equilibrium the reaction may be far to the right (mainly products), far to the left (mainly reactants), or at any point between these extremes.

Water and ice are at equilibrium at the melting point of the ice and the freezing point of the water. Molecules of ice are escaping into the water (melting) and molecules of water are being captured into the surface of the ice (freezing).

Reaching an equilibrium state

A mixture of two solutions of iodine helps to explain what happens when a reversible process reaches a state of equilibrium.

Iodine is slightly soluble in water but is much more soluble in a potassium iodide solution in water. The solution with aqueous potassium iodide is yellow-brown. Iodine is also soluble in organic solvents (such as hexane, a liquid alkane), in which it forms a violet solution. Aqueous potassium iodide and the organic solvent do not mix.

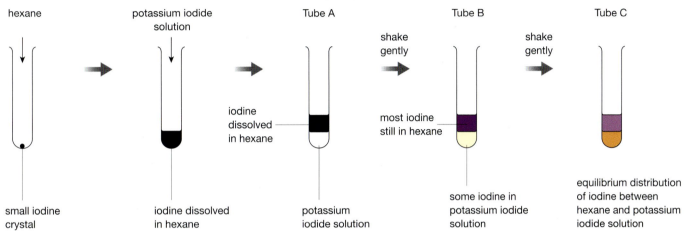

Approaching the equilibrium state starting with iodine in the liquid alkane.

The first graph shows how the iodine concentrations in the two layers change with shaking. In Tube C, the iodine is distributed between the organic and aqueous layers and there is no more change. In this tube there is an equilibrium: $I_2(\text{organic}) \rightleftharpoons I_2(\text{aq})$

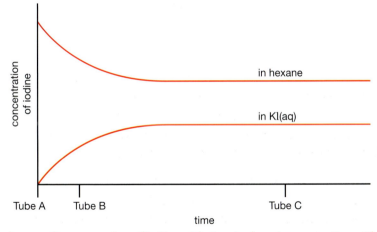

The change of concentration of iodine with time in the mixture, starting with all the iodine in the liquid alkane.

The same equilibrium can be reached starting with all the iodine dissolved in potassium iodide solution rather than hexane.

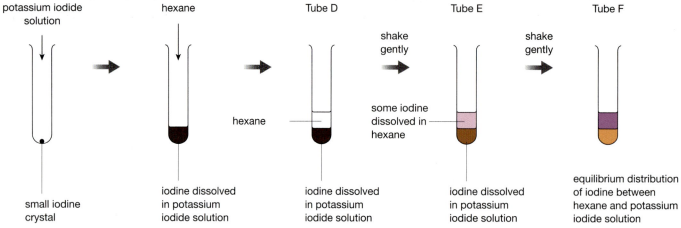

Approaching the equilibrium state starting with all iodine in the aqueous layer.

The second graph shows how the iodine concentration in the two layers changes with shaking.

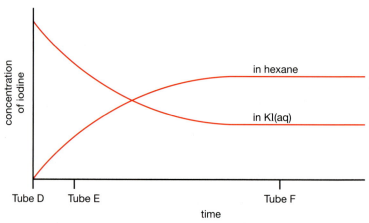

The change of concentration of iodine with time in the mixture, starting with all the iodine in the aqueous layer.

Tube F looks just like Tube C. Tube F is also at equilibrium: equilibrium mixtures in the two tubes are the same. This illustrates two important features of equilibrium processes:

● At equilibrium, the concentrations of reactants and products do not change.

● An equilibrium state can be approached from either the 'reactant side' or the 'product side' of a reaction.

Dynamic equilibrium

The diagram below gives a picture of what happens to the iodine molecules if you shake a solution of iodine in an organic solvent with aqueous potassium iodide (see Tube A in the first graph).

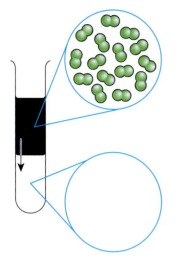 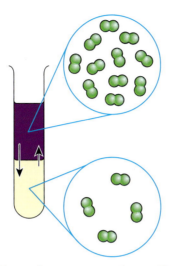 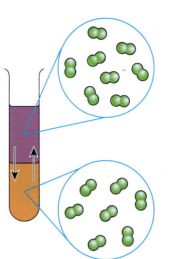

Iodine molecules reaching dynamic equilibrium between two solvents. There would far more molecules than are shown.

All the iodine starts in the upper, organic layer. At first, when the solution is shaken, movement is in one direction (the forward reaction) as some molecules move into the aqueous layer. There is nothing to stop some of these molecules moving back into the organic layer. This backward reaction starts slowly because the concentration in the aqueous layer is low. So to begin with, the overall effect is that iodine moves from the organic to the aqueous layer. This is because the forward reaction is faster than the backward reaction.

As the concentration in the organic layer falls, the rate of the forward reaction goes down. As the iodine concentration in the aqueous layer rises, the rate of the backward reaction goes up. There comes a point at which the two rates are equal. At this point both forward and backward reactions continue, but there is no overall change because each layer is gaining and losing iodine at the same rate. This is **dynamic equilibrium**. It is important to note that when dynamic equilibrium is established there do not have to be equal amounts of reactants and products in the mixture. The position of equilibrium can lie in favour of the reactants, that is, to the left of the equation, or it can lie in favour of the products, that is, to the right of the equation.

$$I_2(\text{organic}) \rightleftharpoons I_2(\text{aq})$$
$$\text{left-hand side} \quad \text{right-hand side}$$

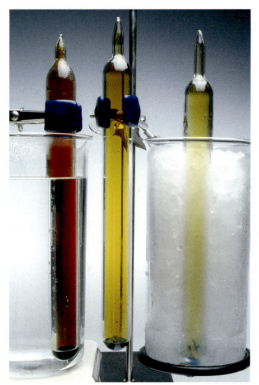

Tubes containing a mixture of the nitrogen compounds at different temperatures.

Factors affecting the position of equilibrium

Data about the yield and rate of chemical processes is used to choose the best condition to make a product. The **position of equilibrium** can be changed by changing the experimental conditions of temperature, concentration, and, where gases are involved, the pressure.

The effect on an equilibrium mixture of changing conditions can be predicted using **Le Chatelier's principle**. It states that:

'When the conditions change, an equilibrium mixture of chemicals responds in a way that tends to counteract the change'.

The presence of a catalyst only affects the rate of reaction and not the yield, as it does not affect the position of equilibrium.

Dinitrogen tetroxide dissociates to nitrogen dioxide in an endothermic reaction. The two gases exist in equilibrium; the proportions of each gas will depend on the conditions. The position of equilibrium can be altered by changing the pressure or temperature of the system.

$$N_2O_4(g) \rightleftharpoons 2NO_2(g)$$
$$\text{colourless} \qquad \text{dark brown}$$

The position of the equilibrium shifts to try and cancel out any changes you introduce. So increasing the pressure moves the equilibrium to the left as this decreases the number of gaseous particles present in the mixture, that is, it favours N_2O_4. Increasing the temperature moves the position of equilibrium to the right as this puts energy into the system, that is, it favours the endothermic production of NO_2.

Changing the temperature of the tubes containing a mixture of the nitrogen compounds determines which way the reaction goes and ultimately the position of equilibrium. When the tube is warmed, more of the brown NO_2 gas forms and the gas in the tube becomes darker, as the position of the equilibrium moves to the right. When the tube is cooled, more of the colourless N_2O_4 gas is formed and the gas in the tube becomes paler, as the position of equilibrium shifts to the left.

Ammonia is produced on an industrial scale in the **Haber process** by the reaction of nitrogen and hydrogen gas.

$$N_2(g) + 3H_2(g) \rightleftharpoons 2NH_3(g)$$

The chemical engineers running the ammonia plant need to choose the reaction conditions carefully to try and make the process as safe and efficient as possible.

From looking at the data presented in the graph opposite, a combination of high pressures and low temperatures produce high yields of ammonia. The chemical engineers must also take into consideration the economic implications of the reaction conditions. Very high pressures are expensive to maintain because the equipment may fail under extreme conditions. If the temperature is too low the rate of reaction will be very slow, thus taking

H a long time to produce the product. So in practice they have to come up with a compromise. The chosen conditions for the Haber process are:

- a pressure between 150 and 300 atmospheres
- a temperature between 400 and 450 °C
- an iron catalyst.

The equilibrium mixture contains about 15% ammonia, which is constantly drawn off by cooling it down to a liquid. The unused gases are recycled.

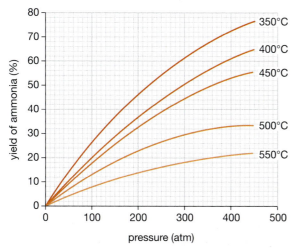

Graph showing how the percentage of ammonia at equilibrium changes with temperature and pressure.

Questions

1 Write a symbol equation to show:
   a water turning into steam
   b steam condensing to water
   c the changes in parts **a** and **b** as a single, reversible change.

2 The pioneering French chemist Lavoisier heated mercury in air and obtained the red solid, mercury oxide. He also heated mercury oxide to form mercury and oxygen.
   a How can both of these statements be true?
   b Why should you not try to repeat the experiment?

3 a Write an equation to show the reversible reaction of carbon monoxide gas with steam to form carbon dioxide and hydrogen.
   b In your equation, what happens in the forward reaction?
   c In your equation, what happens in the backward reaction?

4 Under what conditions are the following in equilibrium?
   a water and ice
   b water and steam
   c salt crystals and a solution of salt in water

5 a Why do iron and steam not reach an equilibrium state when they react as shown in the forward-reaction diagram in C6.3A?
   b Suggest conditions in which a mixture of iron and steam would react to reach an equilibrium state.
   c Which chemicals would be present in an equilibrium mixture of iron and steam?

6 Explain in your own words what is meant by the term 'dynamic equilibrium'.

H 7 When ammonia is manufactured by the Haber process, a yield of only 15% is achieved. Explain why the conditions are not changed to increase this yield.

A: Fertilisers and the chemical industry

Find out about

- essential elements for plant growth
- synthetic and natural fertilisers
- production of ammonia by the Haber process
- the synthesis of ammonium sulfate

Key words

➤ synthetic fertilisers
➤ eutrophication

Essential elements for plant growth

Nitrogen, phosphorus, and potassium are essential plant-nutrient elements. Nitrogen is needed for making protein in the leaves and stalks. Phosphorus speeds up the growth of roots and the ripening of fruits. Potassium protects plants against disease and frost damage, and also promotes seed growth. All these elements are found in the soil naturally as inorganic compounds. However, as crops grow and are eventually harvested, these essential nutrients are lost from the soil and must be replaced by fertilisers.

There are two types of fertiliser: organic fertilisers and inorganic fertilisers. Organic fertilisers are made from rotting and decaying organic matter such as plants and animals. They consist of large organic molecules that are slowly broken down over time. Inorganic fertilisers are made from naturally occurring minerals or manufactured inorganic compounds. These fertilisers contain higher concentrations of the nutrients and work faster than organic fertilisers.

Traditionally, farmers used natural organic fertilisers, including manure and animal waste, but today many farmers use fertilisers manufactured by the chemical industry, that is, **synthetic fertilisers**. This is because:

- natural fertilisers are not available in large enough quantities to meet the world demand for food
- the exact composition of natural fertilisers varies from batch to batch, making land management more difficult
- natural fertilisers are difficult to transport around the country.

Synthetic fertilisers: benefits and costs

Ammonia is one of the most important compounds used to make synthetic fertilisers. The demand for fertilisers (and explosives during the First World War) led to the development of chemical processes to 'fix' nitrogen, to produce nitrogen compounds such as ammonia, nitric acid, and nitrates. The production of ammonia by the Haber process has increased significantly over the past 70 years. More fertilisers, such as ammonium nitrate, are being produced to achieve the increase in food production that is needed to support the world's growing population.

The ability to fix nitrogen has affected society and has also had an impact on the environment. The increased availability of fertilisers has led to changes in land use. Less land is needed to provide food for more people so larger towns and cities can be supported. The fixing of nitrogen also affects the natural nitrogen cycle. For example, the overuse of fertilisers such as ammonium nitrate can lead to excess concentrations of nitrogen compounds being washed into the rivers by rain.

Synoptic link

You can learn more about how the Haber process changed World War I in C7.3 *Are all applications of science ethically acceptable?*

This can lead to increased growth of algae and excessive weed growth, which in turn upsets ecosystems as living organisms die. This process is known as **eutrophication**. Nitrates can also get into drinking water and be harmful to human health.

The label on the fertiliser bag usually shows the ratio of nutrients present. The proportions are always shown in the order nitrogen:phosphorous:potassium (N:P:K). The numbers shown are the ratio of masses the compounds present and not the actual elements. Sometimes two figures appear on the label, giving the ratio of masses of both the compounds and the elements present.

The Haber process

The basis of the Haber process is a reversible reaction between nitrogen and hydrogen gas. Nitrogen is obtained from the air, and the main feedstock for hydrogen is natural gas or methane.

$$\text{nitrogen} + \text{hydrogen} \rightleftharpoons \text{ammonia}$$
$$N_2(g) + 3H_2(g) \rightleftharpoons 2NH_3(g)$$

H The reactant gases are compressed to about 200-times atmospheric pressure, heated to about 450 °C, and passed over an iron catalyst. Haber and Bosch systematically tested about 20 000 catalysts, before finding the right one. Finally they found an iron ore containing traces of alkali metals that worked.

In this reversible reaction, there has to be a compromise between a high yield and a high rate of reaction. These ideas were discussed in the previous section as the supporting data was explored. To get the greatest output as quickly and economically as possible, chemical engineers must consider a number of different factors including:

- sourcing of raw materials
- production of feedstocks
- the rate of reaction
- the position of equilibrium
- separation of ammonia
- recycling of unreacted nitrogen and hydrogen.

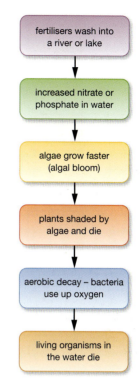

fertilisers wash into a river or lake

↓

increased nitrate or phosphate in water

↓

algae grow faster (algal bloom)

↓

plants shaded by algae and die

↓

aerobic decay – bacteria use up oxygen

↓

living organisms in the water die

Flowchart showing the process of eutrophication.

An algae-covered lake showing signs of eutrophication.

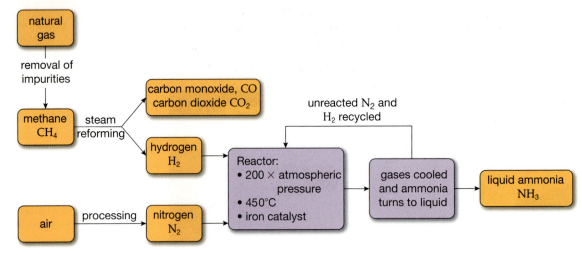

Flowchart showing the synthesis of ammonia.

As shown in the flowchart, in the Haber process the mixture of gases flows continuously through the **reactor**. The gases are only in contact with the catalyst for a short time. This means that the mixture never gets all the way to equilibrium. Unreacted reactants are continuously separated from the ammonia and recycled so that the nitrogen and hydrogen are not wasted.

Getting the conditions right

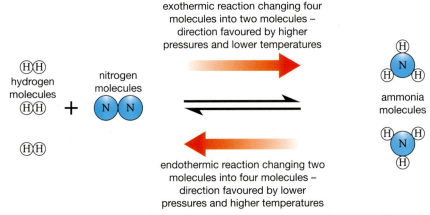

The position of the equilibrium is affected by temperature and pressure. The conditions used for the Haber process are a compromise to achieve a good yield at reasonable cost.

In the reaction to make ammonia there are four molecules of gas on the left-hand side of the equation but only two molecules of gas on the right. An equilibrium mixture of the gases responds to an increase in pressure by changing to make more ammonia and less nitrogen and hydrogen. This is what Le Chatelier's principle predicts because reducing the number of molecules tends to lower the pressure.

The reaction of nitrogen with hydrogen is exothermic. Le Chatelier's principle predicts that if the temperature of an equilibrium mixture of the gases rises, then the equilibrium changes in a way that takes in energy because this tends to lower the temperature. So at a higher temperature there is less ammonia and more nitrogen and hydrogen in the equilibrium mixture.

Key words

➤ reactor
➤ sustainability

H The conditions chosen for the Haber process in industry are a compromise that balances chemical efficiency with cost and safety. The higher the pressure, the higher the yield of ammonia as the gas mixture approaches an equilibrium state. But high-pressure plants are expensive to build and run. They can also be more hazardous for plant operators. The lower the temperature, the higher the possible yield of ammonia, but the reaction becomes too slow to be economic.

Synoptic link

You can learn more about the Haber Process in C6.3 *What factors affect the yield of a reaction?*

Production of ammonium sulfate

Ammonium sulfate is a commonly used fertiliser. Like many synthetic fertilisers it is a salt and can be made in the laboratory by neutralising sulfuric acid with ammonia solution.

sulfuric acid + ammonia solution $\rightarrow$ ammonium sulfate

$$H_2SO_4(aq) + 2NH_3(aq) \rightarrow (NH_4)_2SO_4(aq)$$

On an industrial scale ammonium sulfate is produced by several different processes including:

- by reacting anhydrous ammonia and sulfuric acid
- as a by-product of coke ovens, by reacting ammonia from coke-oven gas with sulfuric acid
- as a by-product of the production of caprolactam.

In order to meet the demand for fertilisers due to the increasing world population, it is necessary for fertilisers to be produced on an industrial scale by several different processes. At the same time, **sustainability** remains an important issue for the chemical industry.

Caprolactam is used in the production of nylon 6 fibres and nylon 6 engineering resins and films. Nylon 6 fibres are used for textiles, carpets, and industrial yarns. The engineering resins are used for film food-packaging products, wire, and cabling, as well as the automotive industry.

So, where possible, scientists try to use by-products or waste products from one process as a reactant or feedstock for another. The following flowchart summarises the industrial processes used in the preparation of ammonium sulfate. A **continuous flow** approach is used as it is more economically viable to keep the plant running rather than producing individual **batches** of the product. Lab processes prepare chemicals in batches; industrial processes are usually continuous.

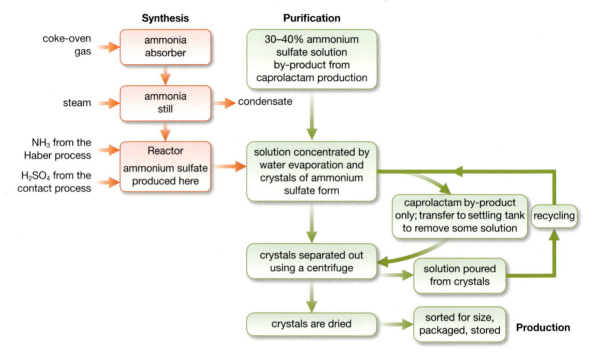

The industrial process for making ammonium sulfate.

Laboratory preparation of ammonium sulfate

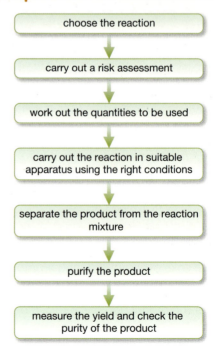

Flowchart summarising the laboratory production of a salt such as ammonium sulfate.

The laboratory process for making a salt is summarised in the flowchart. You will notice that there are many similarities to the industrial process. Preparing a batch of ammonium sulfate in the laboratory follows a similar practical procedure to making magnesium sulfate; the main differences are in the initial steps of the reaction since both reactants are solutions.

Synoptic link

You can learn more about the full procedure for making a salt in C8P *Making a sample of a salt*.

Questions

1 Refer to the flowchart for the industrial process of making ammonium sulfate.
   a What are the main feedstock sources for the production?
   b Identify steps taken to make the yield of the pure salt as large as possible.

2 What are the main differences between the industrial and laboratory production of ammonium sulfate?

3 Why is it necessary for farmers to use fertilisers?

4 Suggest possible consequences for the environment of the large-scale manufacture of ammonia.

H
5 In the Haber process there is a continuous flow of reactants through the reactant chamber, rather than batches of reactants. Explain:
   a why this has an advantage regarding the amount of reactants used
   b how the ammonia is separated from the flow of reactants.

6 Suggest reasons why the Haber process can become uneconomic if the operators try to increase the yield of ammonia by:
   a making the pressure even higher
   b lowering the temperature.

7 At a pressure of 200 times atmospheric pressure and at 450 °C, an equilibrium mixture of nitrogen, hydrogen, and ammonia contains about 40% ammonia. In an industrial plant working under these conditions the mixture of gases leaving the reactor is only about 15% ammonia. Explain why.

B: Green chemistry

Find out about

- atom economy
- **H** green chemistry and sustainability of industrial processes

Key words

➤ atom economy
H ➤ sustainability

Atom economy

In 1998, Barry Trost of Stanford University, US, was awarded a prize for his work in green chemistry. He introduced the term **atom economy** as a measure of the efficiency with which a reaction uses its reactant atoms. This is a theoretical value based on the reaction equation.

$$\text{atom economy} = \frac{\text{mass of atoms in the useful product}}{\text{mass of atoms in the reactants}} \times 100\%$$

In an ideal chemical reaction, all the atoms in the reactants would end up in the useful products, and no atoms would be wasted. If this was the case the atom economy for the reaction would be 100%. The atom economy for the Haber process is in fact 100%, since there are no by-products. From the equation

$$N_2(g) + 3H_2(g) \rightleftharpoons 2NH_3(g)$$

it is easy to see that all the starting atoms end up in the ammonia molecules. For most reactions it is necessary to carry out the full calculation to determine the atom economy of the reaction.

Worked example: Atom economy

What is the atom economy for the thermal decomposition of calcium carbonate, producing calcium oxide (quicklime)?

$$CaCO_3(s) \rightarrow CaO(s) + CO_2(g)$$

Step 1: Use the Periodic Table to find the relative atomic mass for each element.

$Ca = 40, C = 12, O = 16$

Step 2: Calculate the total relative formula mass of the reactants.

$CaCO_3 = (1 \times 40) + (1 \times 12) + (3 \times 16) = 100$

Step 3: Calculate the relative formula mass of the useful product.

CaO is the useful product

$CaO = (1 \times 40) + (1 \times 16) = 56$

Step 4: Use the equation to calculate the atom economy.

Remember that 1 mole of a substance has a mass equal to its relative atomic or formula mass in grams.

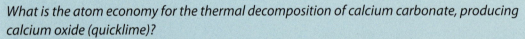

$$\text{atom economy} = \frac{\text{mass of atoms in the useful product}}{\text{mass of atoms in reactant}} \times 100\%$$

$$= \frac{56 \text{ g}}{100 \text{ g}} \times 100\%$$

Answer:

atom economy = 56%

With the very best conditions for the thermal decomposition of calcium carbonate, just over half of the mass of starting materials can end up as product. So this is not a green process. This approach does not take yield into account and does not allow for the fact that many real-world processes use a deliberate excess of reactants. For example, in many neutralisation reactions, such as the reaction between magnesium carbonate and sulfuric acid, the carbonate is in excess. The atom economy is used alongside data about yields and efficiency when processes are evaluated for sustainability.

Sustainability

Modern processes incorporate green chemistry ideas, to provide a sustainable approach to production. Sustainability is a measure of how a process is able to meet current demands without having a long-term impact on the environment. Reactions with a high atom economy are more sustainable that those with a low atom economy. Other issues that affect the sustainability of a process include:

- the nature and amount of by-product or wastes
- the energy inputs or outputs
- whether or not raw materials are renewable
- the impact of other competing uses for the same raw materials.

Can the production of ammonia be more sustainable in the future?

The production of ammonia is a relatively clean process. The only emissions are carbon dioxide and oxides of nitrogen. In a modern plant, both of these gases can be recovered or reduced to very low levels. The atom economy is 100% but the yield is only 15%. The reaction yield is increased by recycling the unreacted nitrogen and hydrogen gas.

Energy use

A major problem with the production of ammonia is the amount of energy needed. More than 1% of all the energy consumed in the world is used for ammonia production. The energy needed to operate the process has decreased over the past 100 years. This is mainly due to the use of more efficient catalysts that allow the reaction to take place at lower temperatures and pressures. Using renewable energy sources for ammonia production would also reduce the amount of greenhouse gases entering the atmosphere.

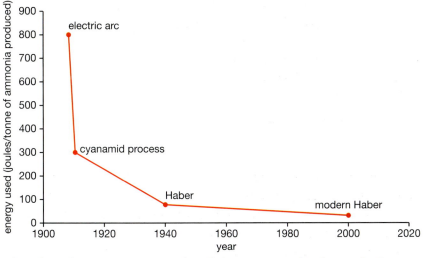

When the Haber process was introduced, the energy used in the production of ammonia dramatically decreased. Application of the principles of green chemistry in more recent years has ensured there is still a downward trend.

Future feedstocks

Currently, hydrogen is extracted from methane by steam re-forming. If fossil-fuel sources of methane run low, alternative ways of producing hydrogen will be needed. The **electrolysis** of water could become a major source of hydrogen. This feedstock would depend on a cheap and renewable electricity supply, such as hydroelectricity or solar power.

New catalysts

The search for new catalysts is an important area of current research and development. The higher the catalytic activity of a catalyst, the more efficient it is at synthesising ammonia.

The Kellogg Advanced Ammonia Process (KAAP)

In 1992, M W Kellogg and the Ocelot Ammonia Company started ammonia production using a new ruthenium catalyst deposited on an active carbon support. With this catalyst a pressure of 40 times atmospheric pressure can be used for ammonia production, instead of 200 times atmospheric pressure.

The ruthenium catalyst is more active than the original iron-based catalyst and yields of about 20% ammonia can be achieved.

This new technology can be fitted to existing ammonia plants, saving money and energy. The newer catalyst is more expensive, but this is outweighed by other cost savings.

Computer-generated image of a nitrogen molecule held on a layer of iron atoms at the surface of the iron catalyst used for making ammonia. Nitrogen and hydrogen molecules react when brought together on the catalyst surface. Replacing iron atoms by ruthenium atoms gives a more effective catalyst.

The Haldor–Topsøe catalyst

In 2000, scientists at a Danish company announced the discovery of a new commercially viable catalyst for the Haber reaction. The new compounds contain iron, molybdenum, nitrogen, nickel, and cobalt. They appear to be two or three times more efficient than the current commercial, iron-based catalysts at the same operating conditions.

H The new catalysts are cheaper than the ruthenium-based catalysts of the KAAP. The same Haldor–Topsøe team have also produced new ruthenium catalysts that are 2.5 times more active than current ruthenium catalysts.

Learning from nature

Chemists are keen to learn about **nitrogen fixation** from nature. Studies of the enzyme nitrogenase have shown that it contains clusters of iron, molybdenum, and sulfur.

Chemists have been successful in making similar artificial clusters that show catalytic activity. By producing and using new catalysts that mimic natural enzymes, it may be possible in the future to produce ammonia at room temperature and pressure. This, of course, would lead to even lower energy use during production.

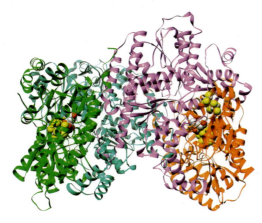

A computer-generated image representing the enzyme nitrogenase. The ribbons represent the enzyme molecule, and the coloured spheres show the position of the iron-molybdenum clusters (in the red-and-pink part of the enzyme on the right) and other non-protein clusters (in the blue-and-green part on the left). These are all important in nitrogen fixation.

When we consider the world-wide use of ammonia it is clear that the search for even more sustainable production methods is worth continuing.

Questions

1 16 g of methane (CH_4) was burned in the air. 32 g of carbon dioxide was collected during the reaction. During the reaction, some sooty deposits were noticed. The equation for the combustion of methane is:

$$CH_4 + 2O_2 \rightarrow CO_2 + 2H_2O$$

H **a** What was the percentage yield of carbon dioxide?
   b Why do you think the percentage yield was so low?
   c Calculate the atom economy for the reaction.

H 2 The enzyme nitrogenase fixes nitrogen at a pressure of 1 atmosphere, the KAAP ruthenium catalyst fixes nitrogen at 40 times atmospheric pressure, and the iron catalyst of the Haber process fixes nitrogen at 200 times atmospheric pressure.
Explain why processes that give a good yield at a lower pressure are more sustainable.

Science explanations

C6 Making useful chemicals

Chemists use their knowledge of chemical reactions to plan and carry out the synthesis of new substances.

You should know:

- the characteristic reactions of acids with metals, metal oxides, metal hydroxides, and metal carbonates that produce salts
- the practical procedures for making salts
- how chemists use indicators and pH meters to detect acids and alkalis and to measure pH
- **H** the correct use of the terms weak, strong, concentrated, and dilute in relation to acids and their reactivity
- how to use the relationship between the pH of a solution and its concentration of hydrogen ions
- how to follow the rate of a change by measuring the disappearance of a reactant or the formation of a product, and how to analyse the results graphically
- that the concentrations of reactants, the particle size of solid reactants, the temperature, the pressure, and the presence of catalysts are factors that affect the rates of reaction
- how collision theory can explain why changing the concentration of reactants or the particle size of solids, affects the rate of a reaction
- how the idea of activation energy can explain the action of catalysts
- how to interpret and process graphs and numerical data for reaction rates
- how enzymes act as biological catalysts
- that reversible reactions can reach a state of dynamic equilibrium
- **H** how changing reaction conditions can change the equilibrium position
- why it is important that there are natural and artificial ways to fix nitrogen **S**
- **H** how and why the conditions for the Haber process are chosen to give the optimum yield
- why it is desirable to find new ways to manufacture ammonia.
- the main stages in the industrial production of fertilisers and compare it with laboratory processes.

Ideas about Science

In the context of laboratory synthesis of new substances, you should be able to:

- suggest appropriate equipment, materials and techniques, justifying the choice with reference to the purity of the sample to be made
- construct a clear and logically sequenced procedure
- identify hazards associated with the procedure.

Science helps us find ways of using natural resources in a more sustainable way. You should be able to:

- identify benefits and costs of making chemicals such as ammonia and ethanol
- suggest reasons why the choice of method for making a chemical depends on the social or economic context
- describe how the principles and practices of green chemistry contribute to sustainable development
- use data, such as atom economies and yields, to compare the sustainability of alternative processes
- explain the importance of regulations to control the chemical industry and the uses of its products.

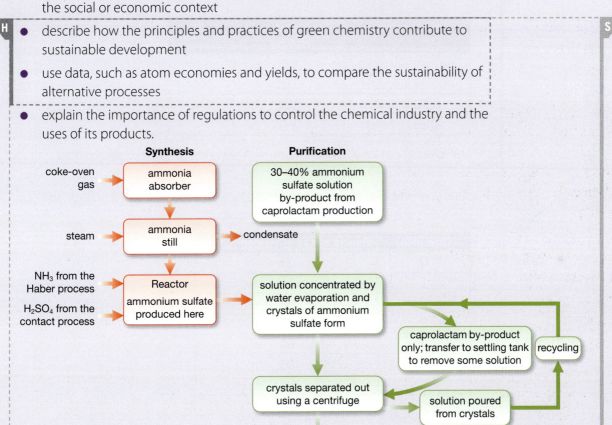

The industrial process for making ammonium sulfate.

C6 Review questions

1. The table provides information about four reactions, A, B, C and D. In each reaction an acid reacts with a metal compound to make a salt.

Reaction	Name of acid	Name of metal compound	Name of salt formed
A	sulfuric acid	sodium hydroxide	
B		copper oxide	copper nitrate
C	hydrochloric acid	calcium carbonate	calcium chloride
D	ethanoic acid		sodium ethanoate

a Copy and complete the table by filling in the empty cells.

b What is the other product of reaction A?

c Write a symbol equation, with state symbols for reaction C.

 2. The table shows the concentration and pH of four acidic solutions.

Acid in solution	Concentration in mol/dm³	pH
hydrochloric acid	0.1	1.0
hydrochloric acid	0.01	2.0
sulfuric acid	0.01	1.7
ethanoic acid	1.0	2.4

Use ideas about acid concentration and acid strength to explain the data in the table.

3. A teacher adds 3 g of zinc granules to dilute hydrochloric acid in a flask. She measures the volume of hydrogen gas given off.

Time (min)	Volume of gas in syringe (cm³)
0	0.0
1	32
2	56
3	74
4	87
5	95
6	95

a Describe one way the teacher could have measured the volume of gas. Draw a diagram to support your answer.

b Draw a line graph of the results.

c i Draw a tangent to the graph at 3.5 minutes.

   Use your tangent to calculate the rate of reaction. Show on the graph what measurements you made and show your working.

   ii How would you expect the rate of reaction at 3.5 minutes to compare to the rate at 2.0 minutes? Explain your reasoning.

d The teacher decides to repeat the experiment with 3 g of zinc powder. She keeps all other factors the same.

   i Draw and label a line on your graph to show the expected results.

   ii Use collision theory to explain the effect of this change.

4 Sulfuric acid is made in a very large scale process.

One reaction involved is the reaction between sulfur dioxide and oxygen to make sulfur trioxide.

$$2SO_2(g) + O_2(g) \rightleftharpoons 2SO_3(g)$$

a This reaction reaches *dynamic equilibrium*.

Describe what is happening to the molecules at dynamic equilibrium.

H b The yield and the rate for this reaction are both improved if high pressures are used.

Explain the disadvantages of using very high pressures.

c Suggest what other conditions could be used to make the rate of reaction as fast as possible.

5 Every year, UK chemical companies produce more than 1 million tonnes of ammonia by the Haber process.

The equation for the Haber process reaction is:

$$N_2(g) + 3H_2(g) \rightleftharpoons 2NH_3(g)$$

a Joe and Eve talk about the Haber Process.

i Explain how the equation shows that this reaction has an atom economy of 100%.

ii Do you agree with Eve's statement? Explain your reasoning.

b The nitrogen and hydrogen for the process are extracted from raw materials.

Gas	Raw material	Extraction process
nitrogen	air	fractional distillation
hydrogen	methane and steam	high temperature reaction

i Suggest the names of **two** by-products that are produced during the extraction of nitrogen.

 ii Evaluate the sustainability of the extraction processes for nitrogen and hydrogen.

c Ammonia factories are very large. They are often positioned near towns.

The speech bubbles show two different views about ammonia factories.

We can't live without ammonia factories and it is energy efficient to put them near towns.

Ammonia factories are unsightly and use a lot of energy. I think we should stop the production of ammonia.

Evaluate the two views and make a judgment about whether you think ammonia production should continue or be stopped.

C7 Ideas about Science

Why study Ideas about Science?

In order to make sense of the scientific ideas that we encounter in everyday life, we need to understand how science explanations are developed, the kinds of evidence and reasoning behind them, their strengths and limitations, and how far we can rely on them.

We also need to think about the impacts of science and technology on society, as well as how we respond to new ideas, artefacts, and processes that science makes possible.

What you already know

- Science explanations are based on evidence, and as new evidence is gathered, explanations may change.
- How to plan and carry out scientific enquiries, choosing the most appropriate techniques and equipment.
- How to collect and analyse data, and draw conclusions.

Case studies

In this chapter some of the Ideas about Science that you have studied across the course are explored in different contexts.

C7.1 How could an increase in carbon dioxide levels affect ocean life?

Find out about how scientists are investigating the possible effects of rising carbon dioxide levels on marine life.

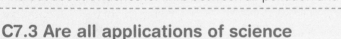

C7.2 How does air quality affect our health?
Find out about evidence for links between air pollution and illness.

C7.3 Are all applications of science ethically acceptable?

Find out about the work of scientist Fritz Haber, and some questions that can be answered using science and some that can't.

C7.1 How could an increase in carbon dioxide levels affect ocean life?

Find out about

- developing an experiment to test the effects of rising carbon dioxide levels on ocean life
- identifying factors that may affect the outcome and factors that need to be controlled
- using data to justify a conclusion
- explaining the extent to which data increases or decreases confidence in a prediction or hypothesis

Key words

➤ carbon sink
➤ ocean acidification

As human activities release more carbon dioxide into the Earth's atmosphere, more carbon dioxide dissolves in the ocean. Without this **carbon sink**, atmospheric carbon dioxide levels – and therefore climate change – would be a lot worse.

However, when carbon dioxide dissolves in the ocean there is an increase in H^+ ions. This decreases the pH of the seawater. This process is referred to as **ocean acidification**.

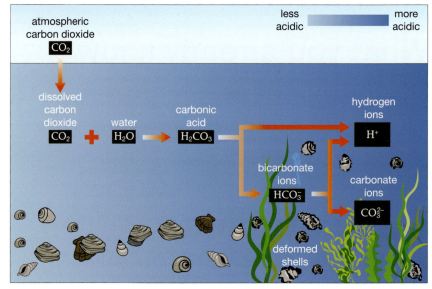

Carbon dioxide dissolves in the ocean, leading to an increase in H^+ ions. This is ocean acidification.

Synoptic link

You can learn more about pH in C6.1B *How strong is an acid*?

Ocean pH is thought to have dropped from 8.4 to 8.3 since the industrial revolution. It is expected to drop by another 0.3–0.4 by the end of this century. A drop of 0.1 may not seem like a big change but remember that if the pH of a solution decreases by 1.0, it is 10-times more acidic.

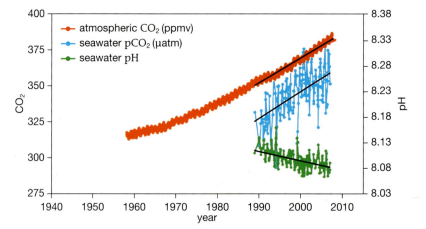

The graph shows a correlation between carbon dioxide (CO_2) in the atmosphere, CO_2 in the seawater, and the pH of the seawater off the coast of Hawaii.

The impact on coral reefs

Underwater ecosystems are just as important as those on land. Coral reefs have been called 'the rainforests of the sea', and support thousands of species.

Even slight changes in ocean pH can affect the marine organisms that live there. A coral reef is an underwater ecosystem, home to thousands of species. The reef is a solid structure made from calcium carbonate. The living coral is found inside.

In order to make the reef, the coral combines calcium ions (Ca^{2+}) and carbonate ions (CO_3^{2-}) to make calcium carbonate. However, H^+ ions can combine with the carbonate ions to make bicarbonate (HCO_3^-). Coral cannot use bicarbonate ions to build the reef. So the growth of coral may be slowed by an increase in H^+ ions. If the pH changes too much, the coral may not survive. If the reef disappears, this will endanger the survival of many species.

Coral and seaweed

Coral and seaweed compete for space and light within a reef ecosystem. Seaweed can attach to coral and each can affect the other's growth. An increase in seaweed growth causes a decrease in coral growth. With less coral to make calcium carbonate, the reef can start to break down.

Seaweed and coral and compete for space and light on the Great Barrier Reef.

Scientists at Heron Island Research Station investigate the many aspects of the marine environment, including the effects of ocean acidification and ocean warming on the reef ecosystem.

Key words

➤ hypothesis
➤ prediction

A hypothesis and predictions

Seaweed uses carbon dioxide in photosynthesis. So we might expect that an increase in carbon dioxide would increase the rate of growth of seaweed. A team of scientists devised a hypothesis to test:

An increase in carbon dioxide will cause an increase in the amount of seaweed relative to coral. This is because seaweed growth will increase due to increased photosynthesis, and coral growth will decrease due to the increase in H+ ions and increased competition from the seaweed.

A **hypothesis** is a tentative explanation of an observation. It can be used to make a **prediction** about how, within a planned experiment, a change in one factor will affect another.

The scientists tested their hypothesis in a laboratory on Heron Island in the Great Barrier Reef, Australia. They planned a variety of measurements and made a prediction for each one.

Measurement	Prediction
Coral survival (percentage of coral surviving each day)	The percentage survival of coral will decrease as the levels of carbon dioxide increase.
Coral growth	The higher the level of carbon dioxide, the less the coral will grow.
Seaweed growth	The higher the level of carbon dioxide, the more the seaweed will grow.

The experiment

The scientists collected samples of a single species of coral and a single species of seaweed from the reef. There were 20 samples of:

● live coral branches without seaweed attached

● live coral with seaweed attached

● dead branches of coral with seaweed attached.

A control was created by attaching artificial seaweed to live coral branches.

The samples were grown in tanks of seawater with different pH levels. The pH of the seawater in the tank was monitored and controlled by adjusting the carbon dioxide input. Different tanks represented different periods in the Earth's history.

Carbon dioxide level (parts per million)	pH	Period
300	8.10–8.20	pre-industrial
400	8.00–8.10	present day
560	7.85–7.95	predicted* mid-century
1140	7.60–7.70	predicted* end of century
*Predictions based on work by the International Panel on Climate Change.		

The light level was one factor that could have an effect on the outcome of the experiment. This was controlled using shading screens to maintain a mean midday level.

Two tanks were used for each carbon dioxide level; each contained numerous samples of coral and seaweed.

The results

Scientists plotted a graph to show the percentage of coral branches that still survived each week. They only tested the control coral (with artificial seaweed attached) at one level of carbon dioxide, but the results were very similar to the coral without seaweed (where almost all the coral survived).

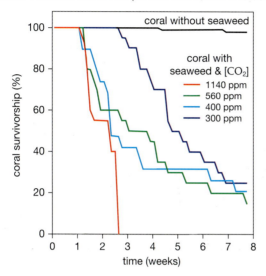

The graph shows how seaweed and carbon dioxide affect the survival of coral over eight weeks. The line for coral without seaweed is the average of the survival rate of isolated corals with no seaweed across all CO_2 treatments.

The scientists used bar graphs to compare the growth of coral with seaweed, seaweed alone, and coral alone at different levels of carbon dioxide.

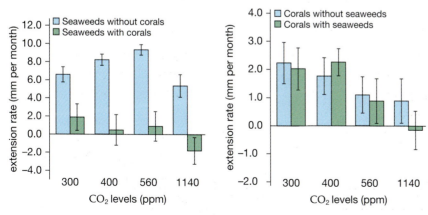

The graphs show how the level of carbon dioxide in the water affected the growth of seaweed and of coral.

The scientists concluded that ocean acidification is a problem for coral reefs because, in the changing conditions, the seaweed grew at the expense of the coral. They acknowledge that they only looked at one species of seaweed and one species of coral, but this study provides a reminder of the interdependence between species in any ecosystem.

Questions

The experiment

1 The scientists made a number of decisions when planning the experiment.

   a Why did they use the chosen range of carbon dioxide levels?

   b Write down one other factor that could affect the outcome, and describe how the scientists controlled it.

   c Why did they use two tanks for each carbon dioxide level, and why did each tank contain numerous coral samples?

The results

2 Look at the graph showing the percentage of coral that survived.

   a Was the coral without seaweed affected by increasing levels of carbon dioxide?

   b Which level of carbon dioxide had the most dramatic effect on the percentage of coral surviving?

   c Describe the difference and similarity shown by the lines plotted for the other levels of carbon dioxide.

   d Did these results agree with the predictions?

3 The scientists concluded the seaweed slows the growth of coral by some sort of biological or chemical means but not physical. Describe and explain the control that the scientists used to help them draw this conclusion.

4 Look at the bar graphs showing the growth of seaweed and coral. The error bars are shown.

   a What can be concluded about the effect of increasing levels of carbon dioxide on:

      i   the growth of seaweed?

      ii the growth of coral?

   b To what extent do these results agree with the predictions?

5 To what extent do the results support the scientists' hypothesis?

6 The scientists made some comments about their research.

   ● Coral reefs are made up of a mixture of different species of coral and seaweed whereas this experiment used only one species of coral and one of seaweed.

   ● The results were consistent with studies by other scientists of a type of red seaweed.

   ● The results were not consistent with studies of a Caribbean species of seaweed.

   Explain how each of these comments would affect your answer to question 5. Would it make you more or less confident about whether the hypothesis was supported?

Is there a link between hayfever and pollen?

Pollen grains under the microscope. Different plants release different types of pollen.

Pollen is released by plants and may travel many kilometres on the wind. Pollen grains are in the air that we breathe.

Hayfever

Do you suffer from a runny nose, sneezing, and itchy eyes in the summer? This could be hay fever.

Hayfever got its name because people noticed that it happens in the summer. This is when grass is being cut to make hay. It is also the time when pollen from plants is at its highest.

To find out what causes hayfever it is important to first look at what factors are linked with hayfever. Factors are variables that may affect the outcome. In this case, hay fever is the outcome. Pollen is a factor that may affect hayfever.

Pollen traps collect pollen grains so that they can be counted using a microscope. This gives the 'pollen count'. Newspapers, radio, and television report the pollen count the summer.

Key words

➤ correlation
➤ causal link
➤ factor
➤ outcome

Is there a correlation?

If an outcome increases (or decreases) as a particular factor increases, this is called a **correlation**. So, do more people have hayfever symptoms when the pollen count increases?

Looking at thousands of people's medical records shows that hayfever is highest in the summer months. This is when most pollen is in the air. It is important to look at a randomly selected sample of medical records to collect data that is representative of the whole population.

This evidence shows that there is a correlation between pollen levels and hay fever symptoms. But does this mean that there is a **causal link** between pollen and hayfever – is pollen the cause of hay fever?

Is there a causal link?

An increase in two things could be caused by a third factor that has not been measured. Or it could be a coincidence that the two things increase at the same time.

Think about ice cream. Most ice cream is sold in the summer months but nobody would say that ice cream causes hayfever. It may be just a coincidence. Or both increases may be caused by some other factor.

To claim that pollen causes hayfever you need some supporting evidence. You need to show that there is a correlation and be able to explain how pollen causes hayfever.

Different types of pollen are released at different times of year. Some people have hayfever in months that correlate with particular types of pollen being released. This is strong extra evidence for the correlation between pollen and hayfever.

Skin-prick tests show that people who suffer from hayfever are allergic to pollen. Hayfever is an allergic reaction caused by pollen. So there is a causal link between hayfever and pollen, because pollen causes hayfever.

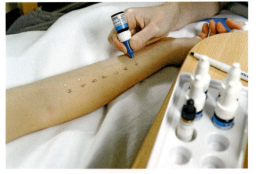

During a skin-prick test, drops of solution are placed on the skin. The skin beneath is pricked. If the patient is allergic to the substance in the solution (e.g., pollen) their skin will turn red and itchy.

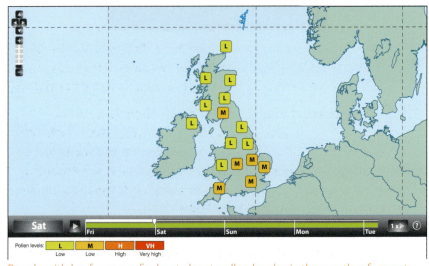

People with hayfever can find out about pollen levels via the weather forecast.

Is there a link between asthma and air quality?

Asthma is a common problem. During an asthma attack, a person's chest feels very tight. They find it difficult to breathe.

Medical evidence shows that asthma attacks are triggered by many different factors. These include:

- tree or grass pollen
- animal skin flakes
- dust mite droppings
- air pollution
- nuts
- shellfish
- food additives

- dusty materials
- strong perfumes
- getting emotional
- stress
- exercise (especially in cold weather)
- colds and flu.

The causes of asthma are not fully understood. There is evidence to show that some people are more likely to have asthma attacks because of their genes. Other evidence indicates that environmental factors, such as air pollution, are also involved.

Nitrogen dioxide is an air pollutant that comes mainly from traffic exhausts. Large-scale studies have shown that if the concentration of nitrogen dioxide stays high for several days, there is an increased risk of people with asthma suffering from asthma attacks. There is a correlation.

People who have asthma have sensitive lungs. Air pollutants may irritate a person's lungs, particularly if their lungs are sensitive. Exposure to nitrogen oxides increases the *chance* of an asthma attack but does not mean that all people with asthma will suffer an attack. Even so, the link between the factor (exposure to polluted air) and the outcome (an asthma attack) is still described as a correlation.

The correlation does not show that nitrogen oxides *cause* asthma. Scientists researching asthma need find a mechanism to explain the link between nitrogen oxides and asthma before we can say there is a causal link.

Using air quality data

There are instruments collecting data about air quality in many cities in the world. Scientists combine that data with other information, such as traffic figures and hospital admissions records to look for patterns. These patterns that might help explain why air quality varies and the effects it has on health.

In China, the rapid development of city areas has led to an increase in air pollution. The levels of air pollution in Beijing, the capital city, are frequently above the World Health Organization guidelines.

Many people in Beijing wear a smog-mask to filter out the particulates in the air.

A team of scientists from the US and China looked at air quality data from Beijing. They investigated a possible link between PM2.5 particles (particles that are less than 2.5 micrometres in size) and cases of influenza (flu). PM2.5 particles are a problem because their small size means that they can penetrate deep into the lungs.

The scientists produced a graph to show patterns in the levels of PM2.5 and cases of flu over a period of four years. They presented the data on the same graph to make it easier to spot any correlation.

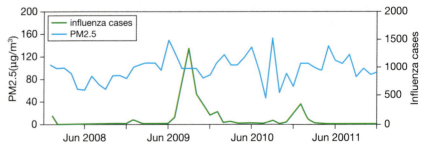

Graph showing levels of PM2.5 and the number of cases of influenza in Beijing between 2008 and 2011. Source: www.ehjournal.net/content/13/1/102

Questions

1 There is a correlation between ice cream sales and the number of people reporting hay fever symptoms. Do you think this is a coincidence? Explain your answer.

2 Suggest why it is useful for the weather forecast to include a pollen forecast.

3 Look at the graph showing levels of PM2.5 and influenza cases in Beijing.

 a Write down when
  - two highest peaks in PM2.5 levels occurred
  - two highest peaks of flu cases occurred.

 b Explain whether the data shows a link between PM2.5 levels and influenza.

4 The Olympic games took place in Beijing in August 2008. Additional measures were used to try to reduce air pollution.

 a Suggest why the city authorities tried to improve air quality during the Olympic Games.

 b Look at the graph showing levels of PM2.5 and influenza cases in Beijing. Discuss whether the data shows whether there was an improvement in air quality.

5 Researchers looked at air quality data from Atlanta, US, from around the time the Olympic Games were held there in 1996. They looked at data for the number of children visiting emergency care for asthma symptoms and at air pollution levels before, during, and after the games.
 They found that during the Olympic Games air pollution dropped and the number of hospital visits for children with asthma symptoms went down. Scientists checked levels of other emergency hospital visits for children for other reasons but these stayed broadly similar.

 a The report of the study describes a correlation between a factor and an outcome. What was the factor? What was the outcome?

 b Suggest two other reasons that could have caused a drop in children's hospital visits for asthma symptoms during the games.

 c What other data helped scientists to decide that the drop in children's visits was connected with the drop in air pollution?

 d More than one child was hospitalized due to severe asthma symptoms during the Olympic Games. Does this mean that the correlation does not show a link?

C7.3 Are all applications of science ethically acceptable?

Find out about

- how the synthesis of ammonia and fertilisers have made a positive difference to our lives
- unintended impacts of fertilisers on the environment
- ethical questions about fertilisers and the work of scientist Fritz Haber
- the difference between questions that can be answered using science and those that can't

Key words

➤ Haber process
➤ synthetic fertiliser
➤ eutrophication

Synoptic link

You can learn more about fertilisers in C6.4A *Fertilisers and the chemicals industry*, and C6.4B *Green chemistry*.

In 1918, German chemist Fritz Haber was awarded the Nobel Prize for his work on ammonia synthesis. The process he developed is now known as the **Haber process**. The process fixes nitrogen from the air to make ammonia, which can be used to make **synthetic fertilisers**. The Haber process has been described as 'the reaction that changed the world' and has enabled us to grow more food than ever before.

Fritz Haber (left) with Albert Einstein in 1915.

Today it is thought that over a third of the world's population relies on food grown with synthetic fertilisers. The continuing challenge of feeding the growing human population means we have become ever more dependent on them. But the use of synthetic fertilisers and the story of Fritz Haber raise ethical questions about the impacts of science on the environment and society.

The importance of fertilisers

Wheat is grown on more land than any other crop. It is ground up to make flour and is an important ingredient in bread, pasta, noodles, couscous, breakfast cereals, pastries, biscuits, and cakes.

Food crops such as wheat, maize, and rice are important in human food security. They are staple foods all over the world. These crops need nitrogen in order to grow; they absorb nitrogen from the soil as nitrates. But in many places, people want to grow crops faster than the nitrates can be replaced by natural processes. To produce enough food, farmers also need to grow crops in places where there is not enough nitrate in the soil to begin with. Synthetic fertilisers are used as a source of nitrogen in these places.

The impact of fertilisers on the environment

Synthetic fertilisers are widely used in agriculture.

The Haber process requires a high energy input, which is primarily produced by burning fossil fuels, creating greenhouse gas emissions. Although synthetic fertilisers are widely used, they can cause problems. Nitrates that run off farmland in rain water can pollute groundwater and drinking water, and they can accumulate in food chains to toxic levels. They can also produce algal blooms, which damage marine ecosystems through **eutrophication**.

An algal bloom in a pond. The algae block sunlight, causing water plants to die because they cannot make food by photosynthesis. Microorganisms break down the dead plants, a process that uses oxygen and leads to low oxygen levels in the water. Animals living in the water soon die because they cannot survive without oxygen. This is eutrophication.

Some organisations, such as Greenpeace, campaign against the use of synthetic fertilisers. They argue that more sustainable farming practices should be promoted.

The development of the Haber process

At the start of the twentieth century, fertilisers were commonly made from sodium nitrate ($NaNO_3$). European countries imported most of the $NaNO_3$ they needed from Chile. Most of this was mined from deposits in the Atacama Desert.

Fritz Haber was working at the Karlsruhe Institute of Technology in Germany. In 1909 he invented a way to fix nitrogen from the air. This involved reacting nitrogen with hydrogen from natural gas at a high temperature and pressure to make ammonia. By 1913, a team led by Carl Bosch at chemical company BASF had turned it into an industrial process. Ammonia can be oxidised into nitric acid, which is used to manufacture nitrate fertilisers.

The Haber process provided Europe with a new source of nitrates, releasing it from dependence on the Chilean deposits. This left mines unprofitable and many unemployed miners in Chile.

Fritz Haber and World War I

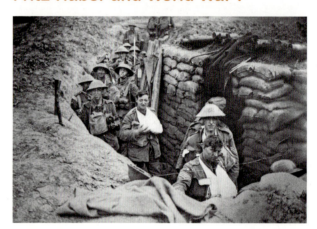

Millions of soldiers fought and died in the battlefield trenches of World War I. Many of them were just teenagers.

At the outbreak of World War I, Fritz Haber was put in charge of the development of the raw materials needed for the war effort. He was keen to do what he thought of as his patriotic duty. As part of this work, he came up with the idea of releasing highly toxic chlorine gas and allowing it to drift across to the enemy trenches.

Haber was married to Clara Immerwahr, a German chemist. She was the first woman in Germany to be awarded a doctorate in chemistry, following her research into the solubility of metal salts. She went on to work as a laboratory assistant, which was the most senior position a woman was allowed to hold at that time. Immerwahr was disgusted by her husband's work on chemical warfare. She pleaded with him to stop but he would not – he regarded her protests as treasonous to her country.

Immerwahr publically protested that the work of her husband was a 'perversion of the ideals of science' and that it was 'a sign of barbarity, corrupting the very discipline which ought to bring new insights into life'.

Clara Immerwahr was a German chemist who married Fritz Haber in 1901. During World War I she strongly disagreed with her husband's work in chemical warfare. In 1991 the German section of the organisation *International Physicians for the Prevention of Nuclear War* presented a prize named in her honour.

In April 1915 the first chlorine gas attack of the First World War was carried out, as a direct result of Haber's work. Just 10 days later, Clara Immerwahr committed suicide.

The Nobel Prize

Haber was awarded the Nobel Prize for Chemistry in 1918, but this award was controversial. The Nobel Prize committee recognised that although he had contributed to the creation of weapons, his development of ammonia synthesis meant that he was a man who had given 'the greatest benefit to mankind'. Not everyone agreed with the decision. It has been reported that the famous physicist Ernest Rutherford refused to shake Haber's hand at the Nobel Prize award ceremony.

Ethical questions

Science and technology are not separate from society. When deciding how scientific knowledge and methods are used, we have to consider social, environmental, political, economic, and **ethical** issues.

Do the benefits of synthetic fertilisers outweigh their damaging impact on the environment? Should scientific knowledge be used to make weapons? Does Haber's work on both ammonia synthesis and chemical warfare amount to 'the greatest benefit to mankind'? These are questions about right and wrong. They are ethical questions, and science cannot provide the answers.

The Nobel Prize is one of the highest and most prestigious awards in science. It is awarded to scientists judged to have made an outstanding contribution to humanity.

Questions

1 The Haber process has been described as 'the reaction that changed the world'.
   a Explain how the Haber process has made a significant positive difference to our lives.
   b Describe how the invention of the Haber process soon had:
      i a positive economic impact in Europe
      ii a negative impact on society in Chile.
   c Describe unintended impacts of the Haber process on the environment.

2 Write down one example of:
   a an action taken by Fritz Haber for political reasons
   b an action taken by Clara Immerwahr for ethical reasons.

3 Some questions cannot be answered using science.
   a The question of whether or not Fritz Haber should have been awarded a Nobel Prize is an ethical question.
      i Summarise views for and against the award of the Nobel Prize to Haber.

   ii Write down one other example of an ethical question from the case study.
   b For each of the following questions, write down whether it is a question that could be answered using science or a question that could not be answered using science.
   Explain your reason in each case.
      i Should scientists help to develop chemical weapons?
      ii What conditions are required to make ammonia in the Haber process?
      iii Is a scientist responsible for everything he or she invents?
      iv Could we reduce the negative environmental impacts of using synthetic fertilisers?
      v Do the benefits of synthetic fertilisers outweigh their damaging impact on the environment?

Ideas about Science

Learning about the Ideas about Science in this course will help you understand how scientific knowledge is obtained, how to respond to science stories and issues in the wider world, and the impacts of scientific knowledge on society.

IaS1: What needs to be considered when investigating a phenomenon scientifically?

The aim of science is to develop good explanations for natural phenomena. There is no single 'scientific method' that leads to explanations, but scientists do have characteristic ways of working. In particular, scientific explanations are based on a cycle of collecting and analysing data.

Usually, developing an explanation begins with proposing a hypothesis. A hypothesis is a tentative explanation for an observed phenomenon ('this happens because…').

The hypothesis is used to make a prediction about how, in a particular experimental context, a change in a factor will affect the outcome. A prediction can be presented in a variety of ways, for example, in words or as a sketch graph.

In order to test a prediction and the hypothesis upon which it is based, it is necessary to plan an experiment that enables data to be collected in a safe, accurate, and repeatable way.

In a given context you should be able to:

- use your scientific knowledge and understanding to develop and justify a hypothesis and prediction

- suggest appropriate apparatus, materials, and techniques, justifying the choice with reference to the precision, accuracy, and validity of the data that will be collected

- explain the importance of accuracy and precision when determining scientific quantities

- use scientific quantities (such as mass, volume, and temperature), and know how they are measured

- identify factors that need to be controlled, and how they could be controlled

- suggest an appropriate sample size and/or range of values to be measured, and justify the suggestion

- plan experiments and describe procedures to make observations, collect data, and test a prediction or hypothesis

- identify hazards associated with the data collection and suggest ways of minimising the risk.

This apparatus could be used to investigate the effect of particle size on the rate of the reaction between calcium carbonate and hydrochloric acid. Can you write a testable hypothesis for this investigation? Which factor would you change during the investigation, and which ones would you need to control?

IaS2: What processes are needed to draw conclusions from data?

The cycle of collecting, presenting, and analysing data usually involves translating data from one form to another, mathematical processing, graphical display, and analysis; only then can we begin to draw conclusions.

A set of repeat measurements can be processed to calculate a range within which the true value probably lies and to give a best estimate of the value (mean).

Displaying data graphically can help to show trends or patterns, and to assess the spread of repeated measurements.

Mathematical comparisons between results and statistical methods can help with further analysis.

When working with data you should be able to:

- produce appropriate tables, graphs, and charts to display the data
- use the appropriate units and be able to convert between units
- use prefixes (from tera to nano) and powers of ten to show orders of magnitude
- use an appropriate number of significant figures.

When displaying data graphically you should be able to:

- select an appropriate graphical form, using appropriate axes and scales
- plot data points correctly, drawing an appropriate line of best fit and indicating uncertainty (e.g., range bars).

When analysing data you should be able to:

- identify patterns or trends
- use statistics (range and mean)
- obtain values from a line on a graph (including gradient, interpolation, and extrapolation).

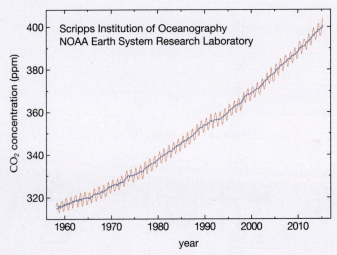

This graph displays data on the concentration of carbon dioxide in the atmosphere in Hawaii since 1958. What patterns can you see in the data? What is the overall trend?

Data can never be relied on completely because observations may be incorrect and all measurements are subject to uncertainty, arising from the limitations of the measuring equipment and the person using it.

Experiments and data obtained must be evaluated before we can make conclusions based on the results. There could be many reasons why the quality (accuracy, precision, repeatability, and reproducibility) of the data could be questioned and a number of ways in which they could be improved. A result that appears to be an outlier should be treated as data, unless there is a reason to reject it (e.g., measurement or recording error).

In a given context you should be able to:

- discuss the accuracy, precision, repeatability, and reproducibility of a set of data
- identify random and systematic errors that are sources of uncertainty in measurements
- explain the decision to discard or retain an outlier
- suggest improvements to an experiment, and explain why they would increase the quality of the data collected
- suggest further investigations that could be done.

A prediction is based on a tentative explanation (a hypothesis). When collected data agrees with the prediction, it increases our confidence that the explanation is correct. But it does not *prove* that the explanation is correct. Disagreement between the data and the prediction indicates that one or other is wrong, and decreases our confidence in the explanation.

In a given context you should be able to:

- use observations and data to make a conclusion
- explain how much the data increases or decreases confidence in a prediction or hypothesis.

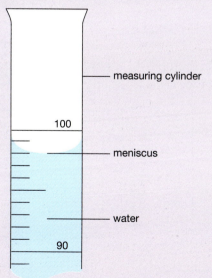

The volume of a liquid can be measured using a measuring cylinder. But there will be uncertainty in the measurement. Can you suggest a source of random error, a source of systematic error, and a mistake that could be made when taking this kind of measurement?

IaS3: How are scientific explanations developed?

Scientists often look for patterns in data to identify correlations that can suggest cause-and-effect links. They then try to explain these links.

The first step is to identify a correlation between a factor and an outcome. The factor may be the cause, or one of the causes, of the outcome. In many situations, a factor may not always lead to the outcome, but increases the chance (or the risk) of it happening. In order to claim that the factor causes the outcome we need to identify a process or mechanism that might account for how it does this.

You should be able to use ideas about correlation and cause to:

- identify a correlation in data presented as text, in a table, or as a graph
- distinguish between a correlation and a cause-and-effect link
- suggest factors that might increase the chance of a particular outcome in a given situation, but do not always lead to it
- explain why individual cases do not provide convincing evidence for or against a correlation
- explain why you would accept or reject a claim that a factor is a cause of an outcome, based on the presence or absence of a causal mechanism.

Scientific explanations and theories do not 'emerge' automatically from data, and are separate from the data. Proposing an explanation involves creative thinking. Collecting sufficient data from which to develop an explanation often relies on technological developments that enable new observations to be made.

As more evidence becomes available, a hypothesis may be modified and may eventually become an accepted explanation or theory.

A scientific theory is a general explanation that applies to a large number of situations or examples (perhaps to all possible ones), which has been tested and used successfully, and is widely accepted by scientists. A scientific explanation of a specific event or phenomenon is often based on applying a scientific theory to the situation in question.

You should be able to:

- describe and explain examples of scientific explanations that have developed over time, and how they were modified when new evidence became available.

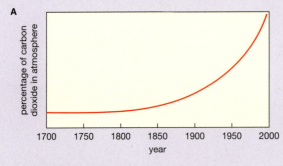
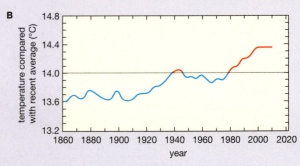

The graphs show how global temperatures and carbon dioxide levels have changed. Can you describe the correlation shown by this pair of graphs?

Findings reported by an individual scientist or group are carefully checked by the scientific community before being accepted as scientific knowledge. Scientists are usually sceptical about claims based on results that cannot be reproduced by anyone else, and about unexpected findings until they have been repeated (by themselves) or reproduced (by someone else).

Two (or more) scientists may legitimately draw different conclusions about the same data. A scientist's personal background, experience, or interests may influence their judgements.

An accepted scientific explanation is rarely abandoned just because new data disagrees with it. It usually survives until a better explanation is available.

You should be able to:

● describe the 'peer review' process, in which new scientific claims are evaluated by other scientists.

Models are used in science to help explain ideas and to test explanations. A model identifies features of a system and rules by which the features interact. It can be used to predict possible outcomes. Representational models use physical analogies or spatial representations to help visualise scientific explanations and mechanisms. Descriptive models are used to explain phenomena. Mathematical models use patterns in data of past events, along with known scientific relationships, to predict behaviour; often the calculations are complex and can be done more quickly by computer.

Models can be used to investigate phenomena quickly and without ethical and practical limitations, but their usefulness is limited by how accurately the model represents the real world.

For a variety of given models (including representational, descriptive, mathematical, computational, and spatial models) you should be able to:

● use the model to explain a scientific idea, solve a problem, or make a prediction
● identify limitations of the model.

The diagram represents the a simple model of the atom, in which the electrons are pictured as occupying energy levels, which give rise to regions of negative charge around a positive nucleus . Can you use it to explain ionic bonding? What are the limitations of this model?

IaS4: How do science and technology impact on society?

Science and technology provide people with many things that they value, and that enhance their quality of life. However, some applications of science can have unintended and undesirable impacts on the quality of life or the environment. Scientists can devise ways of reducing these impacts and of using natural resources in a sustainable way.

You should be able to:

● describe and explain examples of applications of science that have made significant positive differences to people's lives.

Everything we do carries a certain risk of accident or harm. New technologies and processes can introduce new risks. The size of a risk can be assessed by estimating its chance of occurring in a large sample, over a given period of time.

To make a decision about a course of action, we need to take account of both the risks and benefits to the different individuals or groups involved. People are generally more willing to accept the risk associated with something they choose to do than something that is imposed, and to accept risks that have short-lived effects rather than long-lasting ones. People's perception of the size of a particular risk may be different from the statistically estimated risk. People tend to overestimate the risk of unfamiliar things (such as flying as compared with cycling), and of things whose effect is invisible or long-term (such as ionising radiation).

The photograph shows synthetic fertiliser being spread on a field. Can you describe and explain examples of how synthetic chemicals have made a positive difference to people's lives?

You should be able to:

- identify examples of risks that have arisen from a new scientific or technological advance
- for a given situation:
  - identify risks and benefits to the different individuals and groups involved
  - discuss a course of action, taking account of who benefits and who takes the risks
  - suggest reasons for people's willingness to accept the risk
  - **H** distinguish between perceived and calculated risk.

Some forms of scientific research, and some applications of scientific knowledge, have ethical implications. In discussions of ethical issues, a common argument is that the right decision is the one that leads to the best outcome for the greatest number of people.

Where an ethical issue is involved you should be able to:

- suggest reasons why different decisions on the same issue might be appropriate in view of differences in personal, social, economic, or environmental context.
- make a decision and justify it by evaluating the evidence and arguments
- distinguish questions that could be answered using a scientific approach, from those that could not
- state clearly what this issue is and summarise different views that may be held.

Scientists must communicate their work to a range of audiences, including the public, other scientists, and politicians, in ways that can be understood. This enables decision-making based on information about risks, benefits, costs, and ethical issues.

You should be able to:

- explain why scientists should communicate their work to a range of audiences.

Many sunscreens contain nanoparticles. What are the benefits of using a sunscreen? And what are the risks? Who benefits and who is at risk?

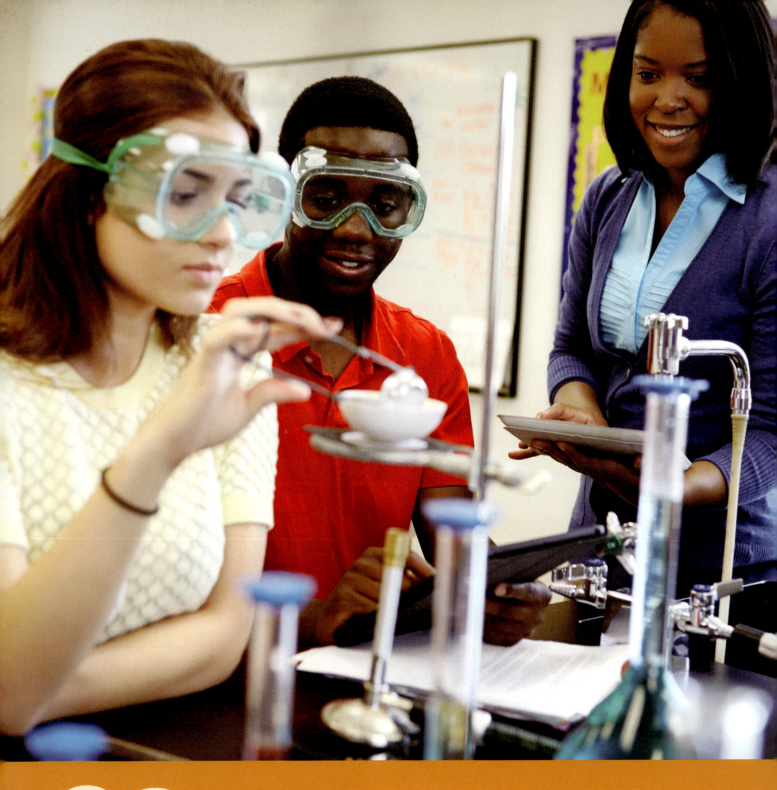

C8 Practical techniques

Why study practical techniques?

Practical work is as essential part of science. It helps us investigate what happens in the world around us. It also helps us explain how and why these things happen.

The aim of science is to develop good explanations for observations of the natural world. Scientific explanations are based on data, which must be collected.

Practising practical techniques helps us to collect data in a safe, ethical and repeatable way, and to improve the accuracy of the data we collect. It is important that different people use the same techniques and standard procedures to collect data. This means data can be compared more easily.

Practical work not only helps us to develop explanations, but also to test explanations proposed by others. Understanding some of the ways scientists collect data helps us to make informed decisions about scientific issues in the news and in our own lives.

Practical techniques

- using appropriate apparatus to make and record a range of measurements accurately, including mass, time, temperature, and volume of liquids and gases
- safely using appropriate heating devices and techniques including a Bunsen burner and a water bath
- using appropriate apparatus and techniques for conducting and monitoring chemical reactions, including appropriate reagents and/or techniques for the measurement of pH
- safely using a range of equipment to purify and/or separate chemical mixtures including evaporation, filtration, crystallisation, chromatography, and distillation
- making and recording appropriate observations during chemical reactions including changes in temperature
- measuring rates of reaction by monitoring the production of gas and colour change
- safely and carefully using gases, liquids and solids, including careful mixing of reagents under controlled conditions, using appropriate apparatus to explore chemical changes and/or products
- using appropriate apparatus and techniques to draw, set up and use electrochemical cells for separation and production of elements and compounds
- using appropriate qualitative reagents and techniques to analyse and identify unknown samples or products including gas tests, flame tests, precipitation reactions, and the determination of concentrations of strong acids and strong alkalis S

A: Measuring temperature and volume

Many measuring instruments have a linear scale. This is a series of equally spaced lines, or **graduations**. You use the scale to read off a value.

You may have to use a linear scale on:

- a thermometer (to measure temperature)
- a measuring cylinder, syringe, pipette or burette (to measure the volume of a liquid or gas)
- an analogue meter.

When a measurement falls between two graduations, it has to be estimated. This means there will be **uncertainty** in the measurement.

Procedure

1 Look at the scale you are reading. Usually, not all of the graduations will be marked with a number. Decide what the distance between two graduations represents.

2 If the reading is between two graduations, decide which graduation it is closest to. Record this as the measurement.

3 Record the uncertainty of the measurement as ± half the smallest graduation.

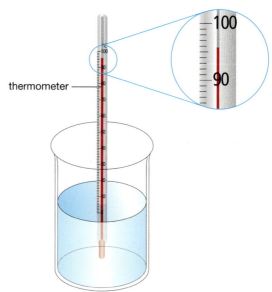

A thermometer has a linear scale. The coloured liquid inside the thermometer expands as the temperature goes up, and rises up the narrow glass tube. Here the reading is between 95 °C and 96 °C, but is closest to 96 °C. We would record the measurement as 96 °C ± 0.5 °C.

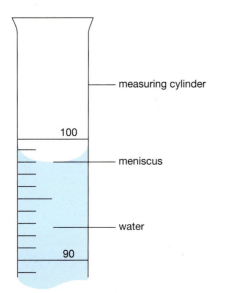

Liquid in a measuring cylinder has a curved top surface. This is called the **meniscus**. The volume is read from the bottom of the meniscus. You should always make the reading at eye level. The volume of water in the diagram is 98 ± 0.5 cm³.

B: Measuring out a specific volume of liquid using a pipette

Glassware for measuring volume is designed to be used in a specific way. Variations in procedure lead to errors, making data less accurate. The errors can be minimised with good practice.

Procedure

1 Check that the pipette is clean, not chipped, and has a clear mark. Attach a suitable filler.

2 Place the tip of the pipette below the liquid surface. Draw up liquid to just above the mark.

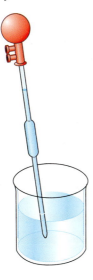

3 Hold the pipette above the liquid. With your eye level with the mark, let liquid out until the bottom of the meniscus is on the mark.

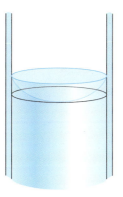

4 Touch the pipette tip against the side of the container to remove any drops.

5 Place the pipette tip just touching the side of the receiving vessel and let the liquid drain out. Allow an extra three seconds after it stops flowing. Do not blow out the last drop.

Apparatus and materials

➤ pipette
➤ conical flask

C: Measuring mass

Find out about

- using appropriate apparatus to accurately measure and record mass

Mass is measured using a balance. Choose a balance with a level of accuracy fit for the purpose of the task – to how many significant figures do you need to measure the mass?

If a very sensitive balance is used, it is necessary to shield the balance from drafts to get an accurate measurement.

Procedures

Weighing direct

1 Check that the balance is clean and reading zero.

2 Place a suitable empty weighing vessel on the balance platform. Set the display to zero. (This is known as taring the balance.)

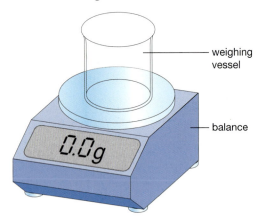

weighing vessel

balance

3 Place the sample in the weighing vessel on the balance platform.

4 The reading on the balance is the mass of the sample.

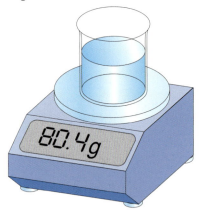

Weighing by difference

1 Check that the balance is clean and reading zero.

2 Place the sample in a suitable clean, dry weighing vessel on the balance platform. Record the mass.

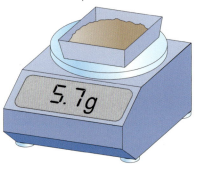

3 Transfer the sample to another container. Weigh the weighing vessel and record its mass.

4 Calculate the mass of the sample from the difference between the two measurements.

Apparatus and materials

➤ balance
➤ weighing vessel

D: Measuring time

Time is measured using a timer such as a stopwatch or stop clock.

You may wish to measure, for example, how long it takes for a chemical reaction to produce a particular volume of gas, or to reach an end point such as a colour change.

Procedure

A digital stopwatch showing a time of 42 s.

A stop clock showing a time of 42 s. This is an example of an analogue meter.

1   At the starting point, start the timer (or if the timer is already running, make a note of the start time).

2   At the end point, stop the timer (or if the timer needs to keep running, make a note of the end time).

3   If the start time was not zero, subtract the start time from the end time.

Uncertainty

A digital stopwatch may show the time to tenths (0.1) or hundredths (0.01) of a second. However, when using a stopwatch, human reaction time can be up to 0.5 s. Therefore, you may need to record your measurement of the time with an uncertainty of ± 0.5 s.

Minutes and seconds

Remember that 1 minute is divided into 60 seconds. This means that 1 minute 50 seconds is *not* the same as 1.50 minutes. (1.50 minutes is one-and-a-half minutes, which is 1 minute 30 seconds).

E: Measuring pH

Find out about

- using appropriate reagents and/or techniques to accurately measure and record pH

Apparatus and materials

➤ test tube
➤ universal indicator solution
➤ dropper
➤ pH test strips
➤ pH meter
➤ colour chart

CHECK SAFETY
Never work unsupervised

wear eye protection

An alternative way to measure pH is to use a digital pH meter. The probe of the pH meter must be cleaned using distilled water between each reading. The probe is very delicate so must be handled with care and not used to stir the solution.

Universal indicator solution contains a mixture of compounds that change colour at different levels of acidity or alkalinity. When universal indicator solution is added to a sample, the colour it turns shows the pH of the sample.

Alternatively, you can use pH test strips and match them to a colour chart, or use a digital pH meter.

Procedure for measuring pH by colour matching

1   Put about 5 cm³ of the sample solution into a test tube.

2   Add three drops of universal indicator solution, and shake the tube gently from side to side to mix. Or dip a pH test strip into the solution.

3   Use the colour chart to estimate the pH value of the sample solution. Compare the colour of the solution or the pH test strip with the chart and decide the nearest match.

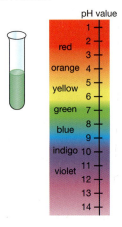

pH value

red 1 2 3
orange 4
yellow 5 6
green 7
blue 8 9
indigo 10
violet 11 12 13 14

F: Heating substances and controlling their temperature

Substances can be heated using a Bunsen burner. The substance is placed in a glass or ceramic container on a tripod and gauze.

It can be difficult to keep a substance at a constant temperature using a Bunsen burner. In this case it is helpful to use a water bath. You may use an electric water bath, or you may make a water bath by immersing the sample container in a large beaker of water at the desired temperature. The temperature of a water bath cannot be raised above the boiling point of water (100 °C).

Procedure for heating with a Bunsen burner

1   Place the tripod and gauze on a flat surface. Ensure that the tripod is stable and will not tip over. Place the container of substance on top.

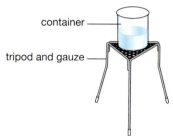

container

tripod and gauze

2   Turn the ring on the Bunsen burner so that the air hole is completely closed. This will produce a yellow safety flame when the Bunsen burner is lit. Turn on the gas, and then light the Bunsen burner.

Bunsen burner

3   Turn the ring on the Bunsen burner so that the air hole is open. The hottest flame (a blue roaring flame) will be produced when the air hole is completely open. Place the Bunsen burner under the tripod.

4   Use a thermometer to measure the temperature of the substance. Hold the thermometer so that the bulb is in the centre of the substance. The bulb should not touch the bottom or sides of the container. Use a clamp to hold the thermometer in place to reduce the risk of spillage when working with higher temperatures, and when monitoring the temperature over a length of time.

Apparatus and materials

- ➤ tripod
- ➤ gauze
- ➤ container
- ➤ Bunsen burner
- ➤ thermometer
- ➤ clamp, boss, and stand
- ➤ water bath

CHECK SAFETY
Never work unsupervised

wear eye protection

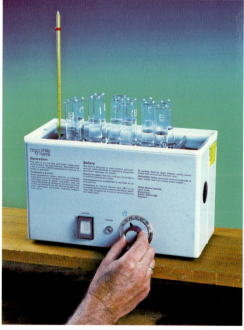

The water in an electric water bath is kept at a constant temperature. The temperature is set using the dial, and is controlled automatically by an electric heater and thermostat. It can be checked using the thermometer. Test tubes containing a substance can be placed in the water to heat the solution and help control its temperature.

G: Separating a mixture using paper chromatography

Find out about

- safely using a range of equipment to separate chemical mixtures including chromatography

Apparatus and materials

- ➤ chromatography tank and lid
- ➤ chromatography paper
- ➤ pencil and ruler
- ➤ fine glass tube
- ➤ sample solution
- ➤ stapler and staples
- ➤ solvent

CHECK SAFETY
Never work unsupervised

Synoptic link

You can learn more about the principles of chromatography in C5.1B *Chromatography*.

This is a technique for qualitative analysis, to find what compounds are present in a solution. Small, concentrated spots of the sample are made on the paper, which is then placed in a **solvent**. In paper chromatography the **stationary phase** is the water trapped in the fibres of the paper; the paper just acts as a support. The solvent is the **mobile phase**. **Reference materials** may be run on the same piece of paper, or R_f values may be calculated and compared with known values.

If you do not have a dedicated chromatography tank, this procedure shows how you can use a glass beaker.

Procedure to run the chromatogram

1 Cut a rectangular piece of chromatography paper to fit a suitable beaker.

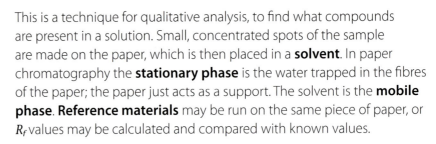

2 Put a 1-cm depth of the chosen solvent into the beaker, and cover it with the lid.

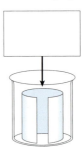

3 Mark the piece of chromatography paper with a pencil as shown.

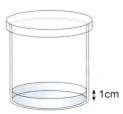

4 With a fine glass tube, place a small spot of the sample solution at S and allow the spot to dry. If the solution is dilute, add further small spots, allowing each to dry before the next is added. Repeat this, using a clean tube, for each reference solution R1, R2,....

5 Roll the paper into a cylinder and staple it as shown, so that the ends do not touch.

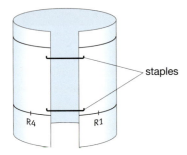

6 Stand the paper in the tank, cover with the lid and leave undisturbed until the **solvent front** gets close to the upper pencil line.

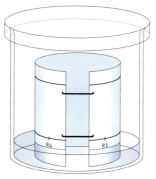

7 Remove the paper from the tank, mark the position of the solvent front with a pencil, and leave it to dry.

Procedure for calculating the retardation factor, R_f

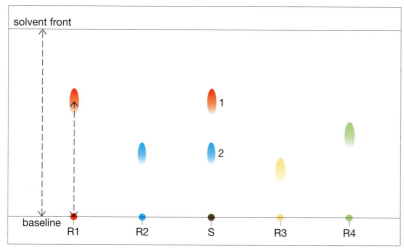

1 If necessary, reveal the positions of colourless spots by developing the chromatogram. (You can develop the chromatogram by spraying it with a locating agent that reacts with the stubstances to form coloured compounds.)

2 Draw around each spot with a pencil. If any samples have more than one spot, number the spots.

3 Prepare a table or spreadsheet to record the positions of the spots.

4 Measure and record the distance between the baseline and the solvent front.

5 Measure and record the distance between the centre of each spot and the baseline.

6 Calculate the R_f value for each spot. See C5.1B for a worked example.

Distance between the baseline and the solvent front = …cm		
Sample and spot number	Distance from baseline (cm)	$R_f = \dfrac{\text{distance spot moved}}{\text{distance solvent moved}}$

H: Separating mixtures and purifying a product

Find out about

● safely using a range of equipment to purify and/or separate chemical mixtures including evaporation, filtration, crystallisation, and distillation

Apparatus and materials

➤ evaporating basin
➤ Bunsen burner
➤ tripod and gauze
➤ glass rod
➤ Petri dish
➤ filter paper
➤ funnel
➤ clamp, boss, and stand
➤ flask
➤ thermometer
➤ condenser
➤ test tube

CHECK SAFETY
Never work unsupervised

wear eye protection

Synoptic link

You can learn more about these procedures in C5.1C *Separation and purification processes*.

It is often necessary to take steps to separate a desired product from the reaction mixture. These steps can also help to purify a product.

Procedures

Evaporation and crystalisation

To separate a soluble solid (such as a salt) from the liquid solvent in which it is dissolved:

1　Pour the solution into an evaporating basin. Heat gently using a Bunsen burner to evaporate some of the liquid. Evaporate until crystals form when a droplet of solution picked up on a glass rod cools.

2　Pour the concentrated solution into a labelled Petri dish and set it aside to cool slowly. The solid will crystallise. The crystals will be larger if they form slowly.

Filtration

To separate an insoluble solid from a liquid, pour the mixture through filter paper in a funnel. The solid will be collected on the filter paper. Collect the liquid filtrate in a collecting vessel such as a beaker of evaporating basin.

Distillation

Use distillation to capture the vapour evaporated from a mixture and collect it as a purified liquid using condensation.

evaporating basin

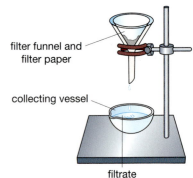

filter funnel and filter paper

collecting vessel

filtrate

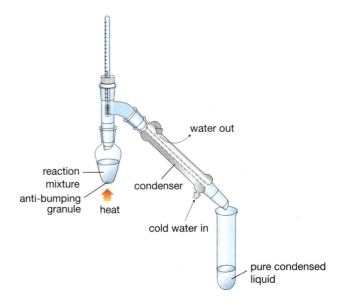

reaction mixture

anti-bumping granule

heat

condenser

water out

cold water in

pure condensed liquid

I: Comparing rates of reactions

We can compare the rate of a chemical reaction in different conditions. For example, is it faster at higher temperature? Or we can compare the rates of two different reactions in the same conditions.

A simple way to do this is to compare how long it takes for each reaction to reach an end point. A reaction that reaches the end point more quickly has a faster rate.

The rate of each reaction could be calculated using the following equation:

$$\text{rate of reaction (/s)} = \frac{1}{\text{time taken to reach end point (s)}}$$

Procedures

Timing how long it takes for an indicator to change colour

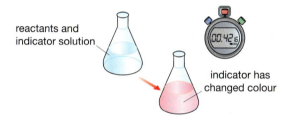

reactants and indicator solution

indicator has changed colour

CHECK SAFETY
Never work unsupervised

wear eye protection

Mix the reactants in a flask with an indicator solution, and start the timer. Stop the timer when the indicator has completely changed colour.

Timing how long it takes for a solution to turn cloudy (for reactions that form a precipitate)

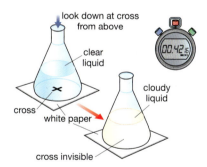

look down at cross from above

clear liquid

cloudy liquid

cross

white paper

cross invisible

Mix the liquids in a flask and start the timer. Stop the timer when you can no longer see the cross.

Timing how long it takes for a solid reactant to dissolve

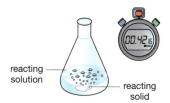

reacting solution

reacting solid

Mix the solid and liquid in a flask and start the timer. Stop the timer when you can no longer see any solid.

J: Measuring the rate of a reaction

Apparatus and materials

➤ conical flask
➤ bung
➤ delivery tube
➤ tray of water
➤ measuring cylinder
➤ clamp, boss, and stand
➤ gas syringe
➤ balance
➤ cotton wool
➤ timer

CHECK SAFETY
Never work unsupervised

wear eye protection

We can measure the rate of a chemical reaction by finding the quantity of product produced or the quantity of reactant used up in a fixed time.

Procedures

Collecting gas in a measuring cylinder or gas syringe

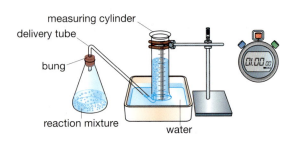

1 Record the volume of gas in the measuring cylinder or gas syringe at the starting point of the reaction.

2 Record the volume of gas in the measuring cylinder or gas syringe 60 seconds later or at the end point of the reaction.

The rate of reaction is calculated using the following equation:

$$\text{rate (cm}^3\text{/s)} = \frac{\text{change in volume of gas (cm}^3\text{)}}{\text{time taken (s)}}$$

Measuring the loss of mass as a gas is formed

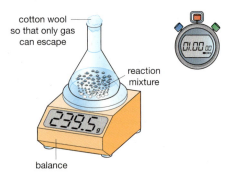

1 Record the mass at the starting point of the reaction.

2 Record the mass 60 seconds later or at the end point of the reaction.

The rate of reaction is calculated using the following equation:

$$\text{average rate (g/s)} = \frac{\text{change in mass (g)}}{\text{time taken (s)}}$$

K: Separating an ionic compound into its elements using electrolysis

An electric current can separate an ionic compound (such as a salt) into its elements, or other products, if it is molten or dissolved in water. This process uses **electrodes** and is called electrolysis.

During electrolysis, the negative electrode attracts the positive ions. The positive electrode attracts the negative ions. When the ions reach the electrodes, they lose their charges and turn into atoms.

Procedure for electrolysis of a salt solution

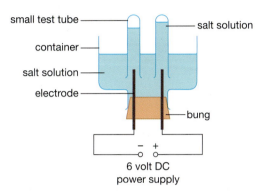

1. Start with the empty container. Make sure the bung is securely pushed into the empty container, and the electrodes are tightly held in place by the bung. The electrodes must not touch one another.

2. Pour a salt solution (such as copper chloride solution) into the container.

3. Make sure you are wearing gloves. Fill a small test tube with the salt solution, then cover the end with your thumb so that solution cannot leak out. Place the test tube upside down over one of the electrodes, as shown in the diagram. Clamp the test tube into place. Repeat this step with another small test tube and the other electrode.

4. Connect each electrode to the power supply using cables and crocodile clips. It can be useful to include a lamp in series in the circuit so you can see that it is working. Switch on the power supply.

5. Look for the formation of product at each electrode. This could be a solid deposit, the appearance of a different colour in the solution, or bubbles of gas that will collect in the test tubes. Record what you see.

6. Switch off the power supply as soon as you have briefly observed the product at each electrode.

7. You may wish to test any gasses that have been collected in the test tubes. The gases formed may be harmful, so gas tests should be done in a fume hood.

Find out about

- using appropriate apparatus and techniques to draw, set up, and use electrochemical cells for separation and production of elements and compounds

Apparatus and materials

- ➤ container
- ➤ bung with two holes
- ➤ electrodes (usually graphite)
- ➤ salt solution
- ➤ small test tubes
- ➤ clamp, boss, and stand
- ➤ cables and crocodile clips
- ➤ 6 V DC power supply

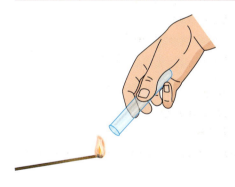

Key word

- ➤ electrode

L: Identifying products using gas tests

Find out about

● using gas tests to identify unknown substances

Apparatus and materials

➤ test tubes
➤ wooden splints
➤ limewater
➤ fume hood
➤ blue litmus paper

CHECK SAFETY
Never work unsupervised

wear eye protection

The gases hydrogen, oxygen, carbon dioxide, and chlorine can be identified using simple tests.

Procedures

Testing for hydrogen gas

Hydrogen gas will make a 'squeaky pop' sound when ignited using a lit wooden splint.

Testing for oxygen gas

Oxygen gas will re-ignite a glowing wooden splint.

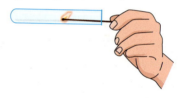

Testing for carbon dioxide gas

Carbon dioxide gas will turn limewater cloudy when bubbled through it.

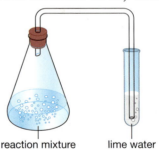

reaction mixture   lime water

Testing for chlorine gas

Chlorine gas is harmful. Always work in a fume hood when handling chlorine gas.

damp-blue litmus paper

gas jar

chlorine gas

Chlorine gas will make damp-blue litmus paper turn red, and will then bleach it white.

M: Identifying products using flame tests

Different substances produce characteristic flame colours when held in the flame of a Bunsen burner.

Procedure

The bright red flame produced by lithium compounds.

1 Light the Bunsen burner and set it to a roaring flame.

2 Clean the wire loop by dipping it into hydrochloric acid and then hold the loop in the clear or blue part of the flame. If a burst of colour is seen, the loop is not clean enough. Repeat this step until the clean loop produces no colour in the flame.

3 Allow the loop to cool, then dip it into the sample to be tested.

4 Hold the loop in the clear or blue part of the flame.

The flame colour produced indicates that particular elements are present in the sample:

Flame colour	Element
bright red	lithium
bright yellow	sodium
lilac	potassium
orange-red	calcium
blue-green	copper

Find out about

● using flame tests to identify unknown substances

Apparatus and materials

➤ wire loop
➤ hydrochloric acid
➤ Bunsen burner

CHECK SAFETY
Never work unsupervised

wear eye protection

N: Identifying products using precipitation reactions

Find out about
● using precipitation reactions to identify unknown substances

Apparatus and materials
➤ test tubes
➤ dilute sodium hydroxide
➤ dilute nitric acid
➤ silver nitrate
➤ barium chloride
➤ barium nitrate

CHECK SAFETY
Never work
unsupervised

wear eye
protection

Iron(III) ions (Fe³⁺) form a red-brown precipitate in dilute sodium hydroxide, which does not dissolve in excess sodium hydroxide.

Different ions produce characteristic precipitates when tested with acid or alkali. This helps us to identify them.

Procedures

Testing for ions with a positive charge

1 Put 2 cm³ of the sample solution in a test tube.

2 Carefully add 1–2 cm³ of dilute sodium hydroxide.

3 Record the colour of any precipitate formed.

4 Add a few more cm³ of dilute sodium hydroxide, and record whether or not the precipitate dissolves.

5 Compare your observations against the table in section C5.2B to identify the positive ion present in the sample.

Testing for ions with a negative charge

1 Put 2 cm³ of the sample solution in a test tube.

2 Carefully add 1–2 cm³ of dilute nitric acid.

3 Carefully add 1–2 cm³ of silver nitrate.

4 Record the colour of any precipitate formed.

5 Compare your observations against the table in section C5.2B to identify the negative ion present in the sample.

O: Titration

A titration is a quantitative technique based on measuring the volumes of solutions that react with each other. Titrations are used to measure concentrations and to investigate the quantities of chemicals involved in reactions.

Procedure

Acid–alkali titration to find the concentration of the acid

1 Use a pipette to transfer 25.0 cm³ of a solution of acid to a conical flask.

2 Fill a burette with a solution of alkali of known concentration. Secure the burette in a clamp.

3 Place a small beaker underneath the burette. Open the tap on the burette to slowly run alkali solution through the burette tip. Stop when the tip is filled with no air bubbles, and the meniscus near the top of the burette is level with the zero graduation.

4 Add a few drops of indicator solution to the acid in the conical flask. Place the flask on a white tile under the burette.

5 Open the tap on the burette to run alkali into the acid a little at a time. Swirl the contents of the flask. Continue adding alkali from the burette until the indicator permanently changes colour. The acid has been neutralised. This is the end point. Record the reading from the burette.

It is useful to do a rough titration first to get an idea where the end point lies, and then to repeat the titration more carefully, adding the acid drop by drop near the end point. Repeat the titration until you have three close results.

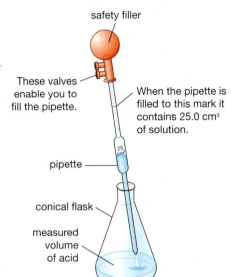

safety filler

These valves enable you to fill the pipette.

When the pipette is filled to this mark it contains 25.0 cm³ of solution.

pipette

conical flask

measured volume of acid

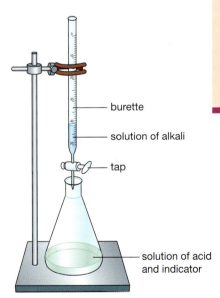

burette

solution of alkali

tap

solution of acid and indicator

Find out about

- using titration to determine the concentration of a strong acid or strong alkali
- using appropriate apparatus and techniques for conducting and monitoring chemical reactions, including appropriate reagents and/or techniques for the measurement of pH
- safely and carefully using liquids, including careful mixing of reagents under controlled conditions, using appropriate apparatus to explore chemical changes

Apparatus and materials

- ➤ pipette
- ➤ solution of acid
- ➤ conical flask
- ➤ clamp, boss, and stand
- ➤ burette
- ➤ solution of alkali
- ➤ small beaker
- ➤ indicator solution
- ➤ white tile

CHECK SAFETY
Never work unsupervised

wear eye protection

P: Making a sample of a salt

Find out about

- safely and carefully using liquids and solids, including careful mixing of reagents under controlled conditions, using appropriate apparatus to make a product
- safely using appropriate heating devices and techniques including a Bunsen burner
- safely using a range of equipment to purify and/or separate chemical mixtures including filtration, evaporation, and crystallisation

Apparatus and materials

- beaker
- glass rod
- filter paper
- funnel
- clamp and clamp stand
- evaporating basin
- Bunsen burner
- tripod and gauze
- Petri dish
- oven
- desiccator
- sample tube
- balance

CHECK SAFETY
Never work
unsupervised

wear eye
protection

The process of making magnesium sulfate (or any other soluble salt) in the laboratory illustrates the practical stages in a chemical synthesis.

Procedure

Mixing reagents under controlled conditions

1 Measure the required volume of acid into a beaker.

2 Add the metal oxide or carbonate bit by bit until no more dissolves in the acid. Stirring with a glass rod makes sure that the magnesium oxide or carbonate and acid mix well. Stirring also helps to prevent the mixture frothing up and out of the beaker.

3 Warm gently until all of the acid has been used up. Make sure that there is a slight excess of solid before moving on to the next stage.

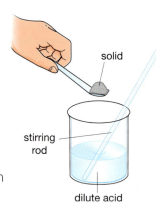

Separating the product from the reaction mixture

4 Filter off the excess solid, collecting the solution of the salt in an evaporating basin. The mixture filters more quickly if the mixture is warm. The residue on the filter paper is the excess solid. The filtrate in the evaporating basin contains the pure salt dissolved in water.

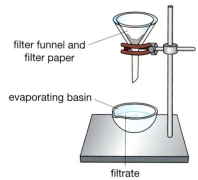

Purifying the product

5 Heat the filtrate gently to evaporate some of the water. Evaporate until crystals form when a droplet of solution picked up on a glass rod crystallises on cooling.

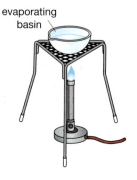

6 Pour the concentrated solution into a labelled glass Petri dish and set it aside to cool slowly so that crystallisation can take place.

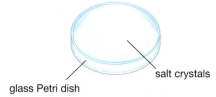

glass Petri dish — salt crystals

7 Complete the drying process in an oven, and then store in a desiccator. A desiccator is a closed container that contains a solid that absorbs water strongly.

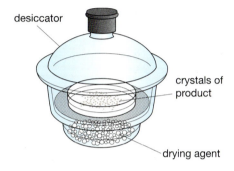

desiccator
crystals of product
drying agent

Measuring the yield and checking the purity of the product

8 Transfer the dry crystals into a weighed sample tube, and then re-weigh it to find the actual yield of crystals.

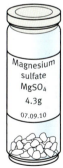

Magnesium sulfate MgSO₄ 4.3g 07.09.10

9 On the sample tube label record the name and formula of the product, the mass of product and the date it was made.

10 Often it is important to carry out tests to check that the product is pure. The appearance of the crystals can give a clue to the purity of the product. A microscope can help if the crystals are small. The crystals of a pure product are often well-formed and even in shape.

Making a salt when both reactants are solutions

Some salts are made from reactants that are both solutions. They can be made using a similar procedure to that shown. For example, to make ammonium sulfate follow the steps given here.

1 Start the procedure by measuring out the required volume of sulfuric acid and pouring it into an evaporating basin.

2 Slowly add ammonia solution to the evaporating basin, stirring the solution as you do.

3 After you have added about two thirds of the ammonia, test the solution with indicator paper or a pH meter to see if it has reached pH 7.

4 If needed, add further amounts of ammonia solution, regularly testing the pH, until it is 7 or above.

5 Slowly evaporate the solution until it is about one-fifth of its original volume.

6 Leave the concentrated solution to cool until crystals form.

Then proceed with steps 7–10 of the procedure for making magnesium sulfate.

Q: Uncertainty in measurements

Find out about

- uncertainty in measurements
- systematic and random errors as sources of uncertainty

Key words

- ➤ variation
- ➤ true value
- ➤ uncertainty
- ➤ random error
- ➤ systematic error
- ➤ outlier
- ➤ accuracy
- ➤ precision

If you measured the pulse rate of 20 people, you wouldn't get the same value for everybody. There would be **variation** in the data – in other words, there would be a range of measured values. This is because there is variation in the population. Everybody is different, and there are factors (such as diet, exercise, and stress) that cause the pulse rate to be different in each person.

If each of the 20 people measured the length of the same piece of paper, you might expect them all to come up with the same length. But you could still see variation in the data. This is due to the measurements themselves. Measured values are usually different to the **true value**, and each time a measurement is taken we can't be certain how close it is to the true value. So it is useful to give an indication of **uncertainty** when recording a measurement.

For example, you could record the volume of a liquid as 15.30 ± 0.05 cm³. From this we can see that:

- the measured value (which could be the mean of several measurements) is 15.30 cm³
- there is uncertainty in the measurement
- the person who took the measurement is confident that the true value is between 15.25 cm³ and 15.35 cm³.

Sources of uncertainty

There are two general sources of uncertainty in measurements: systematic errors and random errors.

Random error causes repeated measurements to give different values. This can happen, for example, when making judgements about the colour change at an end point or when estimating the reading from a thermometer.

A random error is not a mistake made by the person taking the measurement. A random error is a source of variation in measurements that cannot be eliminated, although there are often things we can do to reduce the amount of variation it causes.

Systematic error causes all repeated measurements to be the same amount higher or lower than the true value. This can happen when using an incorrectly calibrated measuring instrument or when taking measurements at a consistent, but wrong, temperature.

An example of how random and systematic errors can occur

The two different types of error are illustrated in the following example. The flask shown in the margin is used to measure out 25 cm³ of a solution.

For the flask in the diagram, the manufacturer states that the line indicates a measured volume of 25.00 ±0.06 cm³. So even if the measured volume of liquid is exactly aligned with the marked line each time, the volume could be consistently larger or smaller than 25.00 cm³ (by up to 0.06 cm³). This is systematic error.

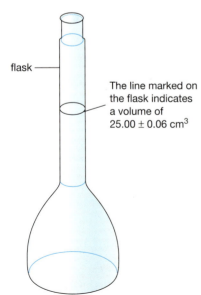

flask —

The line marked on the flask indicates a volume of 25.00 ± 0.06 cm³

This flask is used to measure a particular volume of liquid.

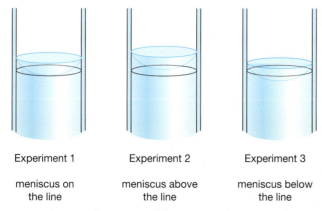

Experiment 1   Experiment 2   Experiment 3

meniscus on   meniscus above   meniscus below
the line   the line   the line

Random errors in the use of a 25.00 cm³ flask.

It is difficult to fill a flask with liquid so that the bottom of the meniscus is aligned exactly with the marked line. In experiments 2 and 3 in the diagram above, the meniscus is not aligned exactly – so the measured amounts are not 25.00 cm³. This is an example of random error. We could reduce the size of the error by aligning the meniscus as closely as possible to the line each time, but it would be very difficult to completely eliminate this source of error.

The difference between 'error' and 'mistake'

Random and systematic errors are not mistakes. Mistakes are failures by the person taking the measurement, such as taking readings from a sensitive balance in a draught. Mistakes of this kind lead to **outliers** in results, and should be avoided by people doing practical work. If it is known that a mistake was made when taking or recording a measurement, the measurement should be taken again.

Accuracy and precision

An analysis or test is often repeated to give a number of measured values, which are then averaged to produce the result.

- **Accuracy** describes how close a result is to the true or 'actual' value.
- **Precision** is a measure of the spread of measured values. A big spread indicates a greater uncertainty than a small spread.

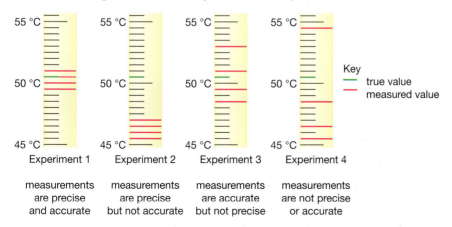

Key
— true value
— measured value

Experiment 1   Experiment 2   Experiment 3   Experiment 4

measurements   measurements   measurements   measurements
are precise   are precise   are accurate   are not precise
and accurate   but not accurate   but not precise   or accurate

Maths skills

The aim of science is to develop good explanations for observations of the natural world. Scientific explanations are based on data. Making sense of the data requires mathematical skills, including making calculations, and presenting and reading graphs and charts. This section of the book will support you in making sense of your own data from experiments and also the data presented by others.

Throughout the book are worked examples to remind you how to apply your mathematical skills in science contexts.

Numbers and units

At the heart of most scientific enquiries are measurements. Measurements are stated as a number and unit. When doing calculations it is important to be consistent in the units you use.

In science we use the SI system of measurements: the base units of kilogram (kg), metre (m), second (s), ampere (A), kelvin (K), and mole (mol) together with derived units including metre per second (m/s), newton (N), metres cubed (m^3), and degrees Celsius (°C).

These are the units you will use in GCSE Chemistry.

Quantity	Unit
mass	kilogram (kg)
length	metre (m)
time	second (s)
area	metre squared (m^2)
volume	metre cubed (m^3), decimetre cubed (dm^3)
energy	joule (J), kilowatt-hour (kWh)
temperature	degrees Celsius (°C), kelvin (K)
amount of substance	mole (mol)
concentration	mol per decimetre cubed (mol/dm^3)
density	kilogram per metre cubed (kg/m^3)
electric charge	coulomb (C)
pressure	pascal (Pa)

Standard form

Sometimes the numbers used in scientific measurements and calculations are very large or very small. For example, the distance from the Earth to the Sun is 146 900 000 000 m.

This is difficult to read and it can be easy to 'lose' one of the zeroes in a calculation. It is more convenient to express the number in **standard form**.

A number written in standard form has two parts:

The power of 10 may be positive or negative.

the multiplier (10 raised to the power needed to give the correct value of the number)

a decimal number more than one, less than ten

1.469×10^{11}

Worked example: Converting large and small numbers to standard form

1 *It is estimated that there are 8 700 000 species on Earth. Write this number in standard form.*

Step 1: Find the decimal number that is more than one and less than 10.

8.7

Step 2: Find how many times you need to *multiply* 8.7 by 10 to get 8 700 000

$8\ 700\ 000 = 8.7 \times 10 \times 10 \times 10 \times 10 \times 10 \times 10$

> Multiply by 10 six times, which is the same as multiplying by 10^6.

Answer:

$8\ 700\ 000 = 8.7 \times 10^6$

2 *The diameter of a white blood cell is about 0.000012 m. Write this quantity in standard form.*

Step 1: Find the decimal number that is more than one and less than 10.

1.2

Step 2: Find how many times you need to *divide* 1.2 by 10 to get 0.000 012.

Remember, dividing by 10 is the same as multiplying by $\frac{1}{10}$

$0.000012 = 1.2 \times \frac{1}{10} \times \frac{1}{10} \times \frac{1}{10} \times \frac{1}{10} \times \frac{1}{10}$

$= 1.2 \times \frac{1}{10^5}$

$= 1.2 \times 10^{-5}$

> Divide by 10 five times, which is the same as multiplying by 10^{-5}.

> The negative power shows the number is *less* than 1.

Answer:

$0.000012\ m = 1.2 \times 10^{-5}\ m$

You should make sure you know how to work with numbers in standard form on your calculator. Different calculators have different labels on buttons for '10 raised to the power of', for example:

 EE   or   **EXP**   or   **x 10ˣ**

Prefixes for units

Sometimes prefixes for units are used instead of writing the quantity in standard form.

For example, the diameter of a white blood cell is 1.2×10^{-5} m, which is 12 µm.

The table shows the prefixes that are used with scientific quantities.

nano	micro	milli	centi	deci	kilo	mega	giga	tera
n	µ	m	c	d	k	M	G	T
0.000 000 001	0.000 001	0.001	0.01	0.1	1000	1 000 000	1 000 000 000	1 000 000 000 000
$\times 10^{-9}$	$\times 10^{-6}$	$\times 10^{-3}$	$\times 10^{-2}$	$\times 10^{-1}$	$\times 10^{3}$	$\times 10^{6}$	$\times 10^{9}$	$\times 10^{12}$

Order of magnitude

Rounding a number to the nearest **order of magnitude** means rounding the number to the nearest power of 10.

Worked example: Order of magnitude

The radius of a carbon atom is measured to be about 0.07 nm.
What is this in metres to the nearest order of magnitude?

Step 1: Write down the length in metres.

$0.07 \text{ nm} = 0.07 \times 10^{-9} \text{ m} = 7 \times 10^{-11} \text{ m}$

> 7 is closer to 10 than to 1

Step 2: Write down the value to the nearest power of 10

$7 \times 10^{-11} \text{ m} \approx 10 \times 10^{-11} \text{ m}$
$10 \times 10^{-11} \text{ m} = 1 \times 10^{-10} = 10^{-10}$

Answer:

$7 \times 10^{-11} \text{ m} \approx 10^{-10} \text{ m}$ (to nearest order of magnitude)

Sometimes when two measurements are compared we simply want to know if they are in the same order of magnitude. If two numbers differ by an order of magnitude, then one number is about 10 times bigger than the other.

Worked example: Comparing orders of magnitude

The radius of a hydrogen atom is measured to be about 0.025 nm. The radius of a lead atom is measured to be about 0.18 nm. Compare the size of hydrogen atoms and lead atoms, and decide if their sizes are the same order of magnitude.

Step 1: Write down what you know, ensuring both measurements are in the same units.

hydrogen: 0.025 nm
lead: 0.18 nm

Step 2: Divide the larger number by the smaller number.

$\dfrac{\text{larger value}}{\text{smaller value}} = \dfrac{0.18 \text{ nm}}{0.025 \text{ nm}} = 7.2$

Answer:

Lead atom is 7.2 times larger, which is less than 10 times larger – this means they are the same order of magnitude.

Significant figures

Data from measurements in scientific experiments should show the **precision** of the measurement. The number of **significant figures** shows the precision that can be claimed for a piece of data.

The first significant figure in a number tells you the approximate size of the number. The first significant figure in a number is the first non-zero digit from the left.

For example, in the number 5437 the first significant figure is the 5; this tells you that the value is more than 5000 and less than 6000. You can round the number to any number of significant figures. So this number is 5000 (to 1 significant figure), 5400 (to 2 significant figures), or 5440 (to 3 significant figures).

The same principle applies to numbers less than one. In the number 0.0342 the first significant figure is 3, so we know the number is between three-hundredths and four-hundredths. The value is 0.03 (to 1 significant figure) or 0.034 (to 2 significant figures).

You may see significant figures abbreviated to 'sig. fig.' or 'S.F.'

Significant figures are a more useful way of expressing precision than decimal places. For example, a length measured with a metre ruler is made to the nearest millimetre (thousandth of a metre). The length of an A4 sheet of paper can be written as 297 mm, 29.7 cm, 0.297 m, and 2.97×10^{-1} m. All of those measurements are given to the same number of significant figures, even if the number of decimal places is different.

A calculated value should be given to the same number of significant figures as the least precise measurement in the data.

Worked example: Significant figures

Calculate the area of a rectangle with sides 26 mm and 13 mm.
Give your answer to 2 significant figures.

Step 1: Write down what you know.

length = 26 mm
breadth = 13 mm

Step 2: Write down the equation you will use.

area of rectangle = length × breadth

Step 3: Substitute in the numbers and calculate the area.

area = 26 mm × 13 mm
= 338 mm²

Step 4: Start at the first non-zero digit and count 2 significant figures.

338 mm²

Step 5: Apply the 5 or more rounding rule to the next digit.

area = 340 mm²

Answer:

area = 340 mm² (2 significant figures)

Making sense of data

Once data has been collected from an experiment it can be processed to show any patterns. This processing often means doing calculations and plotting graphs.

Using percentages

Percentages appear everywhere, for example, in interest rates on loans, discounts in the sales, and, of course, in science. Percentage is a way of describing a fraction of something. 'Per cent' (symbol %) means 'per hundred'. If 20 people out of a group of 100 people are left-handed, then we could say that 20% are left handed – and therefore that 80% are right-handed or ambidextrous.

If one quarter of a population have blonde hair we could say that 25 out of every 100 people have blonde hair – that is 25% have blonde hair.

Percentages can be used to make comparisons between sets of data, to calculate efficiency, or to calculate the yield of a process.

Worked example: Calculating percentages

A student carries out an experiment to find the best way to germinate seeds. One seed tray has 80 seeds and 60 of them germinate to produce plants. A second tray has 90 seeds and 70 of them germinate. Which tray gives the better yield of plants?

Step 1: Write down what you know.

Tray 1:	Tray 2:
number of seeds = 80	number of seeds = 90
number of plants = 60	number of plants = 70

Tray 2 produced more plants, but there were more seeds so percentage yields make it easier to compare.

Step 2: Write down an equation to work out the percentage yield.

$$\text{yield} = \frac{\text{number of plants}}{\text{number of seeds}} \times 100\%$$

Step 3: Substitute in the values to calculate the percentage yield.

$$\text{yield 1} = \frac{60}{80} \times 100\% \qquad \text{yield 2} = \frac{70}{90} \times 100\%$$
$$= 0.75 \times 100\% \qquad\qquad = 0.78 \times 100\%$$
$$= 75\% \qquad\qquad\qquad = 78\%$$

Answer: Tray 2 had a slightly better yield (78%) than tray 1 (75%).

Finding the best estimate

The **range** of a data set describes the spread of data. For example, in a class of students the heights of the students may have a range from 160 cm to 185 cm.

The **average** of a data set is the single number that best represents the data set. There are different ways of representing that average.

The most commonly used average is the **mean value**. The mean value is found by adding up a set of measurements and then dividing by the number of measurements.

$$\text{mean} = \frac{\text{sum of all the measurements}}{\text{number of measurements}}$$

Worked example: Finding the mean

A group of students measure their heights. The measurements are: 160 cm, 165 cm, 167 cm, 168 cm, and 185 cm. Calculate the mean height.

Step 1: Write down what you know.

heights: 160 cm, 165 cm, 167 cm, 168 cm, and 185 cm

number of students: 5

Step 2: Write down an equation to work out the mean height.	mean = $\dfrac{\text{sum of all the measurements}}{\text{number of measurements}}$
Step 3: Substitute in the values to calculate the mean.	mean = $\dfrac{160\ cm + 165\ cm + 167\ cm + 168\ cm + 185\ cm}{5}$
	$= \dfrac{845\ cm}{5}$
	$= 169\ cm$
Answer:	**mean height = 169 cm**

Sometimes the mean is not a good representation of the data. In the example of the group of students above, only one of the students is taller than the mean value. The 185 cm measurement has distorted the calculation because it is so much larger than the others.

A better representative value is sometimes the **median:** the middle value. In this data set, the median is 167 cm.

In a large data set with many values, the **mode** might be the best representative value – that is the value that occurs most often.

Whether you choose the mean, the median or the mode, it is always good to also give the range over which the data was spread.

Making sense of graphs

Scientists use graphs and charts to present data clearly and to look for patterns in the data. You will need to plot graphs or draw charts to present data, and then describe the patterns or trends and suggest explanations for them. Examination questions may also give you a graph and ask you to describe and explain it.

Reading the axes

Look at the two charts in the margin, which both provide data about daily energy use in several countries.

On the first chart the value for China is greater than for the US. But on the second chart the value for the US is much greater than for China. Why are the charts so different if they both represent information about energy use?

Look at the labels on the axes.

One shows the *energy use per person per day*, the other shows the *energy use per day by the whole country*.

The first graph shows that China uses a similar amount of energy to the US. But the population of China is much greater so the energy use per person is much less.

> **First rule of reading graphs:** Read the axes and check the units.

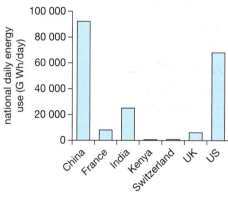

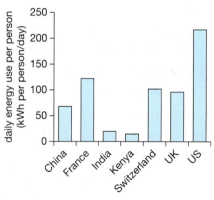

Graphs to show daily energy use in a range of countries, total and per person.

Describing the relationship between variables

The pattern of points plotted on a graph shows whether two factors are related.

Look at the graph of how the mass of a baby changes in the first 12 weeks after birth.

The two variables are age and mass. The graph tells a story about the relationship between these two variables – the baby gets a little lighter in the first two weeks and then heavier. But we can describe the pattern in more detail than that. Between three weeks and nine weeks her mass increases steadily, then increases less quickly up to 12 weeks. The slope of the graph – the **gradient** – is constant between three weeks and nine weeks.

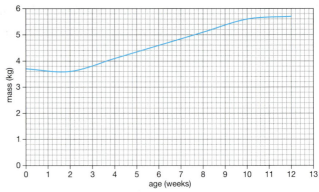

Graph showing how the mass of a baby changes in the first 12 weeks after birth.

Calculating the gradient of a graph

Many graphs show how a quantity changes with time – it might be the size of a population, the height of a plant, the concentration of a substance, or the speed of a moving object. Time is plotted on the *x*-axis and the changing quantity being measured is plotted on the *y*-axis. The gradient of such graphs describes the **rate** of change of the quantity with time.

$$\text{rate of change of quantity} = \frac{\text{change in quantity (}y\text{-axis)}}{\text{time taken to change (}x\text{-axis)}}$$

Worked example: Finding the gradient of a straight-line graph

The graph shows how a baby's mass changed in the weeks after she was born. Use the graph to work out the average rate at which her mass increased between three weeks and nine weeks.

Step 1: Use crosses (**X**) to mark on the line the points the question is asking about.

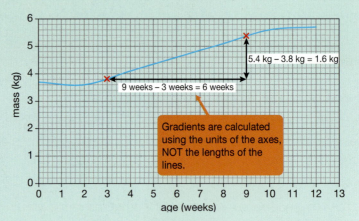

Step 2: Write down the equation to calculate the average rate.

$$\text{rate of change of mass} = \frac{\text{change in mass}}{\text{time taken for change}}$$

Step 3: Substitute in the values from the graph, including the units.

$$\text{rate of change of mass} = \frac{1.6 \text{ kg}}{6 \text{ weeks}} = 0.27 \text{ kg/week}$$

Answer:

average rate of growth = 0.27 kg/week

Sometimes the graph is a curve – the gradient is changing. To calculate the gradient at a point on the curve, you need to find the gradient of the tangent to the curve at that point.

Worked example: Finding the gradient of the tangent to a curve

The graph shows how the population of Africa changed between 1950 and 2015. Use the graph to work out the rate at which the population was growing in 1990.

Step 1: Use crosses (X) to mark on the line the point the question is asking about.

Step 2: Draw a tangent to the line at that point. The tangent just touches the curve and has the same gradient as the curve at that point.

Step 3: Make the tangent the hypotenuse of a right angle triangle large enough for you to calculate the gradient of the line.

Choose the length of the base of the triangle to give an exact quantity; this will improve the accuracy of your calculation.

Step 4: Write down the equation to calculate the rate.

Step 5: Substitute in the values from the graph, including the units.

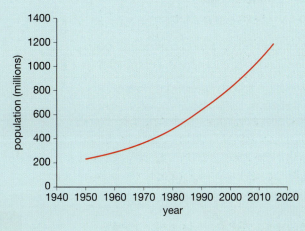

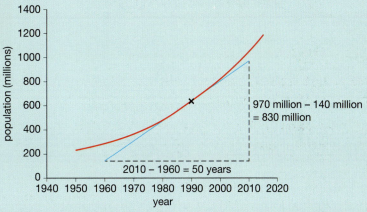

970 million – 140 million = 830 million

2010 – 1960 = 50 years

Graph to show the population of Africa between 1950 and 2015 . Source: United Nations, Department of Economic and Social Affairs, Population Division (2015).

$$\text{rate of change of population} = \frac{\text{change in population}}{\text{time taken for change}}$$

$$\text{rate of change of population} = \frac{830 \text{ million}}{50 \text{ years}}$$

$$= 16.6 \text{ million/year}$$

Answer: **rate of population growth = 17 million/year**

Answer given to 2 significant figures, because that is the precision of the data from the graph.

Second rule of reading graphs: describe each phase of the graph, including ideas about the meaning of the **gradient,** and other **data** including **units.**

Is there a correlation?

Sometimes we are interested in whether one thing changes when another does. If a change in one factor goes together with a change in something else, we say that the two things are correlated.

The two graphs on the right show how global temperatures have changed over time and how levels of carbon dioxide in the atmosphere have changed over time.

Is there a **correlation** between the two sets of data?

Look at the graphs – why is it difficult to decide if there is a correlation?

The two sets of data are over different periods of time, so although both graphs show a rise with time, it is difficult to see if there is a correlation.

It would be easier to identify a correlation if both sets of data were plotted for the same time period and placed one above the other, or on the same axes, like this:

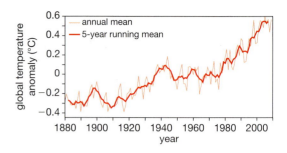

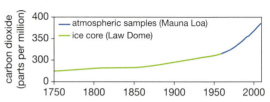

Graphs to show increasing global temperatures and carbon dioxide levels. Source: NASA.

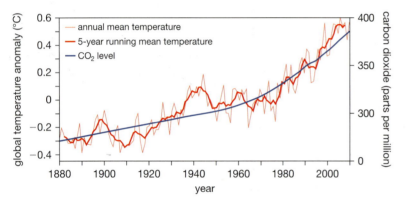

Graph to show the same data as the above two graphs, plotted on one set of axes.

> **Third rule for reading graphs:**
> When looking for a correlation between two sets of data, read the axes carefully.

When there are two sets of data on the same axes, take care to look at which axis relates to which line.

Another way to check for a correlation is to plot the two variables on a scatter graph.

Look at the scatter graph. Is there evidence that the two variables:

- are correlated?
- show a causal relationship?

Explaining graphs

Explaining the patterns or trends shown by a graph is different to describing them. It requires us to use scientific ideas to suggest what could be causing the observed patterns.

When a graph suggests that there is a correlation between two sets of data, scientists try to find out if a change in one factor *causes* a change in the other. They look for a mechanism that explains how one factor affects the other.

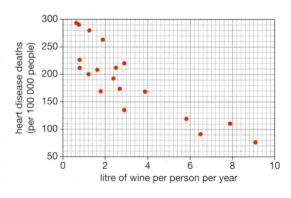

Graph showing that heart disease is less common in people who drink a moderate amount of wine, from a study in over 19 countries.

Displaying frequency data

Frequency data shows the number of times a value occurs. For example, if four students have a pulse rate of 86 beats per minute, then the data value 86 has a frequency of four.

Frequency data is presented using a bar chart or a histogram. It's important to know the difference between these two types of chart and when to use each one.

Bar chart: discontinuous data

Country	Number of nuclear power plants
USA	99
France	58
Japan	43
Russia	34
China	27
UK	16

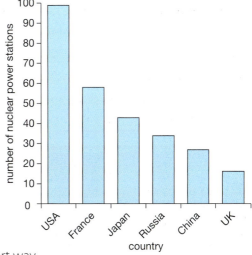

The data in the table has been plotted as a bar chart, showing the number of working nuclear power stations in different countries in June 2015.

Look at the bar chart displaying the number of nuclear power stations per country. Each country is separate, and it's not possible to measure a value part way between any two countries. The data are said to be discontinuous. A bar chart is used to display discontinuous frequency data.

A bar chart would also be used to summarise the number of trees of each species in a wood (species names on the *x*-axis), and the number of students studying chemistry in different schools (school names on the *x*-axis).

In a bar chart all of the bars should be drawn with equal width, and there should be a gap between each bar.

Histogram: continuous data

You may work with frequency data for a **continuous** variable, for example, height. Continuous data can be divided into groups known as class intervals. Collecting data in class intervals can be done by tallying. As a rule of thumb, try to divide the data into at least five class intervals.

Look at the data recording the heights of 31 people:

Height (m)	Tally	Frequency
1.60–1.65	I	1
1.65–1.70	IIII	4
1.70–1.75	IIII IIII II	12
1.75–1.80	IIII III	8
1.80–1.85	IIII	5
1.85–1.90	I	1

Note that the class interval 1.60–1.65 includes all the people with a height that is:

- greater than or equal to 1.60 m
- less than 1.65 m.

This means that a person who is 1.64 m tall is included in the class interval 1.60–1.65, but a person who is 1.65 m tall is included in the next class interval (1.65–1.70).

A histogram is used to display continuous frequency data.

In a histogram the bars are drawn touching, and are labelled at their edges on the *x*-axis.

Rearranging equations

Equations show the relationships between physical variables. They are used to calculate the value of one variable from the values of other known variables. Often you need to rearrange the standard equation to put the quantity you are trying to find on the left-hand side, before the equals sign (this makes it the subject of the equation).

There is a simple rule to follow when rearranging equations:

> Whatever you do to the left hand side of the equation you must always do the same thing to the right hand side – so the two sides remain equal.

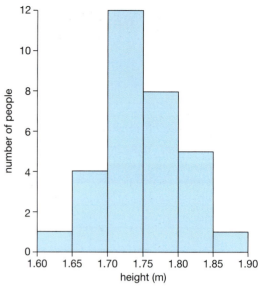

A histogram displaying the frequency of height in a group of 31 people.

Worked example: Changing the subject of the equation

1 *Rearrange the equation for the area of a circle to make the radius the subject of the equation.*

Step 1: Write down what you know.

A = area of a circle

r = radius

$A = \pi r^2$

> Tip: it is much easier to rearrange the equation using algebra symbols than using words.

Step 2: Decide what you need to do to get r^2 on its own. Divide both sides by π.

$$\frac{A}{\pi} = \frac{\pi r^2}{\pi}$$

Step 3: Cancel the π on the right hand side

$$\frac{A}{\pi} = \frac{\cancel{\pi} r^2}{\cancel{\pi}}$$

Step 4: Swap the two sides of the equation so that r^2 is on the left hand side.

$$\frac{A}{\pi} = r^2$$

$$r^2 = \frac{A}{\pi}$$

Step 5: Take the square root of both sides, to give r.

$$\sqrt{r^2} = \sqrt{\frac{A}{\pi}}$$

Answer:

$$r = \sqrt{\frac{A}{\pi}}$$

Often you will be asked to find the value of a variable from other given variables. The worked example below shows the steps to take.

Worked example: Finding the value of a variable from the formula

1 *What is the radius of a circle which has a perimeter of 30 cm?*

Step 1: Write down what you know.　　perimeter = 30 cm

Step 2: Write down the equation for the perimeter of a circle.

perimeter = 2πr

Step 3: Decide what you need to do to get r on its own.

$$\frac{\text{perimeter}}{2\pi} = \frac{2\pi r}{2\pi}$$

Divide both sides by 2 π.

$$\frac{\text{perimeter}}{2\pi} = r$$

Step 4: Swap the two sides of the equation so that r is on the left hand side.

$$r = \frac{\text{perimeter}}{2\pi}$$

Step 5: Substitute what you know into the equation.

$$r = \frac{30 \text{ cm}}{2\pi}$$

Step 6: Work out the value for r.　　$r = 4.77$ cm

Answer:　　$r = 4.8$ cm (to 2 significant figures)

You may find it easier to substitute in the values before rearranging, but the more often you practise rearranging the algebraic equations, the more familiar you will become with the different ways of writing the relationships.

Calculating reacting masses and percentage yields

Worked example: Calculating reacting masses

How much aluminium powder do you need to react with 8.0 g of iron oxide?

Step 1: Write the balanced equation for the reaction.

$$2Al(s) + Fe_2O_3 \longrightarrow Al_2O_3(s) + 2Fe(s)$$

Reactants:

Step 2: Work out the relative formula masses (RFM) by using multiplication and addition.

RFM of $Al = 27$

RFM of $Fe_2O_3 = (2 \times 56) + (3 \times 16) = 160$

Products:

RFM of $Al_2O_3 = (2 \times 27) + (3 \times 16) = 102$

RFM of $Fe = 56$

Step 3: Find the relative masses in this reaction by multiplying the RFM by the numbers in front of each formula in the equation. Then convert the relative masses to reacting masses by including units. These can be g, kg, or tonnes depending on the data in the question. The units must be the same for each of the values.

$$2Al(s) \qquad + \quad Fe_2O_3(s) \quad \longrightarrow \quad Al_2O_3(s) \quad + \quad 2Fe(s)$$
$$(2 \times 27) = 54\ g \qquad 160\ g \qquad\qquad 102\ g \qquad (2 \times 56) = 112\ g$$

Step 4: Scale the reacting masses to the known quantities by using simple ratios. Always include the correct units when substituting values.

$$\frac{\text{mass of aluminium required}}{\text{reacting mass of aluminium}} = \frac{\text{mass of iron oxide used}}{\text{reacting mass of iron oxide}}$$

$$\frac{\text{mass of aluminium required}}{54\ g} = \frac{8\ g}{160\ g}$$

Step 5: Rearrange the equation to find the mass of aluminium.

$$\text{mass of aluminium required} = \frac{(8\ g)}{(160\ g)} \times 54\ g$$
$$= 2.7\ g$$

Answer:

mass of aluminium required = 2.7 g

Worked example: Calculating percentage yield

What is the percentage yield of iron for the same reaction, if 4.9 g of iron is actually produced?

Step 1: Work out the theoretical yield using ratios.

$$\frac{\text{mass of iron yielded}}{\text{reacting mass of iron}} = \frac{\text{mass of iron oxide used}}{\text{reacting mass of iron oxide}}$$

$$\frac{\text{mass of iron yielded}}{112\ g} = \frac{8\ g}{160\ g}$$

Step 2: Rearrange the equation to calculate the theoretical yield of iron.

$$\text{theoretical yield of iron} = \frac{8\ g}{160\ g} \times 112\ g$$

$$= 5.6\ g$$

Step 3: Calculate the percentage yield by substituting the quantities into the equation.

$$\text{percentage yield} = \frac{\text{actual yield (g)}}{\text{theoretical yield (g)}} \times 100\%$$

$$= \frac{4.9\ g}{5.6\ g} \times 110$$

$$= 87.5\%$$

Answer:

percentage yield = 87.5%

Models in science

Find out about

- what a model is
- how models are used to investigate, explain, and predict
- the benefits and limitations of models
- representational, descriptive, and mathematical models

Key words

➤ model
➤ system
➤ representational model>

Most children like to play with toys. A popular toy is a model car. It's called a **model** because it represents something in the real world. It has some of the main features of a real car, such as wheels, doors, a windscreen, and a roof. But some features, such as the engine, are not included. The model shows that the car can move when the wheels turn.

A model car is a simple way to represent the main features of a car, but much of the detail of a real car is not included in the model.

Models are everywhere

People use models all the time, even if they don't realise it. A model isn't always an object, such as a toy car. Models can be words, pictures, and numbers. For example, a map is a model.

Millions of people visit London every year, and they use maps to help them travel around. The usual map of the London Underground is a useful guide for getting from one tube station to another. It's a good model of the city's underground train network because it shows how the stations and lines are connected. It is quick and easy to understand. It can answer questions such as: 'What is the most direct way to get from Bond Street to Westminster?'

This map is a simple model of the London Underground train network. It helps people to work out how to get from one Tube station to another.

We can also use the model to make predictions. For example, we could predict that it would take longer to travel by Tube to Westminster from Camden Road than it would from Bond Street.

However, the London Underground map can't solve all of a traveller's problems. It doesn't show how the Tube stations relate to streets and landmarks on the surface. These features are not included in the model. To answer the question 'What is there to see around Westminster Tube station?' we can use a different, more detailed model – for example, a street map.

This map is another model of London, but it includes features such as streets and local landmarks.

What is a model?

From the everyday examples of model cars and maps, we know that a model:

- is a simpler way of representing something in the real world
- includes some, but not necessarily all, of the features of the thing it represents
- can show how these features are connected or interact
- can be used to explain things, answer questions, and make predictions.

We also know that the usefulness of a model is limited by how closely it represents the real world and that different models are useful in different situations.

Usually, a model represents a collection of interacting parts. Scientists often refer to a collection of interacting parts as a **system**.

Using models in science

Models are useful in science. They help us to explain how things work and interact. They also help us to make predictions and investigate possible outcomes. Scientific models can be words, pictures, objects, numbers, graphs, or equations.

Many different models are used in science, but they can be grouped into three main types. These are described below. You've already used models in your science lessons. An example of a model that you should be familiar with is given for each type.

A number of scientific models are highlighted throughout this book – look out for the red boxes like the ones shown here.

Representational models

A **representational model** uses simple shapes or analogies to represent the interacting parts of a system. One example of a representational model is the particle model – it helps us to visualise the tiny particles (atoms and molecules) that make up substances.

Different models of the same thing can contain different amounts of detail. A model car made of blocks will roll on its wheels from the top of a slope to the bottom. A remote-controlled car contains a motor, so can be driven from the bottom of a slope to the top. The models can be used to demonstrate and investigate different behaviours.

The particle model of matter

All matter is made of very tiny particles with attractive forces between them. The particle model represents the particles as spheres. It helps us explain the arrangement and movement of the particles in different states of matter.

However, this model makes some simplifications. For example, particles of matter are not all identical spheres – they have different shapes, sizes, and masses. This means there are limitations to the model and what we can do with it.

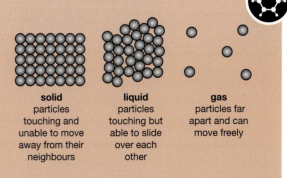

solid
particles touching and unable to move away from their neighbours

liquid
particles touching but able to slide over each other

gas
particles far apart and can move freely

You should already know about some simple models of photosynthesis (the process plant cells use to make food).

A speed camera takes photographs of a car as it travels over markings on the road. The time between the photographs and the distance between the markings are used to calculate the car's speed.

Descriptive models

A **descriptive model** uses words to identify features of a system and describe how they interact. One example of a descriptive model is a simple account of the inputs and outputs of photosynthesis. It helps us to explain how plant cells can make their own food in the form of glucose (a type of sugar).

A simple model of photosynthesis

Photosynthesis is a complicated biological process, involving a number of chemical reactions inside cells. However, a simple description of the inputs and outputs can help us to understand what is going on.

Inputs	Outputs
light	glucose
water	oxygen
carbon dioxide	

This is a very simple way of summing up the process of photosynthesis. We could use this simple model to make a prediction, for example, that increasing the supply of inputs would increase the amount of glucose made.

However, this model misses out much of the detail. It does not tell us anything about the chemical reactions or where they take place.

Mathematical models

A **mathematical model** uses patterns in data of past events, along with known scientific relationships, to predict what will happen to one variable when another is changed. This involves doing calculations. If the calculations are very complicated, they can be done by a computer – this is a **computational model**. But a very simple example of a mathematical model is the relationship between speed, distance, and time.

Modelling the speed of a moving object

There is a simple, scientific relationship between the speed of a moving object, the distance it moves, and the time taken to move that distance. The relationship is:

$$\text{speed} = \frac{\text{distance}}{\text{time}}$$

This is a mathematical model. We can use it to calculate the speed of a moving object if we have values for the other two variables (distance and time). We can also use the model to make a prediction, for example if the car travels for a longer time at the same speed, it will travel further.

But this simple model is limited. It can only be used to calculate an average speed from the total distance and time – it does not include information about changes in speed such as acceleration or deceleration. Also, it does not include any information about the direction the car is travelling.

One type of mathematical model is a **spatial model**. A computer is used to make a model of one or more objects in a three-dimensional space. The model can be used to predict the outcome of changing a variable (e.g., temperature) in a given space (e.g., a landscape and the atmosphere above it). The space can be divided into sections and outcomes predicted for each section. Spatial models are often used to make predictions about weather and climate.

Why use models?

If we tried to describe and explain every unique situation in the world, we would never get finished! This is one reason why models are useful. A model is a general explanation that applies to a large number of situations – perhaps to all possible ones. For example, every cell in your body is different, but we can use a model of an animal cell to describe the main features that animal cells have in common.

A benefit of models is that they enable us to investigate situations that we cannot investigate using practical experiments – perhaps because it is not ethical, is too expensive, or is not possible to do so. For example, mathematical models enable us to investigate the future effects of human activities on the Earth's climate and biodiversity. This helps us to make predictions about the likely outcomes of different courses of action, which can affect the decisions we make.

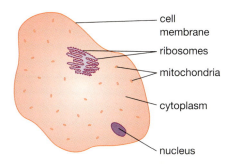

A simple model of an animal cell. Every animal cell is different, but the model includes the main features they have in common.

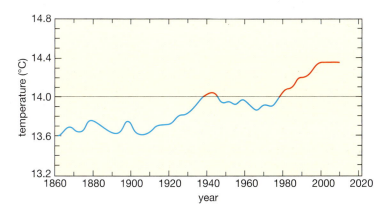

Measurements taken over the past 150 years show that the Earth's temperature is rising. We can use a mathematical model to predict how it may change in the future.

Models are very useful but we must always remember that they are limited. Even a very good model cannot represent the real world exactly, so outcomes in the real world could be different from a model's outcomes. Models help weather forecasters make predictions about the weather, but they don't always get it exactly right. Always think carefully about how much confidence you can have in claims based on a model – be realistic, and don't be too surprised if things turn out a little differently in the real world.

Key words
➤ descriptive model
➤ mathematical model
➤ computational model
➤ spatial model

Appendices

Periodic Table

(1) 1	(2) 2		(3) 13	(4) 14	(5) 15	(6) 16	(7) 17	(0) 18
1 **H** Hydrogen 1.0								2 **He** Helium 4.0

Key

atomic number
symbol
name
relative atomic mass

(1) 1	(2) 2	3	4	5	6	7	8	9	10	11	12	(3) 13	(4) 14	(5) 15	(6) 16	(7) 17	(0) 18
1 **H** Hydrogen 1.0																	2 **He** Helium 4.0
3 **Li** Lithium 6.9	4 **Be** Beryllium 9.0											5 **B** Boron 10.8	6 **C** Carbon 12.0	7 **N** Nitrogen 14.0	8 **O** Oxygen 16.0	9 **F** Fluorine 19.0	10 **Ne** Neon 20.2
11 **Na** Sodium 23.0	12 **Mg** Magnesium 24.3											13 **Al** Aluminium 27.0	14 **Si** Silicon 28.1	15 **P** Phosphorus 31.0	16 **S** Sulfur 32.1	17 **Cl** Chlorine 35.5	18 **Ar** Argon 39.9
19 **K** Potassium 39.1	20 **Ca** Calcium 40.1	21 **Sc** Scandium 45.0	22 **Ti** Titanium 47.9	23 **V** Vanadium 50.9	24 **Cr** Chromium 52.0	25 **Mn** Manganese 54.9	26 **Fe** Iron 55.8	27 **Co** Cobalt 58.9	28 **Ni** Nickel 58.7	29 **Cu** Copper 63.5	30 **Zn** Zinc 65.4	31 **Ga** Gallium 69.7	32 **Ge** Germanium 72.6	33 **As** Arsenic 74.9	34 **Se** Selenium 79.0	35 **Br** Bromine 79.9	36 **Kr** Krypton 83.8
37 **Rb** Rubidium 85.5	38 **Sr** Strontium 87.6	39 **Y** Yttrium 88.9	40 **Zr** Zirconium 91.2	41 **Nb** Niobium 92.9	42 **Mo** Molybdenum 95.9	43 **Tc** Technetium	44 **Ru** Ruthenium 101.1	45 **Rh** Rhodium 102.9	46 **Pd** Palladium 106.4	47 **Ag** Silver 107.9	48 **Cd** Cadmium 112.4	49 **In** Indium 114.8	50 **Sn** Tin 118.7	51 **Sb** Antimony 121.8	52 **Te** Tellurium 127.6	53 **I** Iodine 126.9	54 **Xe** Xenon 131.3
55 **Cs** Caesium 132.9	56 **Ba** Barium 137.3	57–71 Lanthanides	72 **Hf** Hafnium 178.5	73 **Ta** Tantalum 180.9	74 **W** Tungsten 183.8	75 **Re** Rhenium 186.2	76 **Os** Osmium 190.2	77 **Ir** Iridium 192.2	78 **Pt** Platinum 195.1	79 **Au** Gold 197.0	80 **Hg** Mercury 200.6	81 **Tl** Thallium 204.4	82 **Pb** Lead 207.2	83 **Bi** Bismuth 209.0	84 **Po** Polonium	85 **At** Astatine	86 **Rn** Radon
87 **Fr** Francium	88 **Ra** Radium	89–103 Actinides	104 **Rf** Rutherfordium	105 **Db** Dubnium	106 **Sg** Seaborgium	107 **Bh** Bohrium	108 **Hs** Hassium	109 **Mt** Meitnerium	110 **Ds** Darmstadtium	111 **Rg** Roentgenium	112 **Cn** Copernicium 1		114 **Fl** Flerovium		116 **Lv** Livermorium		

Chemical formulae

argon	Ar	methanol	CH_3OH
bromine	Br_2	nitric acid	HNO_3
calcium carbonate	$CaCO_3$	nitrogen	N_2
calcium chloride	$CaCl_2$	nitrogen dioxide	NO_2
calcium sulfate	$CaSO_4$	nitrogen monoxide	NO
carbon dioxide	CO_2	oxygen	O_2
carbon monoxide	CO	potassium bromide	KBr
chlorine	Cl_2	potassium chloride	KCl
ethanoic acid	CH_3COOH	potassium hydroxide	KOH
ethanol	C_2H_5OH	potassium iodide	KI
hydrochloric acid	HCl	sodium bromide	NaBr
hydrogen	H_2	sodium carbonate	Na_2CO_3
iodine	I_2	sodium chloride	NaCl
lithium bromide	LiBr	sodium hydroxide	NaOH
lithium chloride	LiCl	sodium iodide	NaI
lithium hydroxide	LiOH	sodium nitrate	$NaNO_3$
lithium iodide	LiI	sodium sulfate	Na_2SO_4
magnesium carbonate	$MgCO_3$	sulfur dioxide	SO_2
magnesium chloride	$MgCl_2$	sulfuric acid	H_2SO_4
magnesium hydroxide	$Mg(OH)_2$	water	H_2O
magnesium oxide	MgO		
magnesium sulfate	$MgSO_4$		
methanoic acid	HCOOH		

Qualitative analysis data

Tests for negatively charged ions

Ion	Test	Observation
carbonate CO_3^{2-}	add dilute acid	effervesces, and carbon dioxide gas produced (the gas turns lime water milky)
chloride (in solution) Cl^-	acidify with dilute nitric acid, then add silver nitrate solution	white precipitate
bromide (in solution) Br^-	acidify with dilute nitric acid, then add silver nitrate solution	cream precipitate
iodide (in solution) I^-	acidify with dilute nitric acid, then add silver nitrate solution	yellow precipitate
sulfate (in solution) SO_4^{2-}	acidify, then add barium chloride solution or barium nitrate solution	white precipitate

Tests for positively charged ions

Ion	Test	Observation
calcium Ca^{2+}	add sodium hydroxide solution	white precipitate (insoluble in excess)
copper Cu^{2+}	add sodium hydroxide solution	light-blue precipitate (insoluble in excess)
iron(II) Fe^{2+}	add sodium hydroxide solution	green precipitate (insoluble in excess)
iron(III) Fe^{3+}	add sodium hydroxide solution	red-brown precipitate (insoluble in excess)
zinc Zn^{2+}	add sodium hydroxide solution	white precipitate (soluble in excess, giving a colourless solution)

Tests for gases

Gas	Test	Observation
hydrogen	Insert a lit wooden splint into gas.	gas will ignite with a 'squeaky pop' sound
oxygen	Insert a glowing wooden splint into gas.	wooden splint re-ignites
carbon dioxide	Bubble gas through lime water.	lime water turns milky
chlorine (harmful)	Insert damp blue litmus paper into gas. Always work in a fume hood when handling chlorine gas.	damp blue litmus paper turns red, and then is bleached white

Flame tests

Some substances produced characteristic flame colours when held in the flame of a Bunsen burner

Element	Flame colour
calcium	orange-red
copper	blue-green
lithium	bright red
potassium	lilac
sodium	bright yellow

Useful mathematical relationships

You will need to be able to carry out calculations using these mathematical relationships.

Chromatography

$$\text{retardation factor } (R_f) = \frac{\text{distance travelled by solute}}{\text{distance travelled by solvent}}$$

Amount of substance

$$\text{number of moles} = \frac{\text{mass of substance (g)}}{\text{relative formula mass (g)}}$$

$$\text{number of moles of gas} = \frac{\text{volume of gas in sample (dm}^3)}{24\ (\text{dm}^3)}$$

Concentration

$$\text{concentration (g/dm}^3) = \frac{\text{mass of solute (g)}}{\text{volume (dm}^3)}$$

$$\text{concentration (mol/dm}^3) = \frac{\text{number of moles of solute}}{\text{volume (dm}^3)}$$

Yield of reactions

$$\text{atom economy} = \frac{\text{mass of atoms in desired product}}{\text{total mass of atoms in reactants}}$$

$$\text{yield} = \frac{\text{actual yield}}{\text{theoretical yield}} \times 100\%$$

Some physical properties of a range of materials

Material	Melting point (°C)	Electrical conductivity (MS/m)	Strength in tension (MPa)	Strength in compression (MPa)	Stiffness Young's modulus (GPa)	Density (kg/m³)	Mohs hardness
bottle glass	580	10^{-13}			70	2490	5.5
glass fibre		10^{-14}	3445 (when fresh)	1080	72–85	2580	
brick			depends on joints	7–70	7	1500–1800	
porcelain		10^{-18}			70	2500	
polyethene		10^{-20}	15–29		0.15–1.0	920–960	
copper	1085	64	220–430		130	8900	3
iron	1538	11	210		210	7900	4
aluminium	660	41	50–110		70	2700	2.5–3
lead	327	4.55	15		16	11000	1.5
concrete			cracks unless reinforced	6–70	15–40	800–2400	

Glossary

accuracy How close a quantitative result is to the true or 'actual' value.

acid An acid is a compound that produces hydrogen ions when it dissolves in water. Acidic solutions have a pH lower than 7. Acid solutions change the colour of indicators, form salts when they neutralise alkalis, react with carbonates to form carbon dioxide, and give off hydrogen when they react with a metal.

activation energy The minimum energy needed in a collision between molecules if they are to react. The activation energy is the height of the energy barrier between reactants and products in a chemical change.

active site The part of an enzyme where the chemical reaction takes place. The reacting molecules (substrates) fit into the active site.

actual yield The mass of the required chemical obtained after separating and purifying the product of a chemical reaction.

addition polymerisation A chemical reaction in which monomer molecules 'add together' to form a polymer. In addition polymerisation, the only product of the reaction is the polymer. Polyethene is an example of a polymer formed by addition polymerisation.

addition reaction Addition reactions of alkenes involve adding atoms by the breaking of a double bond. Only one product is made.

air pollutants Harmful substances that have been released into the atmosphere.

alcohol An organic compound containing the functional group —OH. Examples inlcude methanol (CH_3OH) and ethanol (C_2H_5OH).

alkali A compound that dissolves in water to give a solution of hydroxide ions which makes the pH higher than 7. An alkali can be neutralised by an acid to form a salt.

alkali metals Elements in Group 1 of the periodic table. Alkali metals react with water to form alkaline solutions of the metal hydroxide.

alkanes Alkanes are hydrocarbons found in crude oil. All the C–C bonds in alkanes are single bonds. Methane is an alkane. It has the formula CH_4.

alkene A hydrocarbon which contains a C=C double bond.

allotropes Structurally different forms of the same element.

alloy A material that is a mixture of two or more elements, where at least one is a metal. The properties of an alloy can be adjusted by adjusting the proportions in which the consituent elements are mixed.

anode The electrode that attracts negative ions during electrolysis.

antibacterial Destroys or slows down the growth of bacteria.

aqueous solution A solution in which water is the solvent.

atmosphere A layer of gas around a planet. The Earth's atmosphere is a mixture of gases (air) containing roughly 78% nitrogen and 21% oxygen, with trace amounts of other gases. Three quarters of the mass of the atmosphere is within the first 11 km of height. The atmosphere protects life on Earth by absorbing ultraviolet solar radiation and reducing temperature extremes between day and night.

atom The smallest particle of an element. The atoms of each element are the same as each other and are different from the atoms of other elements.

atom economy A measure of the efficiency of a chemical process. The atom economy for a process shows the mass of useful product atoms as a percentage of the mass of reactant atoms.

atomic number The number of protons in an atom of an element.

Avogadro constant The number of particles in a mole. Avogadro's constant = 6.0×10^{23}.

balanced equation An equation showing the formulae of the reactants and products. The equation is balanced when there is the same number of each kind of atom on both sides of the equation.

ball-and-stick model A 3-D model of a molecule which uses balls to represent atoms and sticks to represent bonds. It is used to show the shape of the molecule and the bond angles.

barrier method A way of slowing down or preventing corrosion by placing a barrier between the metal, and the oxygen and water in the surrounding environment.

batch A quanity of product produced at one time.

benefits The positive or helpful effects of a product or process.

binary compound A compound that contains atoms of only two types of element.

biocatalyst Substances produced by microorganisms that can alter the rate of reaction (e.g., enzymes).

biodegradable Substances that can be broken down by microorganisms, such as bacteria and fungi. Most paper and wood items are biodegradable, but most synthetic polymers, such as plastics, are not.

bioleaching The use of bacteria to extract metal ions from metal ores or waste material to form a solution of metal ions.

boiling point The temperature at which a pure sample of a liquid all changes into a gas.

bond strength A measure of how much energy is needed to break a covalent bond between two atoms. Bond strength is measured in kilojoules (kJ).

brittle A material that snaps, rather than changing shape, when stressed. Brittle is the opposite of tough.

buckminsterfullerene An allotrope of carbon, made up of 60 carbon atoms held together by covalent bonds in a sphere.

bulk chemical Chemicals made by industry on a scale of thousands or millions of tonnes per year. Examples are sulfuric acid, nitric acid, sodium hydroxide, ethanol, and ethanoic acid.

burette A graduated tube with taps or valves used to measure the volume of liquids or solutions during quantitative investigations such as titrations.

by-product Unwanted product of chemical synthesis. By-products are formed by side-reactions that happen at the same time as the main reaction, thus reducing the yield of the product required.

carbohydrate A substance made of carbon, hydrogen, and oxygen. Carbohydrates include sugars such as glucose, and natural polymers such as starch and cellulose.

carbon capture Technology that can trap carbon dioxide produced by an industrial process and in some cases store it underground.

carbon cycle The natural processes that recycle carbon through the living and non-living parts of an ecosystem.

carbon sink A process by which carbon dioxide is removed from the atmosphere.

carboxylic acid Carboxylic acids are organic compounds containing the functional group –COOH. Methanoic acid is an example. It has the formula CH_3COOH.

catalyst A catalyst is a substance that increases the rate of a chemical reaction, but which is left unchanged by the reaction.

catalytic converter A device fitted to a vehicle exhaust that changes the waste gases into less harmful ones.

cathode The electrode that attracts positive ions during electrolysis.

causal link When a scientific explanation can be given as to how one factor causes a change in another.

cause When a change in a factor produces a particular outcome, and there is a mechanism to explain this link, then the factor is said to cause the outcome.

ceramics Ceramics are solid and brittle materials such as pottery, glass , cement, and brick.

chemical bond An attractive force between atoms that allows the formation of chemical substances that contain two or more atoms.

chemical change/reaction A change that forms a new substance.

chemical equation A summary of a chemical reaction showing the reactants and products with their physical states (see also: balanced equation).

chemical formula A way of describing a substance that uses chemical symbols for atoms. It tells you which elements the substance is made up of, and the relative amounts of each element present in the substance.

chlorination The process of adding chlorine to water in order to kill any microorganisms present.

climate Long-term weather patterns in a region.

climate change A long-term change in the climate of the Earth, or a region of the Earth.

climate model A computer simulation using existing data and scientific understanding that predicts climate patterns in the future.

combustion When a chemical reacts rapidly with oxygen, releasing energy.

complete combustion Carbon compounds burn in a plentiful supply of oxygen to form carbon dioxide and water.

composite A material made by combining two or more materials. The materials which make up a composite material often have very different properties and when combined they work together to give a composite material unique physical properties.

compound A substance made of two or more elements that are chemically combined.

computational model A type of mathematical model in which the calculations are done by a computer.

concentration The amount of solid dissolved in 1 dm³ solution. Concentration may be measured in terms of mass (with units g/dm³) or moles (with units mol/dm³).

condensation The change of state from a gas to a liquid, for example, water vapour in the air condenses to form rain.

condensation polymerisation A chemical reaction in which monomer molecules react to form a polymer. In condensation polymerisation, each time a monomer is added to the chain, a small molecule such as water is also formed. Nylon is an example of a polymer formed by condensation polymerisation.

conservation of atoms All the atoms present at the beginning of a chemical reaction are still there at the end. No new atoms are created and no atoms are destroyed during a chemical reaction.

conservation of mass The total mass of chemicals is the same at the end of a reaction as at the beginning. No atoms are created or destroyed and so no mass is gained or lost.

continuous flow (process) A process used to produce large amounts of product without stopping the production process.

correlation When an outcome happens if a specific factor is present, but does not happen when it is absent, or if a measured outcome increases (or decreases) as the value of a factor increases, there is a correlation between the two. For example, there is a correlation between pollen count and the number of hay fever cases.

corrosion The gradual destruction of metals by their reaction with oxygen and water in the environment. Corrosion results in the metal becoming weaker and brittle.

covalent bond A bond formed between two atoms of non-metal elements by sharing a pair of electrons.

cracking Heat and a catalyst cause chemical changes to molecules in crude oil. Larger molecules are broken down into smaller molecules, some of which have a double bond. Molecules may also change shape.

crude oil A dark, oily liquid found in the Earth, which is a mixture of hydrocarbons.

crystallisation The process of forming crystals, for example, by evaporating the water from a solution of a salt.

denatured When the shape of an enzyme has been changed, usually as a result of high temperature or a pH change. Denatured enzymes no longer work because the shape of the active site has changed.

density The mass per unit volume of a material.

desalination Removal of salt from seawater.

descriptive model A type of scientific model that uses words to identify features of a system and to describe how they interact. One example of a descriptive model is a simple account of the inputs and outputs of photosynthesis. It helps us to explain how plant cells can make their own food in the form of glucose (a type of sugar).

diamond A gemstone. A form of carbon. It has a giant covalent structure and is very hard.

dilute acid An acid which has had water added to it so that the concentration of hydrogen ions becomes very low.

disinfection The process of killing microorganisms.

displacement reaction In a displacement reaction a more reactive element will displace a less reactive reactive element from its compound. The less reactive metal becomes an element and the more reactive element becomes part of a compound. Examples include reactions of halogens (Group 7) and metals.

distillation A method of separating a mixture of two or more substances with different boiling points.

DNA A natural polymer made of nucleotides. DNA carries the genetic code, which controls how an organism develops and functions.

double bond A covalent bond between two atoms involving the sharing of two pairs of electrons.

durable A material is durable if it lasts a long time in use. It does not wear out.

dynamic equilibrium Chemical equilibria are dynamic. At equilibrium the forward and backward reactions are still continuing but at equal rates so that there is no overall change.

efficiency The percentage of energy supplied to a device that is usefully transferred by it.

electrical conductivity A measure of a material's ability to conduct electricity. Metals are good conductors.

electrode A conductor made of a metal or graphite through which a current enters or leaves a chemical during electrolysis.

electrolysis Splitting up an ionic compound when molten or in aqueous solution by passing an electric current through it.

electrolyte A chemical that can be split up by an electric current when molten or in solution is the electrolyte. Ionic compounds are electrolytes.

electron A tiny, negatively charged particle, which is part of an atom. Electrons are found outside the nucleus. Electrons have negligible mass and one unit of charge.

electron shell The electrons in an atom have different energies and are arranged at distinct energy levels. These energy levels are sometimes called shells.

electronic structure The number and arrangement of electrons in an atom of an element.

electroplate Applying a thin layer of a metal onto an object by using the object as the negative electrode during electrolysis.

electrostatic force The force of attraction between objects with opposite electric charges. A positive ion, for example, attracts a negative ion.

element A substance made up of one type of atom.

emission The process in which something is given out by something else, for example, the emission of carbon dioxide from combustion engines.

emission spectroscopy An analytical technique in which the spectrum produced by a chemical sample in a flame is detected and analysed.

empirical formula A type of chemical formula which shows the simplest ratio of atoms in a compound.

end point The point during a titration at which the reaction is just complete. For example, in an acid–alkali titration, the end point is reached when the indicator changes colour. This happens when exactly the right amount of acid has been added to react with all the alkali present at the start.

endothermic An endothermic process uses energy from its surroundings, making them cooler.

energy level The electrons in an atom have different energies and are arranged at distinct energy levels.

energy-level diagram A diagram to show the difference in energy between the reactants and the products of a reaction.

enzyme A protein that catalyses (speeds up) a chemical reaction.

equilibrium A state of balance in a reversible reaction when neither the forward nor the backward reaction is complete. The reaction appears to have stopped. At equilibrium reactants and products are present and their concentrations are not changing.

error The difference between a measured value and the true value.

ethical Concerned with what is right or wrong.

eutrophication A type of environmental damage. Excess nitrate or phosphate in water causes rapid growth of algae, followed by death and decomposition of water plants (due ot lack of light), and a reduction of oxygen levels in the water. This usually leads to the death of water animals (due to lack of oxygen).

evaporation The change of state from a liquid to a gas. For example, when a puddle dries out the water has evaporated.

exothermic An exothermic process transfers energy to its surroundings, making them warmer.

exposure How much of a hazard that a person will come in contact with and in what way. Risk is made up of hazard and exposure.

extract Remove a valuable element (usually a metal) from the natural environment. Extraction may refer to mining the metal ore or to the chemical reaction that produces the metal from the ore.

extrapolation The process of extending the line of a graph to estimate values beyond the original data.

factor A variable that changes and may affect something else (the outcome).

feedstock The chemicals that feed into an industrial process to make a chemical product.

fermentation Anaerobic respiration in yeast cells, which breaks down glucose, and makes ethanol (alcohol) and carbon dioxide.

fibre-reinforced plastics A composite material in which the matrix is a synthetic polymer. The polymer has fibres, usually of glass or carbon, embedded in it.

fibres Long thin threads that make up materials such as wool and polyester. Most fibres used for textiles consist of natural or synthetic polymers.

filtration The process of separating a solid from a liquid by passing it through filter paper.

fine chemical Chemicals made by industry in smaller quantities than bulk chemicals. Fine chemicals are used in products such as food additives, medicines, and pesticides.

flame colour A colour produced when a chemical is held in a flame. Some elements and their compounds give characteristic colours. Sodium and sodium compounds, for example, give bright yellow flames.

flammable Desribes a substance that ignites easily when it is heated or exposed to a flame.

flexible A fllexible material bends easily without breaking or snapping.

formula A representation of the number and types of atom in a compound. Symbols are used to represent atoms.

formulations Mixtures that are prepared according to a specific formula

fossil fuel Coal, gas and oil are fossil fuels. They were formed from the remains of trees and sea creatures over millions of years and are finite, which means they cannot be replaced but will run out.

fraction A mixture of hydrocarbons with similar boiling points that have been separated from crude oil by fractional distillation.

fractional distillation The process of separating a liquid mixture into groups of molecules with similar boiling points.

fuel cell A device that produces electricity by the oxidation of a fuel.

fullerenes A family of carbon nanoparticles in which the carbon atoms are held together by covalent bonds in the shape of a hollow sphere or tube. Buckminsterfullerene is an example of a fullerene.

fully displayed formula A 2-D representation of a molecule using a structural formula that shows every bond, including those in the functional groups.

functional group A reactive group of atoms in an organic molecule. The hydrocarbon chain making up the rest of the molecule is generally unreactive. Examples of functional groups are —OH in alcohols and —COOH in carboxylic acids.

gas scrubbing A process used to remove pollutants from flue gases.

general formula A general formula shows the ratio of the number of carbon to other atoms in molecules of a homologous series. The general formula for the alkanes is C_nH_{2n+2}.

giant covalent structure A giant, three-dimensional arrangement of atoms that are held together by covalent bonds. Diamond is an example of a substance with a giant covalent structure.

giant ionic lattice The structure of a an ionic compound in the solid state. It consists of millions of oppositely charged ions packed closely together in a regular, three-dimensional arrangement.

giant structure A giant, three-dimensional arrangement of atoms or ions. Giant covalent structures contain atoms held together by covalent bonds. Silicon dioxide and diamond have giant covalent structures. Giant ionic structures contain ions. There are no individual molecules, but millions of oppositely charged ions packed closely together in a regular, three-dimensional arrangement.

gradient of a graph The gradient, or slope, of a graph is a measure of its steepness. It is calculated by choosing two points on the graph and calculating: the change in the value of the y-axis variable/the change in the value of the x-axis variable. If the graph is a straight line, you can use any two points on it. If it is curved, you should estimate the gradient of the tangent at the required point.

graduation A line on a container, ruler, or meter that marks a measurement.

graphene A form of carbon consisting of a single sheet, one atom thick.

graphite A form of carbon. It has a giant covalent structure. It is unusual for a non-metal in that it conducts electricity.

green chemistry Green chemistry is sustainable chemistry. It is the design of chemical products and processes that reduce or eliminate the use and generation of hazardous substances

greenhouse effect The atmosphere absorbs infrared radiation from the Earth's surface and radiates some of it back to the surface, making it warmer than it would otherwise be.

greenhouse gas Gases in the atmosphere which absorb radiation emitted by the Earth and thereby contribute to the greenhouse effect (warming). The most potent important greenhouse gases are carbon dioxide, methane and water vapour.

group In the Periodic Table, each vertical column is called a group of elements. All the elements in the same group have the same number of electrons in their outermost shell. Elements in the same group have similar properties.

Haber Process An efficient process of reacting nitrogen and hydrogen gas to make ammonia on an industrial scale.

half equation An equation to show the changes that happen to ions at the electrodes during electrolysis. Half equations show ions gaining or losing electrons.

halogen An element located in Group 7 of the periodic table.

hard A material that is difficult to dent or scratch.

hazard A source of potential harm to health or the environment.

heterogeneous A heterogeneous material varies throughout, eg, it may be composed of layers of different materials. A sample would need to be carefully taken to take account of the different components, in order to be representative of the whole.

homogeneous A homogeneous material is uniform throughout, so a single small sample would be representative of the whole.

homologous series A series of compounds which have similar chemical properties and the same general formula. The alkanes, alkenes, alcohols and carboxylic acids are all examples of homologous series.

hydrocarbon A compound of hydrogen and carbon only. Ethane, C_2H_6, is a hydrocarbon.

hypothesis A tentative explanation for an observation. A hypothesis is used to make a prediction that can be tested.

impurities Small quantities of additional substances in an otherwise pure substance.

incinerator A factory for burning rubbish and generating electricity.

incomplete combustion Occurs when there is insufficient oxygen available during burning to produce carbon dioxide. Carbon monoxide or carbon may be produced instead.

indicator A chemical that shows whether a solution is acidic or alkaline. For example, litmus turns blue in alkalis and red in acids. Universal indicator has a range of colours that show the pH of a solution.

inert An substance that does not readily take part in chemical reactions is inert.

intermolecular forces Weak forces between molecules. Energy is needed to overcome these forces when molecules change state from solid to liquid and from liquid to gas. The bonds inside the molecules which hold the atoms together do not break.

interpolation The process of taking a pair of values from a graph that are in between the data points that were plotted.

ion An electrically charged atom or group of atoms. A positively charged ion has a greater number of protons than electrons. A negatively charged ion has a greater number of electrons than protons.

ionic bonding Very strong attractive forces that hold the ions together in an ionic compound. The forces are electrostic - they come from the attraction between positively and negatively charged ions.

ionic compound Compound formed by the combination of a metal and a non-metal. They contain positively charged metal ions and negatively charged non-metal ions.

ionic equation An ionic equation describes a chemical change by showing only the reacting ions in solution, for example, $Ba^{2+}(aq) + SO_4^{2-}(aq) \rightarrow BaSO_4(s)$

ionisation The process of removing an electron from (or adding an electron to) an atom or group of atoms.

irreversible reaction A chemical change that can only go in one direction, for example, combustion.

isotope Two or more forms of the same element that have different numbers of neutrons in their nuclei, and hence different mass numbers.

landfill Disposing of rubbish in holes in the ground.

law of conservation of mass Matter is not created or destroyed in a chemical reaction. The total mass of the products is equal to the total mass of the reactants.

Le Chatelier's Principle The principle that states the position of an equilibrium will respond to oppose a change in the reaction conditions.

life cycle The full life of a material or product, including production, use, and disposal.

life cycle assessment A way of analysing the production, use and disposal of a material or product to add up the total energy and water used and the effects on the environment.

limiting quantity The substance that is completely used up when a reaction is complete. The theoretical yield is based on the amount of the limiting quantitity that is available for the reaction.

locating agent A chemical used to show up chemical spots on a chromatogram

long-chain molecule Polymers are long-chain molecules. They consist of long chains of atoms held together by covalent bonds.

lubricant A substance that reduces friction between moving parts of a machine.

macroscopic Large enough to be seen without the help of a microscope.

malleable Metals are malleable because they can be hammered into different shapes.

mass number The total number of protons and neutrons in an atom of an element.

mathematical model A type of scientific model that uses patterns in data of past events, along with known scientific relationships, to predict what will happen to one variable when another is changed. A simple example of a mathematical model is the relationship between speed, distance and time.

matrix A component of a composite material. The matrix surrounds and binds together the fibres or particles of the other material making up the composite.

mean value A type of average, found by adding up a set of measurements and then dividing by the number of measurements. You can have more confidence in the mean of a set of measurements than in a single measurement. The mean can be used as the best estimate of the true value.

melting point The temperature at which a pure sample of a solid changes into a liquid.

membrane A thin material which will allow only specific types of particle to pass through.

meniscus The water surface in a narrow tube curves to form a meniscus.

metal Elements that form cations when compounds of it are in solution and oxides of the elements form hydroxides rather than acids in water. Most metals are conductors of electricity, have crystalline solids with a metallic lustre and have a high chemical reactivity and many of these elements are hard and have high physical strength.

metallic bonding Very strong attractive forces that hold metal atoms together in a solid metal. The metal atoms lose their outer electrons and form positive ions. The electrons drift freely around the lattice of positive metal ions and hold the ions together.

microscopic So small that detail cannot be seen without a microscope.

mineral A compound of a valuable element (usually a metal) which is extracted from the Earth.

mitigate Counteract, reduce the effect of.

mixture Two or more difference substances, mixed but not chemically joined together.

model A scientific model is a way of representing something from the real world, such as a system of interacting parts. It includes some, but not necessarily all, of the features of the system it represents. It can show how these features are connected or interact, and can be used to explain scientific ideas, answer questions, and make predictions.

mobile phase The solvent that carries chemicals from a sample through a chromatographic column or sheet

molar mass The mass of one mole of a substance. The molar mass is the relative formula mass of the substance in grams.

molar volume The volume of one mole of any gas. The molar volume at room temperature and pressure is $24\,dm^3$.

mole The mole is the amount of any substance that contains as many particles (e.g., atoms, molecules, ions, electrons) as there are atoms in 12 grams of pure carbon-12 . One mole contains the same number of particles as Avogadro's constant, 6.0×10^{23}.

molecular formula A molecular formula shows the number of each type of atom in a molecule. Molecular formulae are only used for simple covalent compounds. Ionic compounds and giant covalent compounds contain millions of atoms. These are always represented using empirical formulae.

molecule A group of atoms joined together. Most non-metals consist of molecules. Most compounds of non-metals with other non-metals are also molecular.

monomer Small molecules that can be joined to others like it in long chains to make a polymer. Monomers are the repeting units of polymers.

nanometre (nm) A unit of length 1 000 000 000 times smaller than a metre.

nanoparticles A very tiny particle, whose size can be measured in nanometres.

nanotechnology The use and control of matter on a tiny (nanometre) scale.

nanotubes Carbon atoms held together by covalent bonds in the form of a hollow tube.

neutralisation A reaction in which an acid reacts with an alkali to form a salt. During neutralisation reactions, the hydrogen ions in the acid solution react with hydroxide ions in the alkaline solution to make water molecules.

neutron A subatomic particle found in the nucleus with about the same mass as a proton but no electrical charge.

nitrogen fixation The conversion of nitrogen gas into compounds either industrially or by natural means.

noble gases The elements of Group 0 of the periodic table, including helium, neon and argon.

non-aqueous solution A solution in which a liquid other than water is the solvent.

non-biodegradable Non-biodegradable substances cannot be broken down by microorganisms such as bacteria and fungi. Most metals and synthetic polymers are non-biodegradable.

non-metal Elements that do not exhibit such characteristic properties of metals.

nucleus (atom) The tiny central part of an atom (made up of protons and neutrons). Most of the mass of an atom is concentrated in its nucleus.

ocean acidification When carbon dioxide dissolves ocean there is an increase in H^+ ions, which decreases the pH of the seawater.

order of magnitude If two numbers differ by an order of magnitude, then one number is about ten times bigger than the other. A value given to the 'nearest order of magnitude' will be given to the nearest power of ten.

ore A type of rock that contains enough valuable minerals to make it profitable to mine.

organic chemistry The study of carbon compounds. This includes all of the natural carbon compounds from living things and synthetic carbon compounds.

outcome A variable that changes as a result of something else changing.

outlier A measured result that seems very different from other repeat measurements, or from the value you would expect. The measurement should be treated as data, unless there is a reason to reject it (e.g., measurement or recording error).

oxidation A reaction that involves the addition of oxygen and/or the loss of electrons from a chemical.

oxide A compound of a metal with oxygen is a metal oxide, for example copper oxide, CuO.

particulate Tiny bit of a solid.

peer review The process whereby scientists who are experts in their field critically evaluate another scientist's scientific paper or idea before and after publication.

percentage yield A measure of the efficiency of a synthesis reaction.

period In the Periodic Table, the horizontal rows are called periods. From one element to the next the number of electrons increases by one. So from left to right across a period an electron shell is filled up.

Periodic Table A table of the chemical elements arranged in order of atomic number.

PET A type of synthetic polymer. PET is a polyester. It is short for polyethylene terephthalate.

pH A number on a scale from 0-14 that shows the acidity or alkalinity of a solution in water. The number is related to the concentration of hydrogen ions found in the solution.

physical change A change where a substance changes its properties but does not become a different substance.

physical properties Properties of materials such as melting point, density, and electrical conductivity. These are properties that do not involve chemical changes.

phytoextraction An extraction process which uses plants to remove harmful metal ions from the soil by taking them into their roots and storing them in their leaves.

pipette A pipette is used to measure volumes of liquids or solutions accurately. A pipette can be used to deliver the same fixed volume of solution again and again during a series of titrations.

plastic (material) A type of synthetic polymer that can be moulded into shape.

plating Coating a metal with a thin layer of another metal.

plum pudding model of the atom A model of the atom suggested by J. J. Thomson. He said that an atom consists of tiny, negatively charged electrons moving about in a postively-charged sphere.

polar In a polar molecule, the charge is not evenly distributed. One end of the molecule has a small positve charge, and the other end has a small negative charge.

polymerisation A chemical reaction involving the joining together of lots of small molecules called monomers to form a long-chain molecule called a polymer.

polymers A material made of very long molecules formed by joining lots of small molecules, called monomers, together.

position of equilibrium An indication of the proximity of an equilibrium reaction to reactants or to products. The position of equilibrium can be changed by changing the experimental conditions of temperature, concentration, and, where gases are involved, the pressure.

potable Safe to drink

potential difference (p.d.) The difference in potential energy (for each unit of charge flowing) between any two points in an electric circuit. Also called voltage.

precipitate An insoluble solid formed on mixing two solutions. Silver bromide forms as a precipitate on mixing solutions of silver nitrate and potassium bromide.

precision A measure of the spread of quantitative results. If the measurements are precise all the results are very close in value.

prediction A hypothesis is used to make a prediciton about how, in a particular experimental context, a change in a factor will affect the outcome.

pressure The force per unit area acting on a surface. Pressure is measured in Pascals (Pa) or newtons per square metre (N/m^2). $1\ Pa = 1\ N/m^2$.

products The new chemicals formed during a chemical reaction.

proportional Two variables are proportional if there is a constant ratio between them.

protein A natural polymer made from amino acids. Proteins can be structural (such as collagen) or functional (such as enzymes).

proton A subatomic particle found in the nucleus of atoms with a positive electric charge equal in magnitude to the negative charge of an electron

pure substance A substance containing one type of particle; either an element or a compound

qualitative analysis Any method for identifying the chemicals in a sample. Thin-layer chromatography is an example of a qualitative method of analysis.

quantitative analysis Any method for determining the amount of a chemical in a sample. An acid-base titration is an example of quantitative analysis.

random error A measurement error due to results varying in an unpredictable way, for example due to the scientist having to make a judgement about timing or colour.

range Describes the spread between the highest and the lowest of a set of measurements.

rate of reaction A measure of how quickly a reaction happens. Rates can be measured by following the disappearance of a reactant or the formation of a product.

reactants The chemicals on the left-hand side of an equation. These chemicals react to form the products.

reactive metals Metals which react more quickly than other metals, for example with water, oxygen and acids.

reactor Vessel used to carry out a chemical reaction.

recycling A range of methods for making new materials from materials that have already been used.

reducing agent A chemical that removes oxygen from another substance during a reaction For example, carbon acts as a reducing agent when it removes oxygen from a metal oxide. The carbon is oxidised to carbon monoxide during this process.

reduction A reaction that involves the loss of oxygen and/or the gain of electrons by a chemical.

reference material Known chemical used in analysis for comparison with unknown chemicals.

regulation The legal control of applications of science by a government or non-governmental body.

relative atomic mass The mass of an atom of an element compared to the mass of an atom of carbon. The relative atomic mass of carbon is defined as 12.

relative formula mass The combined relative atomic masses of all the atoms in a formula. To find the relative formula mass of a chemical, you just add up the relative atomic masses of the atoms in the formula.

repeatable Data are said to be repeatable when the same investigator finds the same or similar results under the same conditions. We can have more confidence in data that are repeatable.

representative model A type of scientific model that uses simple shapes or analogies to represent the interacting parts of a system. One example is the particle model – it helps us to visualise the tiny particles (atoms and molecules) that make up substances.

representative sample The characteristics of a representative sample are very similar to the characteristics of the whole population.

reproducible Data are said to be reproducible when other investigators have found the same or similar results under similar conditions. We can have more confidence in data that are reproducible.

retardation factor This is a ratio used in paper chromatography. If the conditions are kept the same, each chemical in the mixture will move a fixed fraction of the distance moved by the solvent front. The R_f value is a measure of this fraction.

reversible reaction A chemical change which can go forward or backward depending on the conditions. Many reversible processes can reach a state of equilibrium.

risk An estimate of the probability that an unwanted outcome will happen. The size of the risk can be estimated from the chance of it occurring in a large sample over a given period of time.

risk assessment A check on the hazards involved in a scientific procedure. A full assessment includes the steps to be taken to avoid or reduce the risks from the hazards identified.

sacrificial protection Protecting iron or steel against corrosion by using a more reactive metal.

salt An ionic compound formed when an acid neutralises an alkali or when a metal reacts with a non-metal.

sample It is usually not possible to collect data about a whole population of organisms or other specimens. For this reason a study usually collects data about a proportion of them. This is a sample. Conclusions about a sample can only be applied to the whole population if it is a representative sample.

secondary data Data collected by somebody else, which can be compared to the primary data collected in the lab or field by the person doing the investigation.

simple covalent substance A compound which contains molecules with a small number of atoms joined together by covalent bonds. Oxygen and carbon dioxide are examples of covalent substances.

simple molecule Chemical consisting of small numbers of atoms covalently bonded together. Simple molecular substances have relatively low melting and boiling points.

single bond A covalent bond formed when atoms share one pair of electrons.

solubility The ability of a solute (eg, a salt) to dissolve in a solvent (e.g., water).

solvent A liquid which is used to dissolve other substances. Water and ethanol are two examples of solvents.

solvent front The furthest position reached by the solvent during paper chromatography.

spatial model A type of mathematical model in which a computer is used to make a model of one or more objects in a three-dimensional space. The model can be used to predict the outcome of changing a variable (e.g., temperature) in this space (e.g., a landscape and the atmosphere above it).

spectrometer A machine for recording and measuring a spectrum.

standard form Numbers represented in standard form are given with a decimal point after the first digit multiplied by a power of 10, for example, $150 = 1.5 \times 10^2$.

state Describes whether a substance is a solid, liquid, or gas.

state symbols State symbols are used in equations to show the state of a substance: solid (s), liquid (l), gas (g) or aqueous solution (aq).

stationary phase The medium through which the moblie phase passes in chromatography.

stiff A material that is difficult to bend or stretch.

strong A material that is hard to pull apart or crush.

strong acid A strong acid is fully ionised to produce a high concentration of hydrogen ions when it dissolves in water.

structural formula A structural formula is a 2-D representation of a covalently bonded molecule. It shows the number of bonds around each atom.

subatomic particles The tiny particles that make up atoms. Subatomic particles include protons, neutrons, and electrons.

sublimation The change of state directly from a solid to gas (and the reverse).

surface area The total area of a surface available for a chemical reaction or absorption to take place.

sustainability Using resources and the environment to meet the needs of people today without damaging Earth for people of the future. One way to do this is to use resources at the same rate as they can be replaced.

synthetic fertilisers Substances produced by reactions and applied to the soil to restore fertility and increase crop yields

synthetic polymer Polymers made by a chemical process, not naturally occurring.

systematic error A measurement error that differs from the true value by the same amount each time a measurement is made. A systematic error may be due to the environment, methods of observation or the instrument used.

tangent A straight line touching a curve or the circumference of a circle at a single point. It has the same gradient as the curve at that point.

theoretical yield The amount of product that would be obtained in a reaction if all the reactants were converted to products exactly as described by the balanced chemical equation.

theory A scientific theory is a general explanation that applies to a large number of situations or examples (perhaps to all possible ones). It has been tested and used successfully, and is widely accepted by scientists. An example is Faraday's theory of ions.

titration An analytical technique used to find the exact volumes of solutions that react with each other.

tough A material that absorbs energy and changes shape when stressed. Tough is the opposite of brittle.

transition metals Elements in the block between groups 2 and 3 of the Periodic Table.

true value The actual value.

uncertainty An indication of the confidence a scientist has in the accuracy of a measurement. It can be expressed as a range of values within which the true value must lie.

valid The conclusions from an experiment are valid if the procedures ensure that the effects observed are due to the cause claimed, and if the analysis has taken account of other possible factors.

variation Differences within repeated measurements of a quantity.

viable A viable process is one that is cost effective and does not cause unacceptable damage to the environment.

volatile Physical property of substances that evaporate readily.

weak acid A weak acid is only slightly ionised to produce hydrogen ions when they dissolve in water.

weather The atmospheric conditions at a location including temperature, precipitation, wind speeds etc.

word equation A summary in words describing a chemical reaction.

yield The amount of product obtained during a reaction

Young's modulus A measure of the stiffness of a material. A stiff material has a high Young's modulus. A flexible material has a low Young's modulus.

Index

Cooper/Science Photo Library; **p237(R)**: DK Arts/Shutterstock; **p237(L)**: Stefan Ember/123RF; **p238**: Olaf Speier/Shutterstock; **p243**: Martyn F Chillmaid/Science Photo Library; **p244**: Anton_Ivanov/Shutterstock; **p248**: Charles D Winters/Science Photo Library; **p251(T)**: Martyn F Chillmaid/Science Photo Library; **p251(B)**: Chochowy/iStockphoto; **p253(T)**: Matteo Malavasi/123RF; **p253(B)**: Dima Sobko/Shutterstock; **p259**: Laguna Design/Science Photo Library; **p258**: OUP; **p264**: Pniesen/iStockphoto; **p267(T)**: Jodi Jacobson/iStockphoto; **p267(B)**: Dr Guillermo Diaz-Pulido/Coral Reef Algae Research Lab; **p271(T)**: David Scharf/Science Photo Library; **p271(BL)**: Andy Harmer/Science Photo Library; **p271(BR)**: Burkard Manufacturing Company Limited: the original designers and manufacturers of the Spore Trap; **p272**: Paul Rapson/Science Photo Library; **p273**: Bikeriderlondon/Shutterstock; **p274**: Hung Chung Chih/Shutterstock; **p276(T)**: Emilio Segre Visual Archives/American Institute of Physics/Science Photo Library; **p276(B)**: Igor Strukov/Shutterstock; **p277(T)**: Fotokostic/Shutterstock; **p277(B)**: Sauletas/Shutterstock; **p278(R)**: Everett Historical/Shutterstock; **p278(L)**: Science Photo Library; **p279**: Peter van Evert/Alamy; **p280(TL)**: British Antarctic Survey/Science Photo Library; **p280(TR)**: CERN/Science Photo Library; **p280(CL)**: Alexis Rosenfeld/Science Photo Library; **p280(CR)**: Meunierd/Shutterstock; **p280(BL)**: Javier Trueba/MSF/Science Photo Library; **p280(BR)**: Ashley Cooper/Science Photo Library; **p288**: Monkey Business Images/Shutterstock; **p294**: Trevor Clifford Photography/Science Photo Library; **p295**: Martyn F Chillmaid/Science Photo Library; **p303**: David Taylor/Science Photo Library; **p324(TR)**: Silvano Audisio/Shuttertock; **p324(TL)**: Vladimir Prusakov/Shutterstock; **p324(B)**: Imagedb.com/Shutterstock; **p325(L)**: Thinglass/Shutterstock; **p325(R)**: Andrew J Shearer/iStockphoto; **p326(T)**: Photolinc/Shutterstock; **p326(B)**: Esp_imaging/iStockphoto;

These resources have been developed to support teachers and students undertaking the OCR suite of specifications GCSE Twenty First Century Science. They have been developed from the first and second editions of the resources.

We would like to thank Michelle Spiller, Sarah Old, Naomi Rowe, and Ann Wolstenholme at OCR for their work on the specifications for the Twenty First Century Science course.

We would also like to thank the following contributors to case studies in these resources: Lynda Dunlop (University of York Science Education Group) and Guillermo Diaz-Pulido (Coral Reef Algae Research Lab).

Authors and editors from the first and second editions
We thank the authors and editors of the first and second editions: David Brodie, Anna Grayson, Helen Harden, John Holman, Andrew Hunt, John Lazonby, Ted Lister, Allan Mann, Peter Nicholson, Emma Palmer, Cliff Porter, Mike Shipton, Charles Tracey, Dorothy Warren, and Vicky Wong.

Many people from schools, colleges, universities, industry, and the professions contributed to the production of the first edition of these resources. We also acknowledge the invaluable contribution of the teachers and students in the pilot centres.

A full list of contributors can be found on the website: https://global.oup.com/education/content/secondary/online-products/acknowledgements/?region=uk

The first edition of Twenty First Century Science was developed with support from the Nuffield Foundation, The Salters' Institute, and the Wellcome Trust.

The continued development of Twenty First Century Science is made possible by generous support from The Salters' Institute.

OXFORD
UNIVERSITY PRESS

Great Clarendon Street, Oxford, OX2 6DP, United Kingdom

Oxford University Press is a department of the University of Oxford. It furthers the University's objective of excellence in research, scholarship, and education by publishing worldwide. Oxford is a registered trade mark of Oxford University Press in the UK and in certain other countries

© Salters' Educational Resources Ltd and the Nuffield Foundation

Resources developed by the University of York Science Education Group on behalf of Salters' Educational Resources Ltd

The moral rights of the authors have been asserted

First published in 2016

British Library Cataloguing in Publication Data
Data available

978 0 19 835964 7

10 9 8 7 6 5 4 3 2

Paper used in the production of this book is a natural, recyclable product made from wood grown in sustainable forests. The manufacturing process conforms to the environmental regulations of the country of origin.

Printed in India by Manipal Technologies Ltd

This resource is endorsed by OCR for use with specification J258 OCR GCSE (9–1) in Chemistry B (Twenty First Century Science). In order to gain OCR endorsement, this resource has undergone an independent quality check. Any references to assessment and/or assessment preparation are the publisher's interpretation of the specification requirements and are not endorsed by OCR. OCR recommends that a range of teaching and learning resources are used in preparing learners for assessment. OCR has not paid for the production of this resource, nor does OCR receive any royalties from its sale. For more information about the endorsement process, please visit the OCR website, www.ocr.org.uk.